Graduate Texts in Mathematics

1 TAKEUTI/ZARING. Introduction to Axiomatic Set Theory. 2nd ed.
2 OXTOBY. Measure and Category. 2nd ed.
3 SCHAEFFER. Topological Vector Spaces.
4 HILTON/STAMMBACH. A Course in Homological Algebra.
5 MACLANE. Categories for the Working Mathematician.
6 HUGHES/PIPER. Projective Planes.
7 SERRE. A Course in Arithmetic.
8 TAKEUTI/ZARING. Axiomatic Set Theory.
9 HUMPHREYS. Introduction to Lie Algebras and Representation Theory.
10 COHEN. A Course in Simple Homotopy Theory.
11 CONWAY. Functions of One Complex Variable. 2nd ed.
12 BEALS. Advanced Mathematical Analysis.
13 ANDERSON/FULLER. Rings and Categories of Modules.
14 GOLUBITSKY/GUILLEMIN. Stable Mappings and Their Singularities.
15 BERBERIAN. Lectures in Functional Analysis and Operator Theory.
16 WINTER. The Structure of Fields.
17 ROSENBLATT. Random Processes. 2nd ed.
18 HALMOS. Measure Theory.
19 HALMOS. A Hilbert Space Problem Book. 2nd ed., revised.
20 HUSEMOLLER. Fibre Bundles. 2nd ed.
21 HUMPHREYS. Linear Algebraic Groups.
22 BARNES/MACK. An Algebraic Introduction to Mathematical Logic.
23 GREUB. Linear Algebra. 4th ed.
24 HOLMES. Geometric Functional Analysis and its Applications.
25 HEWITT/STROMBERG. Real and Abstract Analysis.
26 MANES. Algebraic Theories.
27 KELLEY. General Topology.
28 ZARISKI/SAMUEL. Commutative Algebra. Vol. I.
29 ZARISKI/SAMUEL. Commutative Algebra. Vol. II.
30 JACOBSON. Lectures in Abstract Algebra I: Basic Concepts.
31 JACOBSON. Lectures in Abstract Algebra II: Linear Algebra.
32 JACOBSON. Lectures in Abstract Algebra III: Theory of Fields and Galois Theory.
33 HIRSCH. Differential Topology.
34 SPITZER. Principles of Random Walk. 2nd ed.
35 WERMER. Banach Algebras and Several Complex Variables. 2nd ed.
36 KELLEY/NAMIOKA et al. Linear Topological Spaces.
37 MONK. Mathematical Logic.
38 GRAUERT/FRITZSCHE. Several Complex Variables.
39 ARVESON. An Invitation to C^*-Algebras.
40 KEMENY/SNELL/KNAPP. Denumerable Markov Chains. 2nd ed.
41 APOSTOL. Modular Functions and Dirichlet Series in Number Theory.
42 SERRE. Linear Representations of Finite Groups.
43 GILLMAN/JERISON. Rings of Continuous Functions.
44 KENDIG. Elementary Algebraic Geometry.
45 LOÈVE. Probability Theory I. 4th ed.
46 LOÈVE. Probability Theory II. 4th ed.
47 MOISE. Geometric Topology in Dimensions 2 and 3.

continued after Index

Olli Lehto

Univalent Functions and Teichmüller Spaces

With 16 Illustrations

Springer-Verlag
New York Berlin Heidelberg
London Paris Tokyo

Olli Lehto
Department of Mathematics
University of Helsinki
00100 Helsinki
Finland

AMS Subject Classifications: 30-01, 32G15, 30F30, 30F35

Library of Congress Cataloging in Publication Data
Lehto, Olli.
  Univalent functions and Teichmüller spaces.
  (Graduate texts in mathematics; 109)
  Bibliography: p.
  Includes index.
  1. Univalent functions.   2. Teichmüller spaces.
3. Riemann surfaces.   I. Title.   II. Series.
QA331.L427   1986   515   86-4008

Typeset by Asco Trade Typesetting Ltd., Hong Kong.
Printed and bound by R. R. Donnelley & Sons, Harrisonburg, Virginia.
Printed in the United States of America.

9 8 7 6 5 4 3 2 1

ISBN 0-387-96310-3 Springer-Verlag New York Berlin Heidelberg
ISBN 3-540-96310-3 Springer-Verlag Berlin Heidelberg New York

Preface

This monograph grew out of the notes relating to the lecture courses that I gave at the University of Helsinki from 1977 to 1979, at the Eidgenössische Technische Hochschule Zürich in 1980, and at the University of Minnesota in 1982. The book presumably would never have been written without Fred Gehring's continuous encouragement. Thanks to the arrangements made by Edgar Reich and David Storvick, I was able to spend the fall term of 1982 in Minneapolis and do a good part of the writing there. Back in Finland, other commitments delayed the completion of the text.

At the final stages of preparing the manuscript, I was assisted first by Mika Seppälä and then by Jouni Luukkainen, who both had a grant from the Academy of Finland. I am greatly indebted to them for the improvements they made in the text.

I also received valuable advice and criticism from Kari Astala, Richard Fehlmann, Barbara Flinn, Fred Gehring, Pentti Järvi, Irwin Kra, Matti Lehtinen, Ilppo Louhivaara, Bruce Palka, Kurt Strebel, Kalevi Suominen, Pekka Tukia and Kalle Virtanen. To all of them I would like to express my gratitude. Raili Pauninsalo deserves special thanks for her patience and great care in typing the manuscript.

Finally, I thank the editors for accepting my text in Springer-Verlag's well-known series.

Helsinki, Finland
June 1986

Olli Lehto

Contents

Preface   v

Introduction   1

CHAPTER I
Quasiconformal Mappings   4

Introduction to Chapter I   4
1. Conformal Invariants   5
   1.1 Hyperbolic metric   5
   1.2 Module of a quadrilateral   7
   1.3 Length-area method   8
   1.4 Rengel's inequality   9
   1.5 Module of a ring domain   10
   1.6 Module of a path family   11
2. Geometric Definition of Quasiconformal Mappings   12
   2.1 Definitions of quasiconformality   12
   2.2 Normal families of quasiconformal mappings   13
   2.3 Compactness of quasiconformal mappings   14
   2.4 A distortion function   15
   2.5 Circular distortion   17
3. Analytic Definition of Quasiconformal Mappings   18
   3.1 Dilatation quotient   18
   3.2 Quasiconformal diffeomorphisms   19
   3.3 Absolute continuity and differentiability   20
   3.4 Generalized derivatives   21
   3.5 Analytic characterization of quasiconformality   22
4. Beltrami Differential Equation   23
   4.1 Complex dilatation   23

4.2 Quasiconformal mappings and the Beltrami equation   24
4.3 Singular integrals.   25
4.4 Representation of quasiconformal mappings   26
4.5 Existence theorem   27
4.6 Convergence of complex dilatations.   28
4.7 Decomposition of quasiconformal mappings   29
5. The Boundary Value Problem   30
5.1 Boundary function of a quasiconformal mapping.   30
5.2 Quasisymmetric functions   31
5.3 Solution of the boundary value problem   33
5.4 Composition of Beurling–Ahlfors extensions   34
5.5 Quasi-isometry   35
5.6 Smoothness of solutions.   36
5.7 Extremal solutions.   37
6. Quasidiscs   38
6.1 Quasicircles.   38
6.2 Quasiconformal reflections.   39
6.3 Uniform domains.   41
6.4 Linear local connectivity   44
6.5 Arc condition.   45
6.6 Conjugate quadrilaterals   46
6.7 Characterizations of quasidiscs   47

CHAPTER II
Univalent Functions.   50

Introduction to Chapter II   50
1. Schwarzian Derivative   51
1.1 Definition and transformation rules   51
1.2 Existence and uniqueness.   53
1.3 Norm of the Schwarzian derivative   54
1.4 Convergence of Schwarzian derivatives   55
1.5 Area theorem.   58
1.6 Conformal mappings of a disc   59
2. Distance between Simply Connected Domains.   60
2.1 Distance from a disc.   60
2.2 Distance function and coefficient problems   61
2.3 Boundary rotation.   62
2.4 Domains of bounded boundary rotation   63
2.5 Upper estimate for the Schwarzian derivative.   65
2.6 Outer radius of univalence.   66
2.7 Distance between arbitrary domains   67
3. Conformal Mappings with Quasiconformal Extensions   68
3.1 Deviation from Möbius transformations   68
3.2 Dependence of a mapping on its complex dilatation.   69
3.3 Schwarzian derivatives and complex dilatations.   72
3.4 Asymptotic estimates.   73
3.5 Majorant principle.   76
3.6 Coefficient estimates   77

4. Univalence and Quasiconformal Extensibility of Meromorphic
   Functions.. 79
   4.1 Quasiconformal reflections under Möbius transformations 79
   4.2 Quasiconformal extension of conformal mappings............... 81
   4.3 Exhaustion by quasidiscs................................. 83
   4.4 Definition of Schwarzian domains 83
   4.5 Domains not linearly locally connected 84
   4.6 Schwarzian domains and quasidiscs......................... 86
5. Functions Univalent in a Disc 87
   5.1 Quasiconformal extension to the complement of a disc 87
   5.2 Real analytic solutions of the boundary value problem 89
   5.3 Criterion for univalence.................................. 89
   5.4 Parallel strips... 90
   5.5 Continuous extension 91
   5.6 Image of discs 92
   5.7 Homeomorphic extension 94

CHAPTER III
Universal Teichmüller Space 96

Introduction to Chapter III 96
1. Models of the Universal Teichmüller Space 97
   1.1 Equivalent quasiconformal mappings......................... 97
   1.2 Group structures....................................... 98
   1.3 Normalized conformal mappings............................ 99
   1.4 Sewing problem....................................... 100
   1.5 Normalized quasidiscs................................... 101
2. Metric of the Universal Teichmüller Space....................... 103
   2.1 Definition of the Teichmüller distance 103
   2.2 Teichmüller distance and complex dilatation 104
   2.3 Geodesics for the Teichmüller metric........................ 105
   2.4 Completeness of the universal Teichmüller space 106
3. Space of Quasisymmetric Functions............................ 108
   3.1 Distance between quasisymmetric functions................... 108
   3.2 Existence of a section................................... 109
   3.3 Contractibility of the universal Teichmüller space 109
   3.4 Incompatibility of the group structure with the metric 110
4. Space of Schwarzian Derivatives 111
   4.1 Mapping into the space of Schwarzian derivatives 111
   4.2 Comparison of distances 112
   4.3 Imbedding of the universal Teichmüller space 113
   4.4 Schwarzian derivatives of univalent functions. 115
   4.5 Univalent functions and the universal Teichmüller space 116
   4.6 Closure of the universal Teichmüller space.................... 116
5. Inner Radius of Univalence 118
   5.1 Definition of the inner radius of univalence 118
   5.2 Isomorphic Teichmüller spaces 119
   5.3 Inner radius and quasiconformal extensions................... 120

5.4 Inner radius and quasiconformal reflections................... 121
5.5 Inner radius of sectors 122
5.6 Inner radius of ellipses and polygons 125
5.7 General estimates for the inner radius 126

CHAPTER IV
Riemann Surfaces 128

Introduction to Chapter IV 128
1. Manifolds and Their Structures 129
    1.1 Real manifolds... 129
    1.2 Complex analytic manifolds............................... 130
    1.3 Border of a surface...................................... 131
    1.4 Differentials on Riemann surfaces 132
    1.5 Isothermal coordinates 133
    1.6 Riemann surfaces and quasiconformal mappings 134
2. Topology of Covering Surfaces............................... 135
    2.1 Lifting of paths 135
    2.2 Covering surfaces and the fundamental group 136
    2.3 Branched covering surfaces 137
    2.4 Covering groups 138
    2.5 Properly discontinuous groups 140
3. Uniformization of Riemann Surfaces 142
    3.1 Lifted and projected conformal structures 142
    3.2 Riemann mapping theorem 143
    3.3 Representation of Riemann surfaces....................... 144
    3.4 Lifting of continuous mappings 145
    3.5 Homotopic mappings 146
    3.6 Lifting of differentials................................. 147
4. Groups of Möbius Transformations............................ 149
    4.1 Covering groups acting on the plane 149
    4.2 Fuchsian groups 150
    4.3 Elementary groups.. 151
    4.4 Kleinian groups... 152
    4.5 Structure of the limit set.............................. 153
    4.6 Invariant domains 155
5. Compact Riemann Surfaces 157
    5.1 Covering groups over compact surfaces................... 157
    5.2 Genus of a compact surface 158
    5.3 Function theory on compact Riemann surfaces 158
    5.4 Divisors on compact surfaces............................ 159
    5.5 Riemann–Roch theorem 160
6. Trajectories of Quadratic Differentials 161
    6.1 Natural parameters 161
    6.2 Straight lines and trajectories......................... 163
    6.3 Orientation of trajectories 164
    6.4 Trajectories in the large 165

6.5 Periodic trajectories.. 166
6.6 Non-periodic trajectories.................................... 166
7. Geodesics of Quadratic Differentials............................ 168
7.1 Definition of the induced metric............................. 168
7.2 Locally shortest curves..................................... 169
7.3 Geodesic polygons.. 170
7.4 Minimum property of geodesics.............................. 171
7.5 Existence of geodesics 173
7.6 Deformation of horizontal arcs 174

CHAPTER V
Teichmüller Spaces 175

Introduction to Chapter V 175
1. Quasiconformal Mappings of Riemann Surfaces.................... 176
1.1 Complex dilatation on Riemann surfaces...................... 176
1.2 Conformal structures.. 178
1.3 Group isomorphisms induced by quasiconformal mappings........ 178
1.4 Homotopy modulo the boundary............................. 180
1.5 Quasiconformal mappings in homotopy classes................. 181
2. Definitions of Teichmüller Space 182
2.1 Riemann space and Teichmüller space........................ 182
2.2 Teichmüller metric... 183
2.3 Teichmüller space and Beltrami differentials 184
2.4 Teichmüller space and conformal structures.................. 185
2.5 Conformal structures on a compact surface 186
2.6 Isomorphisms of Teichmüller spaces 187
2.7 Modular group 188
3. Teichmüller Space and Lifted Mappings......................... 189
3.1 Equivalent Beltrami differentials 189
3.2 Teichmüller space as a subset of the universal space........... 190
3.3 Completeness of Teichmüller spaces.......................... 190
3.4 Quasi-Fuchsian groups 191
3.5 Quasiconformal reflections compatible with a group 192
3.6 Quasisymmetric functions compatible with a group............. 193
3.7 Unique extremality and Teichmüller metrics 195
4. Teichmüller Space and Schwarzian Derivatives 196
4.1 Schwarzian derivatives and quadratic differentials 196
4.2 Spaces of quadratic differentials............................. 197
4.3 Schwarzian derivatives of univalent functions................. 197
4.4 Connection between Teichmüller spaces and the universal space 198
4.5 Distance to the boundary.................................... 200
4.6 Equivalence of metrics...................................... 201
4.7 Bers imbedding 203
4.8 Quasiconformal extensions compatible with a group 204
5. Complex Structures on Teichmüller Spaces 205
5.1 Holomorphic functions in Banach spaces...................... 205

5.2 Banach manifolds   206
5.3 A holomorphic mapping between Banach spaces   207
5.4 An atlas on the Teichmüller space   208
5.5 Complex analytic structure   209
5.6 Complex structure under quasiconformal mappings.   211
6. Teichmüller Space of a Torus.   212
    6.1 Covering group of a torus   212
    6.2 Generation of group isomorphisms   214
    6.3 Conformal equivalence of tori   215
    6.4 Extremal mappings of tori   216
    6.5 Distance of group isomorphisms from the identity   218
    6.6 Representation of the Teichmüller space of a torus.   219
    6.7 Complex structure of the Teichmüller space of torus   220
7. Extremal Mappings of Riemann Surfaces.   221
    7.1 Dual Banach spaces.   221
    7.2 Space of integrable holomorphic quadratic differentials   222
    7.3 Poincaré theta series   223
    7.4 Infinitesimally trivial differentials.   224
    7.5 Mappings with infinitesimally trivial dilatations.   226
    7.6 Complex dilatations of extremal mappings.   227
    7.7 Teichmüller mappings   229
    7.8 Extremal mappings of compact surfaces.   231
8. Uniqueness of Extremal Mappings of Compact Surfaces   232
    8.1 Teichmüller mappings and quadratic differentials   232
    8.2 Local representation of Teichmüller mappings.   233
    8.3 Stretching function and the Jacobian   235
    8.4 Average stretching.   236
    8.5 Teichmüller's uniqueness theorem   237
9. Teichmüller Spaces of Compact Surfaces   240
    9.1 Teichmüller imbedding   240
    9.2 Teichmüller space as a ball of the euclidean space   241
    9.3 Straight lines in Teichmüller space.   242
    9.4 Composition of Teichmüller mappings.   243
    9.5 Teichmüller discs.   244
    9.6 Complex structure and Teichmüller metric.   245
    9.7 Surfaces of finite type.   247

Bibliography   248

Index.   253

Introduction

The theory of Teichmüller spaces studies the different conformal structures on a Riemann surface. After the introduction of quasiconformal mappings into the subject, the theory can be said to deal with classes consisting of quasiconformal mappings of a Riemann surface which are homotopic modulo conformal mappings.

It was Teichmüller who noticed the deep connection between quasiconformal mappings and function theory. He also discovered that the theory of Teichmüller spaces is intimately connected with quadratic differentials. Teichmüller ([1], [2]) proved that on a compact Riemann surface of genus greater than one, every holomorphic quadratic differential determines a quasiconformal mapping which is a unique extremal in its homotopy class in the sense that it has the smallest deviation from conformal mappings. He also showed that all extremals are obtained in this manner. It follows that the Teichmüller space of a compact Riemann surface of genus $p > 1$ is homeomorphic to the euclidean space $\mathbb{R}^{6p-6}$.

Teichmüller's proofs, often sketchy and intermingled with conjectures, were put on a firm basis by Ahlfors [1], who also introduced a more flexible definition for quasiconformal mappings. The paper of Ahlfors revived interest in Teichmüller's work and gave rise to a systematic study of the general theory of quasiconformal mappings in the plane.

Another approach to the Teichmüller theory, initiated by Bers in the early sixties, leads to quadratic differentials in an entirely different manner. This method is more general, in that it can also be applied to non-compact Riemann surfaces. The quadratic differentials are now Schwarzian derivatives of conformal extensions of quasiconformal mappings considered on the universal covering surface, the extensions being obtained by use of the Beltrami differential equation.

The development of the theory of Teichmüller spaces along these lines gives rise to several interesting problems which belong to the classical theory of univalent analytic functions. Consequently, in the early seventies a special branch of the theory of univalent functions, often studied without any connections to Riemann surfaces, began to take shape.

The interplay between the theory of univalent functions and the theory of Teichmüller spaces is the main theme of this monograph. We do give a proof of the above mentioned classical uniqueness and existence theorems of Teichmüller and discuss their consequences. But the emphasis is on the study of the repercussions of Bers's method, with attention both to univalent functions and to Teichmüller spaces. It follows that even though the topics dealt with provide an introduction to the Teichmüller theory, they leave aside many of its important aspects. Abikoff's monograph [2] and the surveys of Bers [10], [11], Earle [2], Royden [2], and Kra [2] cover material on Teichmüller spaces not treated here, and the more algebraic and differential geometric approaches, studied by Grothendieck, Bers, Earle, Thurston and many others, are not considered.

There is no clearly best way to organize our material. A lot of background knowledge is needed from the theory of quasiconformal mappings and of Riemann surfaces. A particular difficulty is caused by the fact that the interaction between univalent functions and Teichmüller spaces works in both directions.

Chapter I is devoted to an exposition of quasiconformal mappings. We have tried to collect here all the basic results that will be needed later. For detailed proofs we usually refer to the monograph Lehto–Virtanen [1]. The exceptions are cases where a brief proof can be easily presented or where we have preferred to use different arguments or, of course, where no precise reference can be given.

Chapter II deals with problems of univalent functions which have their origin in the Teichmüller theory. The leading theme is the interrelation between the Schwarzian derivative of an analytic function and the complex dilatation of its quasiconformal extension. A large fraction of the results of Chapter II comes into direct use in Chapter III concerning the universal Teichmüller space. This largest and, in many ways, simplest Teichmüller space links univalent analytic functions and general Teichmüller spaces.

A presentation of the material contained in Chapters II and III paralleling the introduction of the Teichmüller space of an arbitrary Riemann surface would perhaps have provided a better motivation for some definitions and theorems in these two chapters. But we hope that the arrangement chosen makes the theory of Teichmüller spaces of Riemann surfaces in Chapter V more transparent, as the required hard analysis has by then largely been dealt with. Also, we obtain a clear division of the book into two parts: Chapters I, II and III concern complex analysis in the plane and form an independent entity even without the rest of the book, while Chapters IV and V are related to Riemann surfaces.

The philosophy of Chapter IV on Riemann surfaces is much the same as that of Chapter I. The results needed later are formulated, and for proofs references are usually made to the standard monographs of Ahlfors–Sario [1], Lehner [1], and Springer [1]. An exception is the rather extensive treatment of holomorphic quadratic differentials, which are needed in the proof of Teichmüller's uniqueness theorem. Here we have largely utilized Strebel's monograph [6].

Finally, after all the preparations in Chapters I–IV, Teichmüller spaces of Riemann surfaces are taken up in Chapter V. We first discuss their various characterizations and, guided by the results of Chapter III, develop their general theory. After this, special attention is paid to Teichmüller spaces of compact surfaces. The torus is first treated separately and then, via the study of extremal quasiconformal mappings, compact surfaces of higher genus are discussed.

Each chapter begins with an introduction which gives a summary of its contents. The chapters are divided into sections which consist of numbered subsections. The references, such as I.2.3, are made with respect to this three-fold division. In references within a chapter, the first number is omitted.

In this book, the approach to the theory of Teichmüller spaces is based on classical complex analysis. We expect the reader to be familiar with the theory of analytic functions at the level of, say, Ahlfors's standard textbook "Complex Analysis". Some basic notions of general topology, measure and integration theory and functional analysis are also used without explanations. Some acquaintance with quasiconformal mappings and Riemann surfaces would be helpful, but is not meant to be a necessary condition for comprehending the text.

CHAPTER I

Quasiconformal Mappings

Introduction to Chapter I

Quasiconformal mappings are an essential part of the contents of this book. They appear in basic definitions and theorems, and serve as a tool over and over again.

Sections 1–4 of Chapter I aim at giving the reader a quick survey of the main features of the theory of quasiconformal mappings in the plane. Complete proofs are usually omitted. For the details, an effort was made to give precise references to the literature, in most cases to the monograph Lehto–Virtanen [1].

Section 1 introduces certain conformal invariants. The Poincaré metric is repeatedly used later, and conformal modules of path families appear in the characterizations of quasiconformality.

In section 2, quasiconformality is defined by means of the maximal dilatation of a homeomorphism. Certain compactness and distortion theorems, closely related to this definition, are considered. Section 3 starts with the classical definition of quasiconformal diffeomorphisms and explains the connections between various geometric and analytic properties of quasiconformal mappings.

Section 4 is concerned with the characterization of quasiconformal mappings as homeomorphic solutions of Beltrami differential equations. Complex dilatation, a central notion throughout our presentation, is introduced, and the basic theorems about the existence, uniqueness and representation of a quasiconformal mapping with prescribed complex dilatation are discussed.

The remaining two sections are more self-contained than sections 1–4, and their contents are more clearly determined by subsequent applications. Section 5 is devoted to the now classical problem of extending a homeomorphic self-mapping of the real axis to a quasiconformal self-mapping of the half-plane. The solution is used later in several contexts.

Section 6 deals with quasidiscs. Along with the complex dilatation and the Schwarzian derivative the notion of a quasidisc is a trademark of this book. For this reason, we have given a fairly comprehensive account of their numerous geometric properties, in most cases with detailed proofs.

1. Conformal Invariants

1.1. Hyperbolic Metric

In the first three chapters of this monograph, we shall be concerned primarily with mappings whose domain and range are subsets of the plane. Unless otherwise stated, we understand by "plane" the Riemann sphere and often use the spherical metric to remove the special position of the point at infinity.

In addition to the euclidean and spherical metrics, we shall repeatedly avail ourselves of a conformally invariant hyperbolic metric. In the unit disc $D = \{z \,|\, |z| < 1\}$ one arrives at this metric by considering Möbius transformations $z \to w$,

$$\frac{w - w_0}{1 - \bar{w}_0 w} = e^{i\theta} \cdot \frac{z - z_0}{1 - \bar{z}_0 z}, \qquad z_0, w_0 \in D,$$

which map D onto itself. By Schwarz's lemma, there are no other conformal self-mappings of D. It follows that the differential

$$\frac{|dz|}{1 - |z|^2}$$

defines a metric which is invariant under the group of conformal mappings of D onto itself.

The shortest curve in this metric joining two points z_1 and z_2 of D is the circular arc which is orthogonal to the unit circle. The hyperbolic distance between z_1 and z_2 is given by the formula

$$h(z_1, z_2) = \frac{1}{2} \log \frac{|1 - \bar{z}_1 z_2| + |z_1 - z_2|}{|1 - \bar{z}_1 z_2| - |z_1 - z_2|}. \tag{1.1}$$

The Riemann mapping theorem says that every simply connected domain A of the plane with more than one boundary point is conformally equivalent to the unit disc. Let $f: A \to D$ be a conformal mapping, and

$$\eta_A(z) = \frac{|f'(z)|}{1 - |f(z)|^2}.$$

Then the differential

$$\eta_A(z) \, |dz|$$

defines the *hyperbolic* (or *Poincaré*) *metric* of A. The function η_A, which is called the *Poincaré density* of A, is well defined, for it does not depend on the particular choice of the mapping f. In the upper half-plane, $\eta_A(z) = 1/(2 \operatorname{Im} z)$. The geodesics, which are preserved under conformal mappings, are called hyperbolic segments.

The Poincaré density is monotonic with respect to the domain: If A_1 is a simply connected subdomain of A and $z \in A_1$, then

$$\eta_A(z) \leq \eta_{A_1}(z). \tag{1.2}$$

For let f and f_1 be conformal maps of A and A_1 onto the unit disc D, both vanishing at z. Then $\eta_A(z) = |f'(z)|$, $\eta_{A_1}(z) = |f_1'(z)|$, and application of Schwarz's lemma to the function $f \circ f_1^{-1}$ yields (1.2).

Similar reasoning gives an upper bound for $\eta_A(z)$ in terms of the euclidean distance $d(z, \partial A)$ from z to the boundary of A. Now we apply Schwarz's lemma to the function $\zeta \to f(z + d(z, \partial A)\zeta)$ and obtain

$$\eta_A(z) \leq \frac{1}{d(z, \partial A)}. \tag{1.3}$$

For domains A not containing ∞ we also have the lower bound

$$\eta_A(z) \geq \frac{1}{4d(z, \partial A)}. \tag{1.4}$$

This is proved by means of the *Koebe one-quarter theorem* (Nehari [2], p. 214): If f is a conformal mapping of the unit disc D with $f(0) = 0$, $f'(0) = 1$, $f(z) \neq \infty$, then $d(0, \partial f(D)) \geq 1/4$. We apply this to the function $w \to (g(w) - z)/g'(0)$, where g is a conformal mapping of D onto A with $g(0) = z$. Because $\eta_A(z) = 1/|g'(0)|$, the inequality (1.4) follows. Both estimates (1.3) and (1.4) are sharp.

There is another lower estimate for the Poincaré density which we shall need later. Let A be a simply connected domain and w_1, w_2 finite points outside A. Then

$$\eta_A(z) \geq \frac{|w_1 - w_2|}{4|z - w_1||z - w_2|} \tag{1.5}$$

for every $z \in A$. To prove (1.5) we observe that $z \to f(z) = (z - w_1)/(z - w_2)$ maps A onto a domain A' which does not contain 0 or ∞. Hence, by the conformal invariance of the hyperbolic metric and by (1.4),

$$\eta_A(z) = \eta_{A'}(f(z))|f'(z)| \geq \frac{1}{4d(f(z), \partial A')} \cdot \frac{|w_1 - w_2|}{|z - w_2|^2}.$$

Since $d(f(z), \partial A') \leq |f(z)|$, the inequality (1.5) follows.

The hyperbolic metric can be transferred by means of conformal mappings to multiply connected plane domains with more than two boundary points and even to most Riemann surfaces. This will be explained in IV.3.6. Finally, in V.9.6 we define the hyperbolic metric on an arbitrary complex analytic manifold.

1.2. Module of a Quadrilateral

A central theme in what follows is to measure in quantitative terms the deviation of a homeomorphism from a conformal mapping. A natural way to do this is to study the change of some conformal invariant under homeomorphisms.

In 1.6 we shall exhibit a general method to produce conformal invariants which are appropriate for this purpose. Hyperbolic distance is not well suited to this objective, whereas two other special invariants, the module of a quadrilateral and that of a ring domain, have turned out to be particularly important. We shall first discuss the case of a quadrilateral.

A Jordan curve is the image of a circle under a homeomorphism of the plane. A domain whose boundary is a Jordan curve is called a Jordan domain.

Let f be a conformal mapping of a disc D onto a domain A. Suppose that A is locally connected at every point z of its boundary ∂A, i.e., that every neighborhood U of z in the plane contains a neighborhood V of z, such that $V \cap A$ is connected. Under this topological condition on A, a standard length-area argument yields the important result that f can be extended to a homeomorphism between the closures of D and A. It follows, in particular, that ∂A is a Jordan curve (Newman [1], p. 173).

Conversely, a Jordan domain is locally connected at every boundary point. We conclude that a conformal mapping of a Jordan domain onto another Jordan domain has a homeomorphic extension to the boundary, and hence to the whole plane. For such a mapping, the images of three boundary points can, modulo orientation, be prescribed arbitrarily on the boundary of the image domain. In contrast, four points on the boundary of a Jordan domain determine a conformal module, an observation we shall now make precise.

A *quadrilateral* $Q(z_1, z_2, z_3, z_4)$ is a Jordan domain and a sequence of four points z_1, z_2, z_3, z_4 on the boundary ∂Q following each other so as to determine a positive orientation of ∂Q with respect to Q. The arcs (z_1, z_2), (z_2, z_3), (z_3, z_4) and (z_4, z_1) are called the sides of the quadrilateral.

Let f be a conformal mapping of Q onto a euclidean rectangle R. If the boundary correspondence is such that f maps the four distinguished points z_1, z_2, z_3, z_4 to the vertices of R, then the mapping f is said to be *canonical*, and R is called a canonical rectangle of $Q(z_1, z_2, z_3, z_4)$. It is not difficult to prove that *every quadrilateral possesses a canonical mapping and that the canonical mapping is uniquely determined up to similarity transformations.*

The existence can be shown if we first map Q conformally onto the upper half-plane, arrange the four distinguished points in pairwise symmetric positions with respect to the origin, and finally perform a conformal mapping by means of a suitable elliptic integral. The uniqueness part follows directly from the reflection principle. (For the details, see Lehto–Virtanen [1]; for this monograph, to which several references will be made in Chapter I, we shall henceforth use the abbreviation [LV].)

Now suppose that $R = \{x + iy | 0 < x < a, 0 < y < b\}$ is a canonical rectangle of $Q(z_1, z_2, z_3, z_4)$ and that the first side (z_1, z_2) corresponds to the line segment $0 \leq x \leq a$. The number a/b, which does not depend on the particular choice of the canonical rectangle, is called the (*conformal*) *module* of the quadrilateral $Q(z_1, z_2, z_3, z_4)$. We shall use the notation

$$M(Q(z_1, z_2, z_3, z_4)) = a/b$$

for the module. It follows from the definition that $M(Q(z_1, z_2, z_3, z_4)) = 1/M(Q(z_2, z_3, z_4, z_1))$.

From the definition it is also clear that the module of a quadrilateral is conformally invariant, i.e., if f is a conformal mapping of a domain A and $Q(z_1, z_2, z_3, z_4)$ is a quadrilateral such that $Q \subset A$ and $f(Q)$ is a Jordan domain, then $M(Q(z_1, z_2, z_3, z_4)) = M(f(Q)(f(z_1), f(z_2), f(z_3), f(z_4)))$.

1.3. Length-Area Method

It is possible to arrive at the notion of the module of a quadrilateral through an extremal problem, by use of a length-area method. This approach has turned out to be extremely useful and it leads to far-reaching generalizations, even beyond complex analysis. In the general situation we shall discuss it in 1.6. In explicit form the idea was announced by A. Beurling in the 1946 Scandinavian Congress of Mathematicians in Copenhagen, and a few years later it was used systematically for the first time by L. Ahlfors and A. Beurling.

In order to arrive at this characterization of the module we consider the canonical mapping f of the quadrilateral $Q(z_1, z_2, z_3, z_4)$ onto the rectangle $R = \{u + iv | 0 < u < a, 0 < v < b\}$. Then

$$\iint_Q |f'(z)|^2 \, dx \, dy = ab.$$

Let Γ be the family of all locally rectifiable Jordan arcs in Q which join the sides (z_1, z_2) and (z_3, z_4). Then

$$\int_\gamma |f'(z)| \, |dz| \geq b$$

for every $\gamma \in \Gamma$, with equality if γ is the inverse image of a vertical line segment of R joining its horizontal sides. Hence

$$M(Q(z_1, z_2, z_3, z_4)) = \frac{\displaystyle\iint_Q |f'(z)|^2 \, dx \, dy}{\left(\displaystyle\inf_{\gamma \in \Gamma} \int_\gamma |f'(z)| \, |dz|\right)^2}. \tag{1.6}$$

We can get rid of the canonical mapping f if we introduce the family P whose elements ρ are non-negative Borel-measurable functions in Q and satisfy the condition $\int_\gamma \rho(z)|dz| \geq 1$ for every $\gamma \in \Gamma$. With the notation

$$m_\rho(Q) = \iint_Q \rho^2 \, dx \, dy,$$

we then have

$$M(Q(z_1, z_2, z_3, z_4)) = \inf_{\rho \in P} m_\rho(Q). \tag{1.7}$$

This basic formula can be proved by a length-area reasoning. Define for every given $\rho \in P$ a function ρ_1 in the canonical rectangle R by $(\rho_1 \circ f)|f'| = \rho$. Then, by Fubini's theorem and Schwarz's inequality,

$$m_\rho(Q) = \iint_R \rho_1^2 \, du \, dv \geq \frac{1}{b} \int_0^a du \left(\int_0^b \rho_1(u + iv) \, dv \right)^2.$$

The last integral at right is taken over a line segment whose preimage is in Γ. Therefore, the integral is ≥ 1, and so $m_\rho(Q) \geq a/b = M(Q(z_1, z_2, z_3, z_4))$. To complete the proof we note that $\rho = |f'|/b$ belongs to P. By (1.6) this is an extremal function for which $m_\rho(Q) = M(Q(z_1, z_2, z_3, z_4))$.

1.4. Rengel's Inequality

The power of the characterization (1.7) is that it yields automatically upper estimates: $M(Q(z_1, z_2, z_3, z_4)) \leq m_\rho(Q)$ for any $\rho \in P$. An important application is obtained if we choose ρ to be the euclidean metric. Let s_1 denote the euclidean distance of the sides (z_1, z_2) and (z_3, z_4) in Q, and m the euclidean area. Then (1.7) gives *Rengel's inequality*

$$M(Q(z_1, z_2, z_3, z_4)) \leq \frac{m(Q)}{s_1^2}. \tag{1.8}$$

It is not difficult to prove that *equality holds if and only if $Q(z_1, z_2, z_3, z_4)$ is a rectangle with its usual vertices* ([LV], p. 22).

Using (1.8) we can easily prove that the module depends continuously on the quadrilateral. For a precise formulation of the result let us consider a sequence of quadrilaterals $Q_n(z_1^n, z_2^n, z_3^n, z_4^n)$, $n = 1, 2, \ldots$. Suppose that this sequence converges to $Q(z_1, z_2, z_3, z_4)$ from inside, i.e., $\bar{Q}_n \subset Q$ for every n and to every $\varepsilon > 0$ there corresponds an n_ε such that for $n \geq n_\varepsilon$ every point of the sides of $Q_n(z_1^n, z_2^n, z_3^n, z_4^n)$ has a spherical distance $< \varepsilon$ from the corresponding sides of $Q(z_1, z_2, z_3, z_4)$. Then

$$\lim_{n \to \infty} M(Q_n(z_1^n, z_2^n, z_3^n, z_4^n)) = M(Q(z_1, z_2, z_3, z_4)).$$

To prove this, we only need to carry out a canonical mapping of $Q(z_1, z_2, z_3, z_4)$ and apply Rengel's inequality to the image of $Q_n(z_1^n, z_2^n, z_3^n, z_4^n)$.

Rengel's inequality also makes it possible to characterize conformality in terms of the modules of quadrilaterals, without any a priori differentiability:

Theorem 1.1. *Let $f: A \to A'$ be a sense-preserving homeomorphism which leaves invariant the modules of the quadrilaterals of the domain A. Then f is conformal.*

We sketch a proof. Map a quadrilateral of A and its image in A' canonically onto identical rectangles R and R' whose sides are parallel to the coordinate axes. Given a point $z = x + iy$ of R, we consider the two rectangles R_1 and R_2 onto which R is divided by the vertical line through z. Since all modules remain invariant, it follows from Rengel's inequality, with regard to the possibility for equality, that the images of R_1 and R_2 in R' are also rectangles (cf. [LV], p. 29). But then the real part of the image of z must be x. A similar argument shows that the imaginary part of z does not change either. Thus the induced mapping of R onto R' is the identity, and the conformality of f follows.

1.5. Module of a Ring Domain

A doubly connected domain in the extended plane is called a *ring domain*. Unlike simply connected domains, which fall into three conformal equivalence classes, ring domains possess infinitely many conformal equivalence classes. A counterpart for Riemann's mapping theorem says that a ring domain can always be mapped conformally onto an annulus $r < |z| < R$, where $r \geq 0$, $R \leq \infty$. It follows that every ring domain B is conformally equivalent to one of the following annuli: $1°$ $0 < |z| < \infty$, $2°$ $1 < |z| < \infty$, $3°$ $1 < |z| < R$, $R < \infty$. In case $3°$ the number R determines the equivalence class, and

$$M(B) = \log R$$

is called the module of B. In cases $1°$ and $2°$ the module of B is said to be infinite. A conformal mapping of B onto an annulus is called a *canonical mapping* of B.

Just as in the case of quadrilaterals, the module of ring domains can also be defined without reference to canonical mappings. Let Γ now be the family of all rectifiable Jordan curves in a ring domain B which separate the boundary components of B. As before, P is the family of all non-negative Borel-measurable functions in B with $\int_\gamma \rho |dz| \geq 1$ for every $\gamma \in \Gamma$. Then

$$M(B) = 2\pi \inf_{\rho \in P} m_\rho(B).$$

By use of this formula, many geometrically more or less obvious statements can be rigorously founded. We list here some applications. The first result says that the module cannot be large if neither of the boundary com-

ponents is small in the spherical metric: *Let B be a ring domain which sepa-*
rates the points a_1, b_1 from the points a_2, b_2. If the spherical distance between
a_i and b_i, $i = 1, 2$, is $\geq \delta$, then

$$M(B) \leq \pi^2/2\delta^2 \tag{1.9}$$

The second estimate shows that if the boundary components are close to
each other and none of them is small, then the module is small. More precisely: *Let B be a ring domain whose boundary components have spherical*
diameters $> \delta$ and a mutual spherical distance $< \varepsilon < \delta$. Then

$$M(B) \leq \pi^2/\log\frac{\tan(\delta/2)}{\tan(\varepsilon/2)}. \tag{1.10}$$

For the proofs of (1.9) and (1.10), see [LV], p. 34.
    The third inequality solves an extremal problem. We introduce the *Grötzsch*
ring domain whose boundary components are the unit circle and the line
segment $\{x | 0 \leq x \leq r\}$, $0 < r < 1$; let $\mu(r)$ denote its module. *If B is a ring*
domain separating the unit circle from the points 0 and r, then

$$M(B) \leq \mu(r).$$

This was proved by Grötzsch in 1928 ([LV], p. 54).
    A simple application of the reflection principle shows that the *Teichmüller*
ring domain B bounded by the line segments $-r_1 \leq x \leq 0$ and $x \geq r_2$ has the
module

$$M(B) = 2\mu((r_1/(r_1 + r_2))^{1/2}). \tag{1.11}$$

This domain is also connected with an extremal problem: *If the ring domain*
B separates the points 0 and z_1 from the points z_2 and ∞, then

$$M(B) \leq 2\mu((|z_1|/(|z_1| + |z_2|))^{1/2}). \tag{1.12}$$

Inequality (1.12) generalizes a result of Teichmüller; for the proof we refer
to [LV], p. 56.

1.6. Module of a Path Family

We showed above that the modules of quadrilaterals and ring domains can
be defined with the aid of certain path families. We shall now consider the
more general situation in which an arbitrary family of paths is given.
    By our terminology, a path is a continuous mapping of an interval into the
plane and a curve the image of the interval under a path. We feel free not to
make a very clear distinction between a path and a curve, if there is no fear
of confusion, e.g., to use the same symbol for a path and its image.
    Let A be a domain and Γ a family of paths in A. We associate with Γ the
class P of non-negative Borel measurable functions ρ in A which satisfy the
condition

$$\int_\gamma \rho|dz| \geq 1$$

for every locally rectifiable γ in Γ; such a ρ is said to be admissible for Γ. The number

$$M(\Gamma) = \inf_{\rho \in P} \int\int_A \rho^2 \, dx \, dy$$

is called the *module of the path family* Γ.

The module of a quadrilateral and of a ring domain are special cases of this notion, whose properties are studied in [LV], pp. 132–136.

If $f: A \to A'$ is a homeomorphism, we define $f(\Gamma) = \{f \circ \gamma | \gamma \in \Gamma\}$. It follows from the definition of the module that if f is conformal, then $M(f(\Gamma)) = M(\Gamma)$, i.e., *the module of a path family is conformally invariant*.

2. Geometric Definition of Quasiconformal Mappings

2.1. Definitions of Quasiconformality

Given a domain A, consider all quadrilaterals $Q(z_1, z_2, z_3, z_4)$ with $\bar{Q} \subset A$. Let $f: A \to A'$ be a sense-preserving homeomorphism. The number

$$\sup_Q \frac{M(f(Q)(f(z_1), f(z_2), f(z_3), f(z_4)))}{M(Q(z_1, z_2, z_3, z_4))}$$

is called the *maximal dilatation* of f. It is always ≥ 1, because the modules of $Q(z_1, z_2, z_3, z_4)$ and $Q(z_2, z_3, z_4, z_1)$ are reciprocals.

Since the module is a conformal invariant, the maximal dilatation of a conformal mapping is 1. By Theorem 1.1, the converse is also true: if the maximal dilatation of f is equal to 1, then f is conformal. From this observation we arrive conveniently at the notion of quasiconformality.

Definition. A sense-preserving homeomorphism with a finite maximal dilatation is quasiconformal. If the maximal dilatation is bounded by a number K, the mapping is said to be K-quasiconformal.

This "geometric" definition of quasiconformality was suggested by Pfluger [1] in 1951, and its first systematic use was by Ahlfors [1] in 1953.

By this terminology, f is 1-quasiconformal if and only if f is conformal. If f is K-quasiconformal, then $M(f(Q)) \geq M(Q)/K$ for every quadrilateral in A. A mapping f and its inverse f^{-1} are simultaneously K-quasiconformal. From the definition it also follows that if $f: A \to B$ is K_1-quasiconformal and $g: B \to C$ is K_2-quasiconformal, then $g \circ f$ is $K_1 K_2$-quasiconformal.

The definition of quasiconformality could equally well have been given in terms of the modules of ring domains: *A sense-preserving homeomorphism f of a domain A is K-quasiconformal if and only if the module condition*

$$M(f(B)) \leq KM(B) \tag{2.1}$$

holds for every ring domain B, $\bar{B} \subset A$.

The necessity of the condition can be established easily if the canonical annulus of B, cut along a line segment joining the boundary components, is transformed to a rectangle with the aid of the logarithm. To prove the sufficiency requires somewhat more elaborate module estimations ([LV], p. 39).

Inequality (2.1) shows that a quasiconformal mapping cannot "blow up" a point, and the well known result on the removability of isolated singularities of conformal mappings can be readily generalized ([LV], p. 41): *A K-quasiconformal mapping of a domain A with an isolated boundary point a can be extended to a K-quasiconformal mapping of $A \cup \{a\}$.*

We mention here another generalized extension theorem (cf. 1.2): *A quasiconformal mapping of a Jordan domain onto another Jordan domain can be extended to a homeomorphism between the closures of the domains.* This can be proved by a modification of the proof for conformal mappings ([LV], p. 42), or deduced directly from the corresponding result for conformal mappings by use of Theorem 4.4 (Existence theorem for Beltrami equations).

The modules of quadrilaterals and ring domains, which we used to characterize quasiconformality, are modules of certain path families. As a matter of fact, the following general result, proved by Väisälä in 1961, is true.

A K-quasiconformal mapping f of a domain A satisfies the inequality

$$M(f(\Gamma)) \leq KM(\Gamma) \tag{2.2}$$

for every path family Γ of A.

Hersch, one of the pioneers in applying curve families to quasiconformal mappings, asked as early as 1955 in his thesis whether (2.2) could be true for all path families. At that time, certain "analytic" properties of quasiconformal mappings, which the proof ([LV], p. 171) seems to require, were not yet known. These properties will be discussed in section 3. The reason we mention the result (2.2) here is that we could have taken it as a definition for K-quasiconformality. In a way, a characterization by means of the general relation (2.2) is more satisfactory than the definition which is based on the special notion of the module of a quadrilateral. However, we have preferred to use quadrilaterals, not merely for historical reasons, but also to remain true to the presentation in the monograph [LV], to which repeated references are being made.

2.2. Normal Families of Quasiconformal Mappings

Let us consider a family whose elements are mappings of a plane domain A into the plane. Such a family is said to be *normal* if every sequence of its elements contains a subsequence which is locally uniformly convergent in A. To cover the possibility that ∞ lies in A or in the range of the mappings, we use the spherical metric. If ∞ is not there, we can of course switch to the topologically equivalent euclidean metric.

If a family is equicontinuous, then it is normal. This is the result on which the proofs for a family to be normal are usually based in complex analysis. For instance, if the family consists of uniformly bounded analytic functions, equicontinuity follows immediately from Cauchy's integral formula, and so normality can be deduced. A generalization of this result, proved by use of the elliptic modular function, says that if the functions are meromorphic and omit the same three values, then the family is normal.

Much less is needed for normality if the functions are assumed to be injective.

Lemma 2.1. *Let F be a family of K-quasiconformal mappings of a domain A. If every $f \in F$ omits two values which have a mutual spherical distance $\geq d > 0$, then F is equicontinuous in A.*

PROOF. Let s denote the spherical distance. Given an ε, $0 < \varepsilon < d$, and a point $z_0 \in A$, we consider a ring domain $B = \{z \mid \delta < s(z, z_0) < r\}$ with $\{z \mid s(z, z_0) < r\} \subset A$, and choose $\delta > 0$ so small that $M(B) > \pi^2 K / 2\varepsilon^2$. Let z_1 be an arbitrary point in the neighborhood $V = \{z \mid s(z, z_0) < \delta\}$ of z_0.

Let us consider an $f \in F$. By assumption, f omits two values a and b with $s(a, b) \geq d$. The ring domain $f(B)$ separates the points $f(z_0)$, $f(z_1)$ from the points a, b. If $\eta = \min(d, s(f(z_0), f(z_1)))$, it follows from formula (1.9) that $M(f(B)) \leq 2\pi^2 / \eta^2$. This yields $\eta < \varepsilon$ and hence the desired estimate $s(f(z_0), f(z)) < \varepsilon$ whenever $z \in V$, for every $f \in F$.   □

Lemma 2.1 yields various criterions for a family to be equicontinuous and hence normal. The following will come into use several times.

Theorem 2.1. *A family F of K-quasiconformal mappings of a domain A is equicontinuous and normal, if for three fixed points z_1, z_2, z_3 of A and for every $f \in F$, the distances $s(f(z_i), f(z_j))$ are uniformly bounded away from zero for $i, j = 1, 2, 3, i \neq j$.*

PROOF. By Lemma 2.1, the family F is equicontinuous in $A \setminus \{z_i, z_j\}$, $i, j = 1, 2, 3, i \neq j$, and hence throughout A.   □

2.3. Compactness of Quasiconformal Mappings

Let (f_n) be a sequence of K-quasiconformal mappings of a domain A which is locally uniformly convergent in A. If the limit function f is not constant, it must take at least three different values, because it is continuous. It follows from Theorem 2.1 that the functions f_n constitute an equicontinuous family.

K-quasiconformal mappings possess the same compactness property as conformal mappings.

Theorem 2.2. *The limit function f of a sequence (f_n) of K-quasiconformal mappings of a domain A, locally uniformly convergent in A, is either a constant or a K-quasiconformal mapping.*

PROOF. If f is a homeomorphism, then it follows easily from the definition of K-quasiconformality and from the continuity of the module of quadrilaterals that f is K-quasiconformal ([LV], p. 29). A continuous injective map of an open set of the plane into the plane is a homeomorphism (Newman [1], p. 122). Therefore, it is sufficient to show that a non-constant limit function f is injective. This we can prove, utilizing the fact that the family $\{f_n\}$ is equicontinuous, with the aid of the module estimate (1.10) ([LV], p. 74). □

In case every f_n maps A onto a fixed domain A', more can be said about the limit function.

Theorem 2.3. *Let A be a domain with at least two boundary points and (f_n) a sequence of K-quasiconformal mappings of A onto a fixed domain A'. If the sequence (f_n) converges in A, then the limit function is either a K-quasiconformal mapping of A onto A', or a mapping of A onto a boundary point of A'.*

Here we need not assume that (f_n) is locally uniformly convergent, because we conclude from Lemma 2.1 that $\{f_n\}$ is a normal family. The theorem follows from equicontinuity and normal family arguments ([LV], p. 78).

In 4.6 we shall study the convergence of K-quasiconformal mappings f_n more closely. It turns out that, even though the mappings f_n tend uniformly towards a K-quasiconformal limit f, the local mapping properties of f and the approximating functions f_n may be quite different.

2.4. A Distortion Function

In later applications we shall often encounter a distortion function which we shall now introduce, starting from its simple geometric interpretation.

Let F be the family of K-quasiconformal mappings of the plane which map the real axis onto itself and fix the points $-1, 0$ and ∞. By Theorem 2.1, F is a normal family, and so

$$\lambda(K) = \max\{f(1)|f \in F\} \tag{2.3}$$

exists. This defines our distortion function λ, for which we shall now derive a more explicit expression.

Consider the quadrilateral $H(-1, 0, 1, \infty)$, where H is the upper half-plane.

The Möbius transformation $z \to (1 + z)/(1 - z)$ maps it onto the quadrilateral $H(0, 1, \infty, -1)$. Hence, these two quadrilaterals have the same module. On the other hand, the modules are reciprocal, and so $M(H(-1, 0, 1, \infty)) = M(H(0, 1, \infty, -1)) = 1$.

Let us choose an $f \in F$ and write $t = f(1)$. We form the Teichmüller ring B bounded by the line segments $-1 \leq x \leq 0$ and $x \geq t$. If B is conformally equivalent to the annulus $A = \{z \mid 1 < |z| < R\}$, then the canonical mapping of B can be so chosen that the upper half of B is conformally equivalent to the upper half of A. By applying the mapping $z \to \log z$ we conclude that

$$M(H(0, t, \infty, -1)) = M(B)/\pi.$$

By formula (1.11), $M(B) = 2\mu((1 + t)^{-1/2})$. Since f is K-quasiconformal, $M(H(0, t, \infty, -1)) \leq KM(H(0, 1, \infty, -1)) = K$. If we combine all these estimates, we obtain

$$t = f(1) \leq (\mu^{-1}(\pi K/2))^{-2} - 1. \tag{2.4}$$

In order to show that there is an f for which equality holds, we first make a general remark: Let f be a homeomorphism of a domain A and I a closed line segment which lies in A with the possible exception of its endpoints. Then f has the same maximal dilatation in A and in $A \backslash I$. This can be proved by means of Rengel's inequality ([LV], p. 45), or by making use of the analytic characterization of quasiconformality which will be given in 3.5. It follows that the *reflection principle* for conformal mappings generalizes as such for K-quasiconformal mappings. In particular, a K-quasiconformal self-mapping of the upper half-plane can be extended by reflection in the real axis to a K-quasiconformal mapping of the plane.

Let us now return to (2.4). Let f_1 be the canonical mapping of the quadrilateral $H(0, 1, \infty, -1)$ onto the square $Q(0, 1, 1 + i, i)$, α the affine stretching $x + iy \to Kx + iy$, and f_2 the canonical mapping of $H(0, t, \infty, -1)$ onto the rectangle $R(0, K, K + i, i)$. Then the mapping f which is equal to $f_2^{-1} \circ \alpha \circ f_1$ in H and its mirror image in the lower half-plane is K-quasiconformal in the plane. For this f, equality holds in (2.4). It follows that

$$\lambda(K) = (\mu^{-1}(\pi K/2))^{-2} - 1.$$

From the obvious result $\lambda(1) = 1$ we conclude that

$$\mu(1/\sqrt{2}) = \pi/2. \tag{2.5}$$

Typically the reasoning goes in the other direction, in that we retrieve information about $\lambda(K)$ by estimating $\mu(r)$. For instance, we obtain in this way

$$\lambda(K) = \tfrac{1}{16}e^{\pi K} - \tfrac{1}{2} + o(1), \tag{2.6}$$

with a positive remainder term $o(1)$ as $K \to \infty$ ([LV], p. 82). Also,

$$\lambda(K) \leq \exp(4.39(K - 1))$$

(Beurling–Ahlfors [1]), which tells about the behavior of $\lambda(K)$ as $K \to 1$. In particular, we see that λ is continuous at $K = 1$; continuity at an arbitrary K follows from the continuity of $r \to \mu(r)$.

2.5. Circular Distortion

A conformal mapping of the plane which fixes 0 and ∞ maps the family of circles centered at the origin onto itself. Under normalized K-quasiconformal mappings the images of these circles have "bounded distortion". To be precise:

Theorem 2.4. *Let f be a K-quasiconformal mapping of the plane fixing 0 and ∞. Then for every $r > 0$,*

$$\frac{\max_\varphi |f(re^{i\varphi})|}{\min_\varphi |f(re^{i\varphi})|} \leq c(K), \tag{2.7}$$

where the constant $c(K)$ depends only on K.

There are many ways to prove this important theorem. A normal family argument shows that a finite bound $c(K)$ must exist. A quantitative estimate is obtained as follows. Let z_1 and z_2 be points on the circle $|z| = r$ at which the minimum and maximum of $|f(z)|$ are attained. For $B' = \{w | \min_\varphi |f(re^{i\varphi})| < |w| < \max_\varphi |f(re^{i\varphi})|\}$, let B be the inverse image of B'. Then B separates the points 0, z_1 from the points z_2, ∞. Hence, by (1.12) and (2.5),

$$M(B) \leq 2\mu((|z_1|/(|z_1| + |z_2|))^{1/2}) = 2\mu(1/\sqrt{2}) = \pi.$$

Consequently, $M(B') \leq KM(B) \leq \pi K$, and it follows that (2.7) holds for $c(K) = e^{\pi K}$.

The sharp bound in (2.7) is $\lambda(K)$ (proved by Lehto–Virtanen–Väisälä in 1959). This follows from the fact that, as a generalization of (2.3), $\lambda(K)$ is the maximum of $|f(z)|$ on the unit circle in the family of K-quasiconformal mappings of the plane which fix $-1, 0$ and ∞ but which are not required to map the real axis onto itself. There seems to be no easy way to prove this result.

Theorem 2.4, in a form in which the value of the sharp bound is not needed, will render us valuable service in section 6 when we study the geometry of quasidiscs. With these applications in mind, we draw here a further conclusion from (2.7).

Let f be a K-quasiconformal mapping of the plane fixing ∞. We infer from (2.7) that if $|z_2 - z_0| \leq |z_1 - z_0|$, then

$$|f(z_2) - f(z_0)| \leq c(K)|f(z_1) - f(z_0)|. \tag{2.8}$$

The following generalization is also readily obtained. If $|z_2 - z_0| \leq n|z_1 - z_0|$, where $n \geq 1$ is an integer, then

$$|f(z_2) - f(z_0)| \leq nc(K)^n |f(z_1) - f(z_0)|. \tag{2.9}$$

In order to prove this, we denote by ζ_k, $k = 0, 1, \ldots, n$, the equidistant points on the ray from z_0 through z_2 for which $|\zeta_k - \zeta_{k-1}| = |z_2 - z_0|/n$; here $\zeta_0 = z_0$, $\zeta_n = z_2$. By (2.8), $|f(\zeta_k) - f(\zeta_{k-1})| \leq c(K)|f(\zeta_{k-1}) - f(\zeta_{k-2})|$ for $k = 2, \ldots, n$. Hence, by the triangle inequality,

$$|f(z_2) - f(z_0)| \leq |f(\zeta_1) - f(z_0)| \sum_{k=0}^{n-1} c(K)^k \leq nc(K)^{n-1}|f(\zeta_1) - f(z_0)|.$$

By (2.8), $|f(\zeta_1) - f(z_0)| \leq c(K)|f(z_1) - f(z_0)|$, and (2.9) follows.

We conclude this section with the remark that quasiconformality can be defined by means of the distortion function H,

$$H(z) = \limsup_{r \to 0} \frac{\max_\varphi |f(z + re^{i\varphi}) - f(z)|}{\min_\varphi |f(z + re^{i\varphi}) - f(z)|},$$

even though we shall not make use of this characterization. *A sense-preserving homeomorphism f of a domain A is K-quasiconformal if and only if H is bounded in $A \backslash \{\infty, f^{-1}(\infty)\}$ and $H(z) \leq K$ almost everywhere in A* ([LV], pp. 177–178).

3. Analytic Definition of Quasiconformal Mappings

3.1. Dilatation Quotient

When the definition of quasiconformality in terms of the modules of quadrilaterals was given in the early fifties, quasiconformal mappings had been studied and successfully applied in complex analysis for more than two decades. Historically, the starting point for generalizing conformal mappings was to consider, not arbitrary sense-preserving homeomorphisms, but *diffeomorphisms,* i.e., homeomorphisms which with their inverses are continuously differentiable. We can then generalize the characteristic property of conformal mappings that the derivative is independent of the direction by requiring that the ratio of the maximum and minimum of the absolute value of the directed derivatives at a point is uniformly bounded.

We shall now show that this classical definition gives precisely those quasiconformal mappings which are diffeomorphic. This local approach using derivatives is often much more convenient than the definition using modules of quadrilaterals when the problem is checking the quasiconformality of a mapping given by an analytic expression.

To make the above remarks precise, we introduce for a sense-preserving diffeomorphism f the complex derivatives

$$\partial f = \tfrac{1}{2}(f_x - if_y), \qquad \bar{\partial} f = \tfrac{1}{2}(f_x + if_y),$$

and the derivative $\partial_\alpha f$ in the direction α:

$$\partial_\alpha f(z) = \lim_{r \to 0} \frac{f(z + re^{i\alpha}) - f(z)}{re^{i\alpha}}.$$

Then $\partial_\alpha f = \partial f + \bar{\partial} f e^{-2i\alpha}$, and so

$$\max_\alpha |\partial_\alpha f(z)| = |\partial f(z)| + |\bar{\partial} f(z)|, \quad \min_\alpha |\partial_\alpha f(z)| = |\partial f(z)| - |\bar{\partial} f(z)|.$$

The difference $|\partial f(z)| - |\bar{\partial} f(z)|$ is positive, because the Jacobian $J_f = |\partial f|^2 - |\bar{\partial} f|^2$ is positive for a sense-preserving diffeomorphism. We conclude that the *dilatation quotient*

$$D_f = \frac{\max_\alpha |\partial_\alpha f|}{\min_\alpha |\partial_\alpha f|} = \frac{|\partial f| + |\bar{\partial} f|}{|\partial f| - |\bar{\partial} f|}$$

is finite.

The mapping f is conformal if and only if $\bar{\partial} f$ vanishes identically. Then $\partial_\alpha f$ is independent of α: we have $\partial_\alpha f = \partial f = f'$. This is equivalent to the dilatation quotient being identically equal to 1.

The dilatation quotient is conformally invariant: If g and h are conformal mappings such that $w = h \circ f \circ g$ is defined, then direct computation shows that $D_f(z) = D_w(g^{-1}(z))$.

3.2. Quasiconformal Diffeomorphisms

For diffeomorphisms quasiconformality can be characterized with the aid of the dilatation quotient.

Theorem 3.1. *Let $f: A \to A'$ be a sense-preserving diffeomorphism with the property*

$$D_f(z) \leq K$$

for every $z \in A$. Then f is a K-quasiconformal mapping.

PROOF. We pick an arbitrary quadrilateral Q of A. Let w be the mapping which is induced from the canonical rectangle $R(0, M, M + i, i)$ of Q onto the canonical rectangle $R'(0, M', M' + i, i)$ of $f(Q)$. Because of the conformal invariance of the dilatation quotient, D_w is also majorized by K. Hence $|w_x|^2 \leq \max |\partial_\alpha w|^2 \leq KJ_w$, and the desired result $M' \leq KM$ follows by use of a customary length-area reasoning:

$$M' = m(R') = \int\int_R J_w(z)\,dx\,dy \ge \frac{1}{K}\int\int_R |w_x(z)|^2\,dx\,dy$$

$$\ge \frac{1}{MK}\int_0^1 dy\left(\int_0^M |w_x(z)|\,dx\right)^2 \ge M'^2/MK. \qquad \square$$

Theorem 3.1 is the classical definition of K-quasiconformality given by Grötzsch [1] in 1928.

The converse to Theorem 3.1 is as follows:

Theorem 3.2. *Let $f: A \to A'$ be a K-quasiconformal mapping. If f is differentiable at $z_0 \in A$, then*

$$\max_\alpha |\partial_\alpha f(z_0)| \le K \min_\alpha |\partial_\alpha f(z_0)|. \qquad (3.1)$$

The idea of the proof is to consider a small square Q centered at z_0 and regard it as a quadrilateral with the vertices at distinguished points. The area and the distance of the sides of $f(Q)$ can be approximated by expressions involving the partial derivatives of f at z_0. Application of Rengel's inequality then yields a lower estimate for $M(f(Q))$ from which the desired inequality (3.1) follows. (For details we refer to [LV], p. 50.)

By combining Theorems 3.1 and 3.2 we obtain the following characterization for quasiconformal diffeomorphisms: *A sense-preserving diffeomorphism f is K-quasiconformal if and only if the dilatation condition $D_f(z) \le K$ holds everywhere.*

The class of K-quasiconformal diffeomorphisms does not possess the compactness property of Theorem 2.2. This is one of the reasons for replacing the classical definition of Grötzsch by the more general one. Another reason will be discussed in 4.5.

3.3. Absolute Continuity and Differentiability

We shall soon see that an arbitrary quasiconformal mapping of a domain A is differentiable almost everywhere in A. From Theorem 3.2 it then follows that the dilatation condition (3.1) is true at almost all points of A. However, the converse is not true, i.e., a sense-preserving homeomorphism f which is differentiable a.e. and satisfies (3.1) a.e. is not necessarily K-quasiconformal. What is required is a notion of absolute continuity.

A continuous real-valued function u is said to be *absolutely continuous on lines* (ACL) in a domain A if for each closed rectangle $\{x + iy | a \le x \le b,$ $c \le y \le d\} \subset A$, the function $x \to u(x + iy)$ is absolutely continuous on $[a, b]$ for almost all $y \in [c, d]$ and $y \to u(x + iy)$ is absolutely continuous on $[c, d]$ for almost all $x \in [a, b]$. A complex valued function is ACL in A if its real and imaginary parts are ACL in A.

It follows from standard theorems of real analysis that a function f which is ACL in A has finite partial derivatives f_x and f_y a.e. in A.

Theorem 3.3. *A quasiconformal mapping is absolutely continuous on lines.*

This result was first established by Strebel (1955) and Mori. A later proof by Pfluger, which uses Rengel's inequality and a minimum of real analysis, is presented in [LV], p. 162.

From Theorem 3.3 we conclude that a quasiconformal mapping has finite partial derivatives a.e. From this we can draw further conclusions by making use of the following result:

Let f be a complex-valued, continuous and open mapping of a plane domain A which has finite partial derivatives a.e. in A. Then f is differentiable a.e. in A.

The proof, which is due to Gehring and Lehto (1959), uses the maximum principle and a standard theorem on the density of point sets ([LV], p. 128). Application to quasiconformal mappings yields, with regard to Theorem 3.2, a basic result:

Theorem 3.4. *A K-quasiconformal mapping f of a domain A is differentiable and satisfies the dilatation condition* (3.1) *almost everywhere in A.*

Differentiability a.e. of quasiconformal mappings was first proved by Mori [1] with the aid of the Rademacher–Stepanoff theorem and Theorem 2.4.

3.4. Generalized Derivatives

The ACL-property, which depends on the coordinate system, becomes much more useful when combined with local integrability of the derivatives. A function f is said to possess (generalized) L^p-*derivatives* in a domain A, $p \geq 1$, if f is ACL in A and if the partial derivatives f_x and f_y of f are L^p-integrable locally in A. It is also customary to say that the function f then belongs to the Sobolev space $W^1_{p,\text{loc}}$. This property is preserved under continuously differentiable changes of coordinates ([LV], pp. 151–152).

Roughly speaking, classical transformation rules of Calculus between curve and surface integrals remain valid for functions with L^p-derivatives. This is one reason for the importance of this class of functions. (For details and more information see, e.g., [LV], pp. 143–154 or Lehto [4], pp. 127–131.)

A quasiconformal mapping has L^2-derivatives. In order to prove this we first note that the dilatation condition (3.1) implies the inequality

$$\max_\alpha |\partial_\alpha f(z)|^2 \leq K J(z).$$

In particular, $|f_x(z)|^2 \leq K J(z), |f_y(z)|^2 \leq K J(z)$ a.e. The Jacobian of an almost everywhere differentiable homeomorphism is locally integrable ([LV], p. 131). Consequently, f_x and f_y are locally L^2-integrable.

A homeomorphism with L^2-derivatives is absolutely continuous with respect to two-dimensional Lebesgue measure ([LV], p. 150). Thus quasiconformal mappings have this property. They carry sets measurable with respect to two-dimensional measure onto other sets in this class. The formula

$$\int_E J = m(f(E)) \tag{3.2}$$

holds for every quasiconformal mapping f of a domain A and for every measurable set $E \subset A$. If we apply (3.2) to the inverse mapping f^{-1}, we deduce that for a quasiconformal mapping $J(z) > 0$ almost everywhere.

The considerations in 4.4 will show that every quasiconformal mapping has not only L^2-derivatives but actually L^p-derivatives for some $p > 2$. This is a much deeper result than the existence of L^2-derivatives.

3.5. Analytic Characterization of Quasiconformality

A simple counterexample, constructed with the help of Cantor's function, shows that a homeomorphism need not be quasiconformal even though it is differentiable a.e., satisfies (3.1) a.e. with $K = 1$, has bounded partial derivatives, and is area preserving ([LV], p. 167). What is required is the ACL-property, but once this is assumed it together with (3.1) guarantees quasiconformality.

Theorem 3.5. *A sense-preserving homeomorphism f of a domain A is K-quasiconformal if*

1° *f is ACL in A;*
2° *$\max_\alpha |\partial_\alpha f(z)| \leq K \min_\alpha |\partial_\alpha f(z)|$ a.e. in A.*

PROOF. We first note that being ACL, the mapping f has partial derivatives a.e. and, as a homeomorphism, is therefore differentiable a.e. Thus condition 2° makes sense. As above, we conclude that f has L^2-derivatives. After this, we can follow the proof of Theorem 3.1, apart from obvious modifications. ☐

Theorem 3.5 is called the *analytic definition* of quasiconformality. Apparently different from the equivalent geometric definition, it sheds new light on the connection with the classical Grötzsch mappings. In the next section we shall show that the analytic definition can be written in the form of a differential equation. This leads to essentially new problems and results for quasiconformal mappings.

Under the additional hypotheses that f is differentiable a.e. and has

L^1-derivatives, Theorem 3.5 was first established by Yûjôbô in 1955. Later Bers and Pfluger relaxed the a priori requirements. Under the above minimal conditions, the theorem was proved by Gehring and Lehto in 1959. (For references and more details, see [LV], p. 169.)

4. Beltrami Differential Equation

4.1. Complex Dilatation

Inequality (3.1) is a basic property of quasiconformal mappings. The very natural step to express therein more explicitly the maximum and minimum of $|\partial_\alpha f(z)|$ leads to the important notion of complex dilatation and reveals a connection between the theories of quasiconformal mappings and partial differential equations.

Let $f: A \to A'$ be a K-quasiconformal mapping and $z \in A$ a point at which f is differentiable. Since $\max|\partial_\alpha f| = |\partial f| + |\bar{\partial} f|$, $\min|\partial_\alpha f| = |\partial f| - |\bar{\partial} f|$, the dilatation condition (3.1) is equivalent to the inequality

$$|\bar{\partial} f(z)| \leq \frac{K-1}{K+1}|\partial f(z)|. \tag{4.1}$$

Suppose, in addition, that $J_f(z) > 0$. Then $\partial f(z) \neq 0$, and we can form the quotient

$$\mu(z) = \frac{\bar{\partial} f(z)}{\partial f(z)}.$$

The function μ, so defined a.e. in A, is called the *complex dilatation* of f. Since f is continuous, μ is a Borel-measurable function, and from (4.1) we see that

$$|\mu(z)| \leq \frac{K-1}{K+1} < 1. \tag{4.2}$$

Complex dilatation will play a very central role in our representation. It has a simple geometric interpretation. At a point z at which μ is defined, the mapping

$$\zeta \to f(z) + \partial f(z)(\zeta - z) + \bar{\partial} f(z)(\bar{\zeta} - \bar{z})$$

is a non-degenerate affine transformation which maps circles centered at z onto ellipses centered at $f(z)$. The ratio of the major axis to the minor axis of the image ellipses is equal to $(1 + |\mu(z)|)/(1 - |\mu(z)|)$. We see that the smaller $|\mu(z)|$ is, the less the mapping f deviates from a conformal mapping at the point z. If $\mu(z) \neq 0$, the argument of $\mu(z)$ determines the direction of maximal stretching: $|\partial_\alpha f(z)|$ assumes its maximum when $\alpha = \arg \mu(z)/2$.

4.2. Quasiconformal Mappings and the Beltrami Equation

The definition of complex dilatation leads us to consider differential equations

$$\bar{\partial}f = \mu \partial f. \tag{4.3}$$

An equation (4.3), where μ is measurable and $\|\mu\|_\infty < 1$, is called a *Beltrami equation*. If f is conformal, μ vanishes identically, and (4.3) becomes the Cauchy–Riemann equation $\bar{\partial}f = 0$.

A function f is said to be an L^p-solution of (4.3) in a domain A if f has L^p-derivatives and (4.3) holds a.e. in A.

Theorem 4.1. *A homeomorphism f is K-quasiconformal if and only if f is an L^2-solution of an equation $\bar{\partial}f = \mu \partial f$, where μ satisfies (4.2) for almost all z.*

Proof. The necessity follows from Theorem 3.4 and the sufficiency from Theorem 3.5, when we note that $\|\mu\|_\infty < 1$ implies that f is sense-preserving. $\square$

The Beltrami equation has a long history. With a smooth coefficient μ, it was considered in the 1820's by Gauss in connection with the problem of finding isothermal coordinates for a given surface (cf. IV.1.6). As early as 1938, Morrey [1] systematically studied homeomorphic L^2-solutions of the equation (4.3). But it took almost twenty years until in 1957 Bers [1] observed that these solutions are quasiconformal mappings.

In 4.5 it will become apparent that (4.3) always has homeomorphic solutions, i.e., that the complex dilatation of a quasiconformal mapping can be prescribed almost everywhere. This is a deep result. It is much easier to handle the question of the uniqueness of the solutions of (4.3).

Let f and g be quasiconformal mappings of a domain A with complex dilatations μ_f and μ_g. Direct computation yields the transformation formula

$$\mu_{f \circ g^{-1}}(\zeta) = \frac{\mu_f(z) - \mu_g(z)}{1 - \mu_f(z)\overline{\mu_g(z)}} \left(\frac{\partial g(z)}{|\partial g(z)|} \right)^2, \qquad \zeta = g(z), \tag{4.4}$$

valid for almost all $z \in A$, and hence for almost all $\zeta \in g(A)$.

Theorem 4.2 (Uniqueness Theorem). *Let f and g be quasiconformal mappings of a domain A whose complex dilatations agree a.e. in A. Then $f \circ g^{-1}$ is a conformal mapping.*

Proof. By (4.4), the complex dilatation of $f \circ g^{-1}$ vanishes a.e. From Theorem 3.5 we deduce that $f \circ g^{-1}$ is 1-quasiconformal. Hence, by Theorem 1.1, it is conformal. $\square$

Conversely, if $f \circ g^{-1}$ is conformal, we conclude from (4.4) that f and g have the same complex dilatation.

4.3. Singular Integrals

The Uniqueness theorem says that a quasiconformal mapping of the plane is determined by its complex dilatation μ up to an arbitrary Möbius transformation. It follows that a suitably normalized mapping is uniquely determined by μ. We shall now show that, by use of singular integrals, it is possible to derive a formula which gives the values of a normalized quasiconformal mapping in terms of μ.

Let f be a function with L^1-derivatives in a domain A of the $\zeta = \xi + i\eta$-plane and $D, \bar{D} \subset A$, a Jordan domain with a rectifiable boundary curve. Application of Green's formula yields a generalized Cauchy integral formula ([LV], p. 155)

$$f(z) = \frac{1}{2\pi i} \int_{\partial D} \frac{f(\zeta)}{\zeta - z} d\zeta - \frac{1}{\pi} \iint_D \frac{\bar{\partial} f(\zeta)}{\zeta - z} d\xi \, d\eta, \qquad z \in D. \tag{4.5}$$

The first term on the right, a Cauchy integral, defines an analytic function in D. We conclude, in passing, that *a function f with L^1-derivatives in A is analytic if $\bar{\partial} f = 0$ a.e. in A*. The second term on the right in (4.5) is to be understood as a Cauchy principal value.

Suppose that f has L^1-derivatives in the complex plane $\mathbb{C}$ and that $f(z) \to 0$ as $z \to \infty$. If we take $D = \{\zeta | |\zeta| < R\}$ and let $R \to \infty$, the first term on the right-hand side of (4.5) tends to zero for every fixed z. With the notation

$$T\omega(z) = -\frac{1}{\pi} \iint_{\mathbb{C}} \frac{\omega(\zeta)}{\zeta - z} d\xi \, d\eta, \tag{4.6}$$

we then obtain from (4.5)

$$f = T\bar{\partial} f. \tag{4.7}$$

Assume, for a moment, that ω in (4.6) belongs to class C_0^∞ in the complex plane, i.e., ω is infinitely many times differentiable and has a bounded support. Straightforward computation then shows that

$$\partial T\omega = H\omega \tag{4.8}$$

([LV], pp. 155–157), where

$$H\omega(z) = -\frac{1}{\pi} \iint_{\mathbb{C}} \frac{\omega(\zeta)}{(\zeta - z)^2} d\xi \, d\eta,$$

the integral again being defined as a Cauchy principal value. The linear operator H is called the *Hilbert transformation*. We also see that

$$\bar{\partial} T\omega = \omega,$$

that the operators ∂ and $\bar{\partial}$ commute with T and H, and that $T\omega$ and $H\omega$ belong to C^∞ and are analytic outside the support of ω ([LV], p. 157).

The Hilbert transformation can be extended as a bounded operator to L^p, $1 < p < \infty$. One first proves that if $\omega \in C_0^\infty$, there exists a constant A_p, not depending on ω, such that

$$\|H\omega\|_p \le A_p \|\omega\|_p. \tag{4.9}$$

This is called the Calderón–Zygmund inequality; proofs are given, apart from the original paper by Calderón and Zygmund (1952), in Vekua [1], Ahlfors [5] and Stein [1]. Since C_0^∞ is dense in L^p and L^p is complete, we can use (4.9) to extend the Hilbert transformation to the whole space L^p. Inequality (4.9) then holds for every $\omega \in L^p$ (cf. [LV], p. 159).

Using (4.9) we deduce that (4.8) holds a.e. for every $\omega \in L^p$ (cf. [LV], p. 160). For applications it is also important to note that the norm

$$\|H\|_p = \sup\{\|H\omega\|_p \mid \|\omega\|_p = 1\}$$

depends continuously on p (Ahlfors [5], Dunford–Schwartz [1]).

The special case $p = 2$ is much easier to handle than a general p. A rather elementary integration shows that Hilbert transformation is an isometry in L^2 ([LV], p. 157). In particular, $\|H\|_2 = 1$.

4.4. Representation of Quasiconformal Mappings

We shall now apply the results of 4.3 to quasiconformal mappings. Let f be a quasiconformal mapping of the plane whose complex dilatation μ has a bounded support. Wishing to represent f by means of μ, we introduce a normalization so that μ determines f uniquely.

We first require that $f(\infty) = \infty$. Near infinity, where f is conformal, we then have $f(z) = Az + B +$ negative powers of z. If we set $A = 1$, $B = 0$, then f is uniquely determined by μ.

In a neighborhood of ∞ we thus have

$$f(z) = z + \sum_{n=1}^{\infty} b_n z^{-n}.$$

It follows that the partial derivatives of the function $z \to f(z) - z$, which are locally in L^2, are L^2-integrable over the plane. We conclude from (4.7) and from the generalized formula (4.8) that $\partial f = 1 + H \bar\partial f$ a.e. Since $\bar\partial f = \mu \partial f$ a.e. we thus have

$$\bar\partial f = \mu + \mu H \bar\partial f \quad \text{a.e.} \tag{4.10}$$

This integral equation can be solved by the customary iteration procedure. The Neumann series obtained converges in L^2, but it also converges in L^p, $p > 2$, if p satisfies the condition

$$\|\mu\|_\infty \|H\|_p < 1. \tag{4.11}$$

More explicitly, suppose that $\mu(z) = 0$ if $|z| > R$, and define inductively

$$\varphi_1 = \mu, \qquad \varphi_n = \mu H \varphi_{n-1}, \qquad n = 2, 3, \ldots. \tag{4.12}$$

Then

$$\|\varphi_i\|_p \le (\pi R^2)^{1/p} \|H\|_p^{i-1} (\|\mu\|_\infty)^i. \tag{4.13}$$

Hence, under condition (4.11),

$$\lim_{n\to\infty} \sum_{i=1}^n \varphi_i = \bar{\partial} f, \tag{4.14}$$

and it is a function in L^p.

This solution gives the desired representation formula for $f(z)$, first established by Bojarski in 1955.

Theorem 4.3. *Let f be a quasiconformal mapping of the plane whose complex dilatation μ has a bounded support and which satisfies the condition $\lim_{z\to\infty} (f(z) - z) = 0$. Then*

$$f(z) = z + \sum_{i=1}^\infty T\varphi_i(z),$$

where φ_i is defined by (4.12). The series is absolutely and uniformly convergent in the plane.

PROOF. By (4.7) we have $f(z) = z + T\bar{\partial} f(z)$. By (4.14), $T\bar{\partial} f(z) = (T \sum \varphi_i)(z)$. For $p > 2$, it follows from Hölder's inequality that $|T\varphi_i(z)| \le c_p \|\varphi_i\|_p$, where the constant c_p depends only on p and R. Therefore, by (4.13),

$$|T\varphi_i(z)| \le c_p' (\|H\|_p \|\mu\|_\infty)^i, \tag{4.15}$$

where c_p' depends only on p and R. (For this crucial estimate, (4.10) must be solved in a space L^p with $p > 2$.) We conclude from (4.15) that

$$\left(T \sum_{i=1}^\infty \varphi_i\right)(z) = \sum_{i=1}^\infty T\varphi_i(z)$$

and that the series on the right is absolutely and uniformly convergent.   □

We proved above that under condition (4.11), $\bar{\partial} f \in L^p$ locally. From $\partial f = 1 + H\bar{\partial} f$ and (4.9) we see that the same holds for ∂f. It follows that the partial derivatives of a quasiconformal mapping are locally in L^p for same $p > 2$ ([LV], p. 215). The p will, of course, depend on $\|\mu\|_\infty$.

4.5. Existence Theorem

In proving Theorem 4.3 we started from a quasiconformal mapping which gave the function μ. The following result, fundamental in the theory of quasiconformal mappings, shows that we could equally well have started from a measurable function μ.

Theorem 4.4 (Existence Theorem). *Let μ be a measurable function in a domain A with $\|\mu\|_\infty < 1$. Then there is a quasiconformal mapping of A whose complex dilatation agrees with μ a.e.*

The proof can be divided into three parts. One first shows that if $\mu \in C_0^\infty$, the Beltrami equation $\bar{\partial} w = \mu \partial w$ has a locally injective solution. This can be so constructed that a topological argument shows it to be in fact globally injective. Another way to obtain from locally injective solutions a globally injective one is to use the general uniformization theorem for Riemann surfaces (Theorem IV.3.3). Finally, we get a general solution by approximating the given μ with C_0^∞-functions (cf. Theorem 4.5 below). For details of the proof and historical remarks we refer the reader to Lehto [4], p. 136. The proof in [LV], p. 191, employs step functions, while Vekua [1] makes use of the explicit expression in Theorem 4.3.

For continuous μ, the solutions of the Beltrami equation are not necessarily continuously differentiable. In other words, for a diffeomorphic quasiconformal mapping its complex dilatation, which is continuous, cannot be prescribed as an arbitrary continuous function μ with $\|\mu\|_\infty < 1$. This is one more reason to generalize the classical Grötzsch definition of quasiconformality (cf. the remark made at the end of 3.2).

If μ is a little more regular than just continuous, we are back in the classical situation. For instance, *a quasiconformal mapping whose complex dilatation is locally Hölder continuous is a diffeomorphism.* (See [LV], p. 235, where this conclusion is drawn from a still weaker condition on μ.)

Theorem 4.4 gives immediately a striking generalization of the Riemann mapping theorem: *Let A and B be simply connected domains in the extended plane whose boundaries consist of more than one point, and let μ be a measurable function in A with $\|\mu\|_\infty < 1$. Then there is a quasiconformal mapping of A onto B whose complex dilatation agrees with μ a.e.*

In fact, by Theorem 4.4 there exists a quasiconformal mapping f of A with complex dilatation equal to μ a.e. The boundary of the simply connected domain $f(A)$ consists of more than one point. Hence, by Riemann's mapping theorem, there is a conformal map g of $f(A)$ onto B. Then $g \circ f$ has the desired properties.

4.6. Convergence of Complex Dilatations

It is important in proving Theorem 4.4 that we can initially consider a smooth μ and then obtain the general result by approximation. Let us now study more closely what relations there are between the convergence of mappings and that of their complex dilatations.

We first remark that convergence of mappings need not imply convergence of their complex dilatations. More precisely, let (f_n) be a sequence of quasiconformal mappings of a domain A. We suppose that the complex dilatations

μ_n of f_n satisfy the condition $\|\mu_n\|_\infty \le k < 1$ and that f_n converges locally uniformly in A towards a quasiconformal mapping f with complex dilatation μ. By Theorem 2.2, $\|\mu\|_\infty \le k$, but otherwise there need be no connection between the functions μ_n and μ, i.e., the local mapping properties of f_n and f may be quite different (see [LV], p. 186).

The situation changes if the functions μ_n converge.

Theorem 4.5. *Let* (f_n) *be a sequence of K-quasiconformal mappings of A which converges locally uniformly to a quasiconformal mapping f with complex dilatation μ. If the complex dilatations μ_n of f_n tend to a limit a.e., then* $\lim \mu_n(z) = \mu(z)$ *a.e.*

This result ([LV], p. 187) is needed to take the third step in the proof of Theorem 4.4 sketched above. It can also be used to prove that an arbitrary quasiconformal mapping can be approximated by smooth quasiconformal mappings (cf. [LV], p. 207).

The following complement to Theorem 4.5 shows that convergence of complex dilatations implies convergence of the corresponding normalized mappings.

Theorem 4.6. *Let μ and μ_n, $n = 1, 2, \ldots$, be measurable functions in the plane such that $\|\mu_n\|_\infty \le k < 1$ and $\lim \mu_n(z) = \mu(z)$ a.e. If f and f_n are the quasiconformal mappings of the plane which fix the points 0, 1 and ∞ and have the complex dilatations μ and μ_n, then $f(z) = \lim f_n(z)$ uniformly in the plane in the spherical metric.*

PROOF. By Theorems 4.4 and 4.2, the mappings f and f_n exist and are uniquely determined. By Theorem 2.1, $\{f_n\}$ is a normal family. By Theorems 4.5 and 4.2, every convergent subsequence (f_{n_i}) tends to f. Then the sequence (f_n) itself has the limit f. $\qquad\square$

4.7. Decomposition of Quasiconformal Mappings

Let f be a quasiconformal mapping with maximal dilatation K, and assume that $f = f_2 \circ f_1$, where f_1 and f_2 are K_1- and K_2-quasiconformal. We then have trivially $K \le K_1 K_2$. Using Theorem 4.4 we shall now show that for any given $K_1 \le K$, a "minimal" decomposition $f = f_2 \circ f_1$ always exists with $K = K_1 K_2$.

Theorem 4.7. *Let f be a quasiconformal mapping with maximal dilatation K, and $0 < t < 1$. Then $f = f_2 \circ f_1$, where f_1 is K^t-quasiconformal and f_2 is K^{1-t}-quasiconformal.*

PROOF. Let μ denote the complex dilatation of f. We choose the complex dilatation μ_1 of f_1 as follows: $\mu_1(z)$ is the point on the line segment from 0 to

$\mu(z)$ for which $h(0, \mu_1(z)) = th(0, \mu(z))$, where h is the hyperbolic distance in the unit disc. It then follows from formula (1.1) that

$$\frac{1 + |\mu_1(z)|}{1 - |\mu_1(z)|} = \left(\frac{1 + |\mu(z)|}{1 - |\mu(z)|}\right)^t. \tag{4.16}$$

From this we see that f_1 is K^t-quasiconformal.

If μ_2 denotes the complex dilatation of $f_2 = f \circ f_1^{-1}$, then by formula (4.4),

$$|\mu_2(\zeta)| = \left|\frac{\mu(z) - \mu_1(z)}{1 - \overline{\mu_1(z)}\mu(z)}\right|, \qquad \zeta = f_1(z).$$

Hence, again by (1.1),

$$\log\frac{1 + |\mu_2(\zeta)|}{1 - |\mu_2(\zeta)|} = 2h(\mu_1(z), \mu(z)) = 2(1 - t)h(0, \mu(z)) = (1 - t)\log\frac{1 + |\mu(z)|}{1 - |\mu(z)|}.$$

We conclude that f_2 is K^{1-t}-quasiconformal. $\qquad\square$

It follows from Theorem 4.7 that if f is a K-quasiconformal mapping and $\varepsilon > 0$ is given, we can always write

$$f = f_n \circ \cdots \circ f_2 \circ f_1, \tag{4.17}$$

where each mapping f_i, $i = 1, 2, \ldots, n$, is $(1 + \varepsilon)$-quasiconformal.

5. The Boundary Value Problem

5.1. Boundary Function of a Quasiconformal Mapping

A quasiconformal mapping f of a Jordan domain A onto another Jordan domain B can always be extended to a homeomorphism between the closures of A and B (cf. 2.1). Thus we can speak of the boundary function of f, and it is a homeomorphism of ∂A onto ∂B.

Now let $h\colon \partial A \to \partial B$ be a given homeomorphism under which positive orientations of the boundaries with respect to the Jordan domains A and B correspond to each other. The boundary value problem is to find necessary and sufficient conditions for h to be the boundary function of a quasiconformal mapping $f\colon A \to B$.

We restrict ourselves here to studying the normalized case in which $A = B =$ the upper half-plane, which we denote by H. Then the given mapping h is a homeomorphism of the one point compactification $\overline{\mathbb{R}}$ of the real axis onto itself.

Let x_1, x_2, x_3, x_4 be a sequence of points of $\overline{\mathbb{R}}$ determining the positive orientation with respect to H. We call

$$\sup \frac{M(H(h(x_1), h(x_2), h(x_3), h(x_4)))}{M(H(x_1, x_2, x_3, x_4))},$$

where the supremum is taken over all such sequences x_1, x_2, x_3, x_4, the *maximal dilatation* of h. If it is finite, h is said to be *quasiconformal*, and if it is $\leq K$, the mapping h is K-quasiconformal. From the definition it is clear that these one-dimensional quasiconformal mappings have the customary properties of quasiconformal mappings in the plane: If h is K-quasiconformal, then so is its inverse h^{-1}, and if h_i is K_i-quasiconformal, $i = 1, 2$, then $h_2 \circ h_1$ is $K_1 K_2$-quasiconformal.

Now suppose that h is the boundary function of a K-quasiconformal mapping $f: H \to H$. Then clearly h itself must be K-quasiconformal, and we have found a necessary condition for h, albeit an implicit one.

This necessary condition becomes much more explicit if we choose the points x_1, x_2, x_3, x_4 in a special manner and introduce the normalization $h(\infty) = \infty$. The normalization means that h is a strictly increasing continuous function on the real axis, growing from $-\infty$ to $+\infty$.

Theorem 5.1. *The boundary values h of a K-quasiconformal self-mapping f of the upper half-plane, $f(\infty) = \infty$, satisfy the double inequality*

$$1/\lambda(K) \leq \frac{h(x + t) - h(x)}{h(x) - h(x - t)} \leq \lambda(K) \tag{5.1}$$

for all x and all $t > 0$. Here λ is the distortion function of 2.4. The inequality is sharp for any given K, x and t.

PROOF. Choose $x_1 = x - t$, $x_2 = x$, $x_3 = x + t$, $x_4 = \infty$, and denote the middle term in (5.1) by α. By the considerations in 2.4, we then have $M(H(x_1, x_2, x_3, x_4)) = 1$ and

$$M(H(h(x_1), h(x_2), h(x_3), \infty)) = \left(\frac{2}{\pi} \mu((1 + \alpha^{-s})^{-1/2}) \right)^s$$

for $s = 1$ and $s = -1$. Thus (5.1) follows from the fact that h is K-quasiconformal. From the characterization of λ as an extremal function we deduce that (5.1) is sharp. $\qquad\square$

5.2. Quasisymmetric Functions

An increasing homeomorphism $h: \bar{\mathbb{R}} \to \bar{\mathbb{R}}$ with $h(\infty) = \infty$ is said to be k-*quasisymmetric* if

$$\frac{1}{k} \leq \frac{h(x + t) - h(x)}{h(x) - h(x - t)} \leq k$$

for all $x \in \mathbb{R}$ and all $t > 0$. A function is quasisymmetric if it is k-quasi-symmetric for some k. The smallest possible k is called the quasisymmetry constant of h. From the proof of Theorem 5.1 it follows that *if h is K-quasiconformal and $h(\infty) = \infty$, then h is $\lambda(K)$-quasisymmetric.*

Lemma 5.1. *The family of k-quasisymmetric functions h which keep 0 and 1 fixed is equicontinuous at every point of the real axis.*

PROOF. We first conclude from $h(2^{-n+1}) - h(2^{-n}) \geq h(2^{-n})/k$ that

$$h(2^{-n}) \leq \left(\frac{k}{k+1}\right)^n \tag{5.2}$$

for every non-negative integer n. In looking for a bound for $h(a + x) - h(a)$ we may assume that a and x are non-negative. Then, if $x \leq 2^{-n}$, it follows from (5.2) that

$$0 \leq h(a + x) - h(a) \leq (h(a + 1) - h(a))\left(\frac{k}{k+1}\right)^n. \tag{5.3}$$

If $m \leq a < m + 1$, where m is an integer, then

$$h(a + 1) - h(a) \leq k^m(h(a + 1 - m) - h(a - m)) \leq k^m h(2) \leq k^m(k + 1).$$

Hence, (5.3) implies equicontinuity at a.   □

A quasisymmetric function which fixes 0 and 1 is said to be normalized. We conclude that every infinite sequence (h_n) of normalized k-quasisymmetric functions contains a subsequence which is locally uniformly convergent on the real axis. The limit is also a normalized k-quasisymmetric function. This result allows the following conclusion.

Lemma 5.2. *Let $[a, b]$ be a closed interval on the real axis, and $\varepsilon > 0$. Then there is a $\delta > 0$ such that for a normalized quasisymmetric function h,*

$$|h(x) - x| < \varepsilon, \qquad x \in [a, b],$$

whenever h is $(1 + \delta)$-quasisymmetric.

PROOF. If the lemma is not true, there is an $\varepsilon > 0$ and a sequence of normalized $(1 + 1/n)$-quasisymmetric functions h_n, $n = 1, 2, \ldots$, such that

$$\sup_{a \leq x \leq b} |h_n(x) - x| \geq \varepsilon$$

for every n. Since $\{h_n\}$ is a normal family, there is a subsequence which converges uniformly on $[a, b]$. The limit is 1-quasisymmetric and hence the identity. This is a contradiction.   □

5.3. Solution of the Boundary Value Problem

By Theorem 5.1, quasisymmetry is a necessary condition for h to be the boundary function of a quasiconformal self-mapping of H fixing ∞. The condition is also sufficient (Beurling and Ahlfors [1]):

Theorem 5.2. Let h be k-quasisymmetric. Then there exists a quasiconformal self-mapping of the upper half-plane which has the boundary values h and whose maximal dilatation is bounded by a number $K(k)$ which depends only on k and tends to 1 as $k \to 1$.

PROOF. Let f be defined in the closure of H by the formula

$$f(x + iy) = \tfrac{1}{2} \int_0^1 (h(x + ty) + h(x - ty)) \, dt + i \int_0^1 (h(x + ty) - h(x - ty)) \, dt.$$

$$(5.4)$$

Clearly $f = h$ on the real axis. We call f the *Beurling–Ahlfors extension* of h and prove that it has the desired properties.

Set

$$\alpha(x, y) = \int_0^1 h(x + ty) \, dt = \frac{1}{y} \int_x^{x+y} h(t) \, dt,$$

$$(5.5)$$

$$\beta(x, y) = \int_0^1 h(x - ty) \, dt = \frac{1}{y} \int_{x-y}^x h(t) \, dt.$$

We see that α and β are continuously differentiable in H, and an easy calculation shows that the Jacobian $J = \alpha_y \beta_x - \alpha_x \beta_y$ is positive throughout H. More than that, we deduce from (5.5) that f is a continuously differentiable bijective self-mapping of H (cf. [LV], p. 84). This conclusion can be drawn from the fact that h is a homeomorphism of $\mathbb{R}$ onto itself, quasisymmetry is not needed here.

In estimating the maximal dilatation of (5.4) we make use of linear functions $z \to A_j(z) = a_j z + b_j, j = 1, 2$, with real coefficients $a_j > 0$ and b_j. If f is the Beurling–Ahlfors extension of h, then $A_2 \circ f \circ A_1$ is the Beurling–Ahlfors extension of $A_2 \circ h \circ A_1 | \mathbb{R}$. This can be verified directly from (5.4). Moreover, the maximal dilatation of f and the quasisymmetry constant of h do not change under such a transformation.

Suppose now that there is not a number $K(k)$ bounding the maximal dilatation of an extension (5.4). Then there are k-quasisymmetric functions h_n and points $z_n \in H$ such that the dilatation quotients D_n of the Beurling–Ahlfors extensions f_n of h_n have the property $D_n(z_n) \to \infty$ (cf. Theorem 3.1).

Application of suitable linear transformations A_j makes it possible to assume that the boundary functions h_n are normalized and that $z_n = i$ for every n. By Lemma 5.1, the functions h_n then constitute a normal family. We may thus suppose that the functions h_n converge to a k-quasisymmetric function h, uniformly on bounded intervals of the real axis. Let f denote the Beurling–Ahlfors extension of this limit function h.

Let α_n, β_n be the functions (5.5) for h_n. From $(\alpha_n)_x(i) = 1$, $(\beta_n)_x(i) = -h_n(-1)$, and

$$(\alpha_n)_y(i) = 1 - \int_0^1 h_n(t)\,dt, \qquad (\beta_n)_y(i) = h_n(-1) - \int_{-1}^0 h_n(t)\,dt,$$

we conclude that the partial derivatives of α_n and β_n at i converge to the partials of α and β at i. Because $J(D + 1/D) = (5/4)(\alpha_x^2 + \alpha_y^2 + \beta_x^2 + \beta_y^2) - (3/2)(\alpha_x\beta_x + \alpha_y\beta_y)$, it follows that $D_n(i)$ tends to the dilatation quotient of f at i. Hence, $D_n(i) \to \infty$ is impossible, and the existence of a finite bound $K(k)$ follows.

The above reasoning can also be used to proving the existence of a bound $K(k)$ with the additional property $K(k) \to 1$ as $k \to 1$. If no such $K(k)$ exists, there is a sequence of normalized functions h_n whose quasisymmetry constants tend to 1 while $D_n(i)$ does not converge to 1. From Lemma 5.2 we conclude that $\lim h_n(x) = h(x) = x$. By (5.4), the Beurling–Ahlfors extension of the limit function is then the identity mapping. Thus $\lim D_n(i) = D(i) = 1$, and we have arrived at a contradiction. $\qquad\square$

For all our applications it is sufficient just to know the existence of a bound $K(k)$ which is finite and tends to 1 as $k \to 1$. For the sake of completeness, we mention here that $K(k)$ can be estimated. Beurling and Ahlfors [1] show, after rather laborious computation, that if the imaginary part of (5.4) is multiplied by an appropriate positive constant, the maximal dilatation K of the modified extension satisfies the inequality $K \le k^2$. For the required calculations, see also Lehtinen [1].

The smallest upper bound known at present for the maximal dilatation in the class of quasiconformal mappings with k-quasisymmetric boundary values is $\min(k^{3/2}, 2k - 1)$ (Lehtinen [3]).

5.4. Composition of Beurling–Ahlfors Extensions

Let h_1 and h_2 be k-quasisymmetric functions and f_1 and f_2 their Beurling–Ahlfors extensions. If the quasisymmetry constant of $h_2 \circ h_1^{-1}$ tends to 1, it follows from the proof of Theorem 5.2 that the maximal dilatation of the Beurling–Ahlfors extension of $h_2 \circ h_1^{-1}$ converges to 1. However, $f_2 \circ f_1^{-1}$ is not necessarily the Beurling–Ahlfors extension of $h_2 \circ h_1^{-1}$. A small modification for the proof of Theorem 5.2 is therefore required to show that the maximal dilatation of $f_2 \circ f_1^{-1}$ then also tends to 1. (Cf. Earle and Eells [1].) This is a result which we shall need in III.3.2–3.

Lemma 5.3. *Let h_1 and h_2 be k-quasisymmetric functions and f_1 and f_2 their Beurling–Ahlfors extensions. If the quasisymmetry constant of $h_2 \circ h_1^{-1}$ tends to 1, then the maximal dilatation of $f_2 \circ f_1^{-1}$ converges to 1.*

PROOF. Suppose the lemma is not true. Making use again of the linear transformations A_j, appealing to Lemmas 5.1 and 5.2, and reasoning as in the proof of Theorem 5.2, we would arrive at the following situation. There exist normalized k-quasisymmetric mappings h_{n1} and h_{n2}, $n = 1, 2, \ldots$, which converge locally uniformly on the real axis to k-quasisymmetric mappings h_1 and h_2 and for which $h_{n2} \circ h_{n1}^{-1}$ converges locally uniformly to the identity mapping. On the other hand, if f_{n1} and f_{n2} are the Beurling–Ahlfors extensions of h_{n1} and h_{n2}, the dilatation quotient of $f_{n2} \circ f_{n1}^{-1}$ at the point i is bounded away from 1.

Let f_1 and f_2 be the Beurling–Ahlfors extensions of h_1 and h_2. Arguing as in Theorem 5.2 we obtain the result

$$\lim D_{f_{n2} \circ f_{n1}^{-1}}(i) = D_{f_2 \circ f_1^{-1}}(i). \tag{5.6}$$

But from $(h_{n2} \circ h_{n1}^{-1})(x) \to x$ it follows that $h_1 = h_2$. Hence $f_1 = f_2$, and so $D_{f_2 \circ f_1^{-1}}(i) = 1$. Thus (5.6) is a contradiction. □

5.5. Quasi-Isometry

In a later application, the following property of the mapping (5.4) will be needed (Ahlfors [4]).

Lemma 5.4. *The Beurling–Ahlfors extension of a quasisymmetric function is a quasi-isometry in the hyperbolic metric of the upper half-plane.*

PROOF. Let h be a k-quasisymmetric function and f its Beurling–Ahlfors extension (5.4). We have to prove the existence of a constant c depending only on k, such that

$$\frac{1}{c} \frac{|dz|}{\operatorname{Im} z} \leq \frac{|df(z)|}{\operatorname{Im} f(z)} \leq c \frac{|dz|}{\operatorname{Im} z}. \tag{5.7}$$

The mapping f is K-quasiconformal for a K which depends only on k. We have $|df(z)|/|dz| \leq \max|\partial_\alpha f(z)|$ and $\max|\partial_\alpha f(z)|^2 \leq KJ_f(z)$, where J_f is the Jacobian of f. Hence, the right-hand inequality (5.7) follows if we prove that

$$y^2 K J_f(z) \leq c^2 (\operatorname{Im} f(z))^2 \tag{5.8}$$

with $y = \operatorname{Im} z$.

Suppose that (5.8) is not true, and apply again the same method of reasoning used in the proofs of Theorem 5.2 and Lemma 5.3. We conclude that there are normalized k-quasisymmetric functions h_n, which converge locally uniformly to a k-quasisymmetric function h, such that for the Beurling–Ahlfors extensions f_n of h_n,

$$J_{f_n}(i)/(\operatorname{Im} f_n(i))^2 \to \infty. \tag{5.9}$$

But if f is the Beurling–Ahlfors extension of h, then $J_{f_n}(i) \to J_f(i)$, $\operatorname{Im} f_n(i) \to \operatorname{Im} f(i)$, as we showed in the proof of Theorem 5.2. It follows that (5.9) is impossible.

Similar reasoning yields the lower inequality (5.7).   □

5.6. Smoothness of Solutions

In the 1950's it was a famous open problem whether the boundary function of a quasiconformal self-mapping of H is absolutely continuous. A direct construction (Beurling–Ahlfors [1]) combined with Theorem 5.2 gives an entirely negative answer:

For every $k > 1$, there is a k-quasisymmetric function h which is singular, i.e., for which $h'(x) = 0$ a.e.

Hence, a quasiconformal mapping, while absolutely continuous on lines, need not be absolutely continuous on every closed line segment in its domain of definition. Singular quasisymmetric functions were first regarded as a curiosity. But now we know that, except for linear mappings, all boundary functions encountered in the classical theory of Teichmüller spaces are singular (see V.3.6).

In spite of the fact that a given boundary function may be singular, it always admits extensions which are very smooth in the upper half-plane. We remarked already that the solution (5.4) is continuously differentiable in H, and very much more is true:

Theorem 5.3. *For every quasisymmetric function, the boundary value problem has a real-analytic solution.*

PROOF. The result can be proved with the aid of the decomposition formula (4.17) in 4.7 which makes it possible to express a k-quasisymmetric function as a composition of $(1 + \varepsilon)$-quasisymmetric functions. The proof based on this method will be given in II.5.2. A more direct proof (Lehtinen [1]) is obtained by a modification of the formula (5.4). We can write in (5.4)

$$\operatorname{Re} f(z) = \tfrac{1}{2} \int_{-\infty}^{\infty} \kappa(t)(h(x + ty) + h(x - ty))\, dt,$$

where $\kappa(t) = 1$ on $[0, 1]$ and vanishes elsewhere, and a similar expression obtains for $\operatorname{Im} f(z)$. If κ is replaced by a suitable exponential, the corresponding f turns out to be a real-analytic solution.   □

5.7. Extremal Solutions

Let F_h be the family of all quasiconformal self-mappings of H whose boundary values agree with a given quasisymmetric h. Then F_h is a countable union of normal families, and each of these contains its limits under locally uniform convergence. This allows an important conclusion:

In the class F_h, there always exists an extremal mapping which has the smallest maximal dilatation in F_h.

While the existence of an extremal mapping can be deduced immediately, it is much more difficult to study its uniqueness. It was in fact an open problem for quite a long time whether the extremal mapping in F_h is always unique, until Strebel in 1962 gave an example which shows that F_h can contain more than one extremal.

In Strebel's example one considers the domain A which is the union of the lower half-plane and the "chimney" $\{z|\operatorname{Im} z \geq 0\} \cap \{z|0 < \operatorname{Re} z < 1\}$ (Fig. 1). In A we set $f_1(x + iy) = x + iKy$, $K > 1$, and define another mapping f_2 so that f_2 agrees with f_1 in the chimney and is the identity in the lower half-plane. Then f_1 and f_2 are K-quasiconformal in A, they agree on the boundary of A, and both are extremal for their boundary values (Strebel [1]). By using a suitable conformal mapping of A onto H we can transform f_1 and f_2 to normalized self-mappings of H.

More examples of boundary values with non-unique extremals will be obtained in V.3.7. On the other hand, there are important cases in which uniqueness can be proved (V.8.5). The question of unique extremality has been systematically studied by Reich and Strebel; see, for instance, Strebel [1], Reich and Strebel [1] and Reich [1].

In certain cases the Beurling–Ahlfors solution (5.4) is far from extremal. Lehtinen [3] proved that if h has the quasisymmetry constant k, the maximal dilatation of (5.4) is always $\geq k$. Now let h be the restriction to $\mathbb{R}$ of the K-quasiconformal extremal mapping described in 2.4. Then h has the quasisymmetry constant $\lambda(K)$. In this case the minimal maximal dilatation in F_h is equal to K, whereas by (2.6), the maximal dilatation of the Beurling–Ahlfors extension is $\geq e^{\pi K}/16 - 1/2$.

The preceding considerations make it possible to compare the maximal dilatation of the boundary function h with the extremal maximal dilatation for F_h.

Figure 1. Strebel's chimney.

Lemma 5.5. *Let K^* be the maximal dilatation of the quasisymmetric function h and K the minimal maximal dilatation of the quasiconformal self-mappings of H with boundary values h. Then $K \to 1$ as $K^* \to 1$.*

PROOF. The function h is $\lambda(K^*)$-quasisymmetric by the remark in 5.2 preceding Lemma 5.1. Here $\lambda(K^*) \to 1$ as $K^* \to 1$, as we noted at the end of 2.4. By Theorem 5.2, the maximal dilatation of the Beurling–Ahlfors extension of h tends to 1 as $K^* \to 1$. A fortiori, the same is true of the minimal maximal dilatation K.   □

The bounds at the end of 5.3 for the maximal dilatation of the quasiconformal extensions of h yield quantitative estimates between K and K^*. For instance, we have

$$K^* \le K \le \lambda(K^*)^{3/2}. \tag{5.10}$$

Here the lower estimate is trivial, and the upper estimate follows if we use Lehtinen's bound for the maximal dilatation of the modified extension (5.4). For our later applications, the qualitative result in Lemma 5.5 is sufficient.

6. Quasidiscs

6.1. Quasicircles

A Jordan curve can be defined as the image of a circle under a homeomorphism of the plane. If the homeomorphism is conformal, then the image is a circle. Between the topological and proper circles, quasicircles form a class of curves which will come to frequent use in Chapters II, III and V.

A *quasicircle* in the extended plane is the image of a circle under a quasiconformal mapping of the plane. If the mapping can be taken to be K-quasiconformal, the image curve is called a K-quasicircle. A domain bounded by a quasicircle is called a *quasidisc*.

Let f be a quasiconformal mapping of a domain A and F a compact subset of A. Then there exists a quasiconformal mapping of the plane whose restriction to F agrees with f ([LV], p. 96). It follows that f maps circles in A onto quasicircles.

Since a quasiconformal mapping preserves sets of area zero, a quasicircle has zero area. On the other hand, it is possible that all non-empty subarcs of a given quasicircle are non-rectifiable; concrete examples are provided in [LV], p. 104. Gehring and Väisälä [1] have proved the striking result that, while the Hausdorff dimension of a quasicircle is always less than 2, it can take any value λ, $1 \le \lambda < 2$. We remark that quasicircles with Hausdorff dimension greater than 1 play a role in the modern theory of iteration of polynomials in the plane.

It follows from what we said in 2.4 that a homeomorphism of the plane which is K-quasiconformal in the complement of a straight line is K-quasiconformal everywhere. From the definition of quasicircles we obtain immediately a generalization, which we shall need later several times.

Lemma 6.1. *Let C be a quasicircle and f a homeomorphism of the plane which is K-quasiconformal in the complement of C. Then f is a K-quasiconformal mapping of the plane.*

PROOF. It follows from the definition of a quasicircle that there is a quasiconformal mapping w of the plane which maps the real axis onto C. Then $f \circ w$ is quasiconformal in the plane, as is also $f = (f \circ w) \circ w^{-1}$. Since the area of C is zero, we conclude from Theorem 3.5 (Analytic definition) that f is K-quasiconformal.   □

6.2. Quasiconformal Reflections

Let C be a Jordan curve bounding the domains A_1 and A_2. A sense-reversing K-quasiconformal involution φ of the plane which maps A_1 onto A_2 is a K-*quasiconformal reflection* in C if φ keeps every point of C fixed.

Theorem 6.1. *A Jordan curve admits a quasiconformal reflection if and only if it is a quasicircle.*

PROOF. Suppose first that C is a quasicircle. Let f be a quasiconformal mapping of the plane which maps A_1 onto the upper half-plane H. Then the mapping $\varphi = f^{-1} \circ j \circ f$, where $j(z) = \bar{z}$, is a quasiconformal reflection in C.

Conversely, let φ be a quasiconformal reflection in a Jordan curve C. Let h map H conformally onto A_1. Define f by $f(z) = h(z)$ in the closure of H, and $f(z) = \varphi(h(\bar{z}))$ in the lower half-plane. Then f is a homeomorphism of the plane which is quasiconformal off the real axis, which it maps onto C. By Lemma 6.1, f is quasiconformal in the plane, and so C is a quasicircle.   □

We can draw certain additional conclusions from the above proof. First, if C admits a K-quasiconformal reflection, then C is a K-quasicircle. In the opposite direction we deduce that a K-quasicircle always admits a K^2-quasiconformal reflection.

In the second part of the proof, the required quasiconformal mapping of the plane is conformal in a half-plane. For the sake of later reference, we wish to express certain connections between conformal mappings and quasicircles explicitly.

Lemma 6.2. *A K-quasidisc A has the following properties:*

1° *Every quasiconformal reflection in ∂A is of the form $f \circ j \circ f^{-1}$, where f is a quasiconformal mapping of the plane which maps the upper half-plane H conformally onto A, and j denotes the reflection $z \to \bar{z}$.*

2° *A is the image of H under a K^2-quasiconformal mapping f of the plane which is conformal in H.*

3° *Every conformal mapping $f: H \to A$ has a K^2-quasiconformal extension to the plane.*

PROOF. Let φ be an arbitrary quasiconformal reflection in ∂A. If f is constructed as in the second part of the proof of Theorem 6.1, with A replacing A_1, we obtain 1°. Since ∂A is a K-quasicircle, there exists a K^2-quasiconformal reflection φ. The corresponding f is also K^2-quasiconformal, and 2° follows. We use this same φ in 3° and conclude that $\varphi \circ f \circ j$ is a desired extension of f.   □

We return to our previous notation and denote by C a Jordan curve which bounds the domains A_1 and A_2. In what follows, $c(K)$, $c_1(K)$, ... denote constants which depend only on K.

Lemma 6.3. *Let φ be a K-quasiconformal reflection in C which passes through ∞. Then*

$$|\varphi(z) - z| \leq \frac{c_1(K)}{\eta_2(\varphi(z))}, \qquad z \in A_1, \tag{6.1}$$

where η_2 is the Poincaré density of A_2.

PROOF. Let $h: H \to A_1$ be a conformal mapping, $h(\infty) = \infty$, and set again $f = h$ in the closure of H and $f = \varphi \circ h \circ j$ in the lower half-plane. Then f is a K-quasiconformal mapping of the plane.

Fix $z \in A_1$ and $z_0 \in C$. The function f maps the circle $\{w \mid |w - h^{-1}(z_0)| = |h^{-1}(z) - h^{-1}(z_0)|\}$ onto a curve which passes through the points z and $\varphi(z)$. By Theorem 2.4,

$$|z - z_0|/c(K) \leq |\varphi(z) - z_0| \leq c(K)|z - z_0|. \tag{6.2}$$

If we choose z_0 such that $|\varphi(z) - z_0|$ is equal to the distance $d(\varphi(z), C)$, then

$$|\varphi(z) - z| \leq |\varphi(z) - z_0| + |z - z_0| \leq (1 + c(K))d(\varphi(z), C).$$

From this we obtain (6.1) by using the inequality (1.3).   □

We remark that (6.2) yields the estimate

$$d(z, C)/c(K) \leq d(\varphi(z), C) \leq c(K)d(z, C). \tag{6.3}$$

Here we can take $c(K) = \lambda(K)$ (see 2.5).

For our later applications (see II.4.2) it is important to know that there exist quasiconformal reflections which are quasi-isometries in the euclidean metric (Ahlfors [4]). We also call such reflections Lipschitz-continuous.

Lemma 6.4. *Let C be a K-quasicircle bounding the domains A_1 and A_2 and passing through ∞. Then there exists a $c_2(K)$-quasiconformal reflection φ in C,* *continuously differentiable in A_1 and A_2, such that*

$$|d\varphi(z)| \le c_3(K)|dz| \qquad (6.4)$$

at every point $z \in A_1$.

PROOF. Let h_1 be a conformal mapping of A_1 onto the upper half-plane H and h_2 a conformal mapping of A_2 onto the lower half-plane, both fixing ∞. Since h_1 and h_2 have homeomorphic extensions to the boundary, we can form the function $h_2 \circ h_1^{-1}$ on the real axis. It is quasisymmetric. For if ψ is a K^2-quasiconformal reflection in C and j again denotes the reflection $z \to \bar{z}$, then $j \circ h_2 \circ \psi \circ h_1^{-1}$ is a quasiconformal self-mapping of H with boundary values $h_2 \circ h_1^{-1}$.

By using the Beurling–Ahlfors extension (5.4), we construct a quasi-conformal diffeomorphism $f: H \to H$ with boundary values $h_2 \circ h_1^{-1}$. Set $\varphi = \varphi_1 = h_2^{-1} \circ j \circ f \circ h_1$ in the closure of A_1 and $\varphi = \varphi_1^{-1}$ in A_2. Then φ is a $c_2(K)$-quasiconformal reflection in C, which is continuously differentiable outside C.

Since the Poincaré metric is conformally invariant, it follows from formula (5.7) that

$$\eta_2(\varphi(z))|d\varphi(z)| \le c_4(K)\eta_1(z)|dz|.$$

From this we obtain (6.4), in view of (1.3), (1.4) and (6.3).   □

6.3. Uniform Domains

Let us now start studying the geometry of quasidiscs. We shall prove a chain of theorems which, when put together, give several characterizations for quasidiscs and shed light on their geometric properties. In subsections 6.3–6.5, we follow the presentation of Gehring [4]. A summary will be given at the end of subsection 6.7.

Throughout this subsection we assume that A is a simply connected proper subdomain of the complex plane. The domain A is said to be *uniform* if there are constants a and b such that each pair of points $z_1, z_2 \in A$ can be joined by an arc α in A with the following properties:

$1°$ The euclidean length of α satisfies the inequality

$$l(\alpha) \le a|z_1 - z_2|. \qquad (6.5)$$

$2°$ For every $z \in \alpha$,

$$\min(l(\alpha_1), l(\alpha_2)) \le bd(z, \partial A), \qquad (6.6)$$

where α_1 and α_2 are the components of $\alpha \backslash \{z\}$.

It will appear that a domain A is uniform if and only if it is a quasidisc. We first prove directly that a quasidisc is uniform. This requires a fairly lengthy

argument, and we begin with a lemma in which, as before, $c(K)$, $c_1(K)$, ...
denote constants depending only on K. These are not necessarily the same
constants that arose in the previous subsection.

Lemma 6.5. *Let A be a K-quasidisc with $\infty \in \partial A$, and $f: H \to A$ a conformal
mapping satisfying $f(\infty) = \infty$. Then*

$$\int_0^y |f'(it)|\, dt \le c_1(K)d(f(iy), \partial A)$$

for $0 < y < \infty$.

PROOF. By Lemma 6.2, f has a K^2-quasiconformal extension to the plane. We
may assume without loss of generality that $f(0) = 0$.

For $y > 0$, the mapping f satisfies the inequality

$$|f'(iy)| \le \frac{4d(f(iy), \partial A)}{y}. \tag{6.7}$$

This follows directly from the Koebe one-quarter theorem cited in 1.1, if we
apply it to the function $\zeta \to (f(iy + y\zeta) - f(iy))/yf'(iy)$.

In order to estimate the distance $d(f(iy), \partial A)$ we fix y. After this, we
choose a sequence (y_j) so that $0 < y_{j+1} < y_j \le y$ and that $|f(iy_j)| = c^{-j}|f(iy)|$,
$j = 0, 1, \ldots$, where $c = c(K^2)$ is a constant for which Theorem 2.4 holds for f.
(We can take, for instance, $c = e^{\pi K^2}$.)

Let $y_{j+1} \le t \le y_j$. Because $d(f(it), \partial A) \le |f(it) - f(0)|$, we obtain from (2.8)

$$d(f(it), \partial A) \le c|f(iy_j) - f(0)| = c^{-j+1}|f(iy)|.$$

Hence, by (6.7)

$$\int_{y_{j+1}}^{y_j} |f'(it)|\, dt \le 4c^{-j+1}|f(iy)| \log \frac{y_j}{y_{j+1}}.$$

The logarithm can be estimated by aid of (2.9). Let n be the smallest integer
for which $c \le n$. Then $|f(iy_j) - f(0)| \le n|f(iy_{j+1}) - f(0)|$, and so by (2.9),

$$y_j = |iy_j - 0| \le nc^n|iy_{j+1} - 0| = c_2(K)y_{j+1}.$$

It follows that

$$\int_0^y |f'(it)|\, dt \le 4\log c_2(K)|f(iy)| \sum_{j=0}^{\infty} c^{-j+1} = c_3(K)|f(iy)|.$$

Finally, if $x \in \partial H$, then by applying (2.8) again we infer that $|f(iy)| \le
c|f(iy) - f(x)|$. Thus $|f(iy)| \le cd(f(iy), \partial A)$, and the proof is completed. $\square$

Let z_1 and z_2 be points of A and α the hyperbolic segment of A joining z_1
and z_2. If A is a disc or a half-plane, it is easy to show that (6.5) and (6.6) hold
for $a = b = \pi/2$. This property generalizes to quasidiscs.

Theorem 6.2. *Let A be a K-quasidisc and α a hyperbolic segment in A with endpoints z_1 and z_2. Then α satisfies the conditions (6.5) and (6.6) with constants a and b which depend only on K.*

PROOF. By Lemma 6.2, there is a K^2-quasiconformal mapping f of the plane which maps A conformally onto the unit disc D. We can choose f so that $f(z_1)$ and $f(z_2)$ are real. Let D' be the open disc in D which has the line segment with endpoints $f(z_1)$ and $f(z_2)$ as a diameter. Then $f^{-1}(D')$ is a bounded K^2-quasidisc, in which α is a hyperbolic line. Since $d(z, f^{-1}(\partial D')) \le d(z, \partial A)$ for $z \in f^{-1}(D')$, we may assume, therefore, that A itself is bounded and α is a hyperbolic line in A, i.e., $z_1, z_2 \in \partial A$.

Let A' and α' be the images of A and α under the Möbius transformation $z \to g(z) = (z - z_1)/(z - z_2)$. Then $(g^{-1})'(w) = (z_1 - z_2)(w - 1)^{-2}$, and so

$$l(\alpha) = |z_1 - z_2| \int_{\alpha'} \frac{|dw|}{|w - 1|^2}. \tag{6.8}$$

In order to estimate the integral, we use the arc length representation $s \to w(s)$ for α'. Let $s_0 = c_1(K)/(c_1(K) + 1)$, where $c_1(K)$ is the constant of Lemma 6.5. If $0 < s \le s_0$, then $|w(s) - 1| \ge 1 - |w(s)| \ge 1 - s \ge 1/(c_1(K) + 1)$. For $s > s_0$, we apply Lemma 6.5 to a conformal mapping $f \colon H \to A'$ fixing 0 and ∞. Then α' is the image of the positive imaginary axis and so by Lemma 6.5,

$$s = \int_0^y |f'(it)|\, dt \le c_1(K) d(w(s), \partial A')$$

for $iy = f^{-1}(w(s))$. From $1 = g(\infty) \notin A'$ we further conclude that $d(w(s), \partial A') \le |w(s) - 1|$. It follows that

$$\int_{\alpha'} \frac{|dw|}{|w - 1|^2} \le \int_0^{s_0} (c_1(K) + 1)^2\, ds + \int_{s_0}^\infty \frac{c_1(K)^2}{s^2}\, ds = 2c_1(K)(c_1(K) + 1).$$

Thus (6.5) is obtained from (6.8) with $a = 2c_1(K)(c_1(K) + 1)$.

In order to establish (6.6), we consider a K^2-quasiconformal mapping f of the plane, $f(\infty) = \infty$, which maps A conformally onto the unit disc D. Fix $z \in \alpha$ and choose $z_0 \in \partial A$ so that $|z - z_0| = d(z, \partial A)$. Since $f(\alpha)$ is a hyperbolic line in D,

$$\min_{j=1,2} |f(z) - f(z_j)| \le 2d(f(z), \partial D) \le 2|f(z) - f(z_0)|.$$

By formula (2.9),

$$\min_{j=1,2} |z - z_j| \le 2c(K^2)^2 |z - z_0| = 2c(K^2)^2 d(z, \partial A).$$

Since $l(\alpha_j) \le a|z - z_j|$ by (6.5), we obtain (6.6) with $b = 4c(K^2)^2 c_1(K)(c_1(K) + 1)$. □

The following result follows immediately from Theorem 6.2.

Theorem 6.3. *A quasidisc is a uniform domain.*

Theorem 6.3 is the first link in a closed chain of four theorems. Once these have all been proved we get the converse to Theorem 6.3.

6.4. Linear Local Connectivity

A set E is *linearly locally connected* if there is a constant c such that the following two conditions hold for every finite z and every $r > 0$, where $D(z, r) = \{w \mid |w - z| < r\}$:

1° Any two points of the set $E \cap \overline{D(z, r)}$ can be joined by an arc in $E \cap \overline{D(z, cr)}$.
2° Any two points of the set $E \backslash D(z, r)$ can be joined by an arc in $E \backslash D(z, r/c)$.

In order to illustrate this notion, we let E be the parallel strip $\{x + iy \mid -1 < y < 1\}$. The points 0 and $2r > 0$ of $E \backslash D(r, r)$ can be joined in $E \backslash D(r, r/c)$ only if $c > r$. Letting $r \to \infty$ we conclude that E is not linearly locally connected.

If a simply connected domain A with more than one boundary point is linearly locally connected, then A is a Jordan domain. To prove this, we consider a finite boundary point z of A. Let U be an arbitrary neighborhood of z, and choose $r > 0$ such that the closure of the disc $D(z, cr)$ lies in U. Then $V = D(z, r)$ is a neighborhood of z such that $A \cap V$ lies in a component of $A \cap U$. It follows that A is locally connected at z. A similar reasoning, based on the use of condition 2°, shows that if $\infty \in \partial A$, then A is locally connected at ∞. We conclude that A is a Jordan domain (see 1.2 or Newman [1], pp. 167 and 161).

We shall see after completing our chain of theorems that in fact, a simply connected domain with more than one boundary point is linearly locally connected if and only if it is a quasidisc. We shall now establish the second link of the chain.

Theorem 6.4. *A uniform domain A is linearly locally connected.*

PROOF. Fix a finite z_0 and $r > 0$, and suppose that $z_1, z_2 \in A \cap \overline{D(z_0, r)}$. Since A is uniform, there exists an arc α joining z_1 and z_2 in A such that $l(\alpha) \le a|z_1 - z_2| \le 2ar$. If $z \in \alpha$, we thus have

$$|z - z_0| \le |z - z_1| + |z_1 - z_0| \le l(\alpha) + r \le (2a + 1)r.$$

It follows that α joins z_1 and z_2 in $A \cap \overline{D(z_0, cr)}$ if $c = 2a + 1$.

Next assume that $z_1, z_2 \in A \backslash D(z_0, r)$. We consider an arc α joining z_1 and z_2 in A with $\min l(\alpha_j) \le b d(z, \partial A)$ for every $z \in \alpha$. Set $c = 2b + 1$. If α joins z_1 and z_2 in $A \backslash D(z_0, r/c)$, the theorem is proved. Now suppose that α does not join z_1 and z_2 in $A \backslash D(z_0, r/c)$. We prove that, nonetheless, $A \backslash D(z_0, r/c)$ is connected.

By our hypothesis, there is a point $z \in \alpha$ for which $|z - z_0| < r/c$. For $j = 1, 2$ we have

$$l(\alpha_j) \geq |z_j - z| \geq |z_j - z_0| - |z - z_0| \geq r - r/c.$$

Thus

$$d(z, \partial A) \geq \frac{r(1 - 1/c)}{b} = \frac{2r}{c} > |z - z_0| + \frac{r}{c}.$$

We conclude that $\overline{D(z_0, r/c)} \subset A$. It follows that z_1 and z_2 can be connected in $A \backslash D(z_0, r/c)$. □

From the proof we see that A is linearly locally connected with the constant $c = 2 \max(a, b) + 1$.

6.5. Arc Condition

Let C be a Jordan curve and z_1, z_2 finite points of C. They divide C into two arcs, and we consider the one with the smaller euclidean diameter. The curve C is said to satisfy the *arc condition* if the ratio of this diameter to the distance $|z_1 - z_2|$ is bounded by a fixed number k for all finite $z_1, z_2 \in C$.

A circle satisfies the arc condition for the constant $k = 1$. An example in the opposite direction is obtained if we consider the curve $C = \{x + iy \,|\, x \geq 0, y = \pm x^2\}$. If $z_1 = x + ix^2$, $z_2 = x - ix^2$, then $|z_1 - z_2| = 2x^2$, and the smaller diameter in the above definition is $\geq x$. Hence $d/|z_1 - z_2| \to \infty$ as $x \to 0$, so that this curve does not satisfy the arc condition.

We shall now establish the third link of our chain.

Theorem 6.5. *Let A be a simply connected domain whose boundary contains more than one point. If A is linearly locally connected, then ∂A is a Jordan curve which satisfies the arc condition.*

PROOF. In 6.4, after defining the notion of linear local connectivity, we proved that ∂A is a Jordan curve.

Choose two finite points $z_1, z_2 \in \partial A$ and set $z_0 = (z_1 + z_2)/2$, $r = |z_1 - z_2|/2$. The theorem follows if we prove that at least one of the arcs α_1, α_2 into which the points z_1, z_2 divide ∂A lies in the closure of the disc $D(z_0, c^2 r)$.

The proof is indirect. Suppose there is a $t > r$ and points $w_i \in \alpha_i \backslash D(z_0, c^2 t)$, $i = 1, 2$. Let $r < s_1 < s_2 < t$; then z_1 and z_2 belong to the set $\partial A \cap D(z_0, s_1)$. Since A is a Jordan domain, its boundary points are accessible. We can thus find points $z'_i \in A \cap D(z_0, s_1)$ and arcs β_i joining z'_i to z_i in $A \cap D(z_0, s_1)$. By the linear local connectivity of A, the points z'_1 and z'_2 can be joined by an arc β_3 in $A \cap D(z_0, cs_2)$.

The points w_1 and w_2 lie in $\partial A \backslash \overline{D}(z_0, c^2 s_2)$. Therefore, we can find an arc γ

joining w_1 and w_2 in $A \backslash D(z_0, cs_2)$. But then the cross-cut γ does not meet the cross-cut $\beta_1 \cup \beta_2 \cup \beta_3$. This is a contradiction, because their endpoints are in the order z_1, w_1, z_2, w_2 on ∂A.   □

The proof indicates that if A is linearly locally connected with a constant c, then ∂A satisfies the arc condition with the constant c^2.

6.6. Conjugate Quadrilaterals

We shall now close our chain of theorems by proving that if the boundary of a Jordan domain A satisfies the arc condition, then A is a quasidisc. This is a difficult step for which no simple proof seems to exist. As a preparatory result, we give a characterization of quasicircles in terms of quadrilaterals.

Let C be a Jordan curve bounding the domains A_1 and A_2. Take a sequence of four points $z_1, z_2, z_3, z_4 \in C$ such that $A_1(z_1, z_2, z_3, z_4)$ is a quadrilateral. Then $A_2(z_4, z_3, z_2, z_1)$ is also a quadrilateral, and these two quadrilaterals are said to be *conjugate*.

Lemma 6.6. *Let C be a Jordan curve such that for all conjugate quadrilaterals A_1, A_2 with $M(A_1) = 1$ we have $M(A_2) \leq K$. Then C is a $c(K)$-quasicircle, where $c(K)$ depends only on K.*

PROOF. Let $g_1: A_1 \to H$ and $g_2: A_2 \to H'$ be conformal mappings, where H' is the lower half-plane. Consider the increasing homeomorphism $x \to h(x) = g_2(g_1^{-1}(x))$ of the real axis. For all quadrilaterals $H(z_1, z_2, z_3, z_4)$ with module 1 we then have

$$1/K \leq M(H(h(z_1), h(z_2), h(z_3), h(z_4))) \leq K.$$

We proved in section 5 that the validity of this module inequality is a sufficient condition for the existence of a $c(K)$-quasiconformal mapping $f: H \to H$ with boundary values h. Then $f \circ g_1$ extended by g_2 is a $c(K)$-quasiconformal mapping of the plane carrying A_1 onto H. Thus C is a $c(K)$-quasicircle.   □

In order to utilize Lemma 6.6, we need a result about the geometry of conformal squares.

Lemma 6.7. *Let $Q(z_1, z_2, z_3, z_4)$ be a quadrilateral with module 1, and let s_1 and s_2 denote the euclidean distances in Q between the sides (z_1, z_2), (z_3, z_4) and (z_2, z_3), (z_4, z_1), respectively. Then*

$$s_1/s_2 > 10^{-3}.$$

PROOF. We may assume that among the arcs which join the sides (z_2, z_3) and (z_4, z_1) in Q there is a γ_0 of length s_2. Let z_0 be the point which divides γ_0 into two parts of length $s_2/2$. Set $\rho(z) = 2/s_2$ if $|z - z_0| \leq s_2/2$, $\rho(z) = 1/|z - z_0|$ if $s_2/2 < |z - z_0| \leq s_1 + s_2/2$, and $\rho(z) = 0$ elsewhere. The area $m_\rho(Q)$ of Q in this ρ-metric then satisfies the inequality

$$m_\rho(Q) \leq \pi(1 + 2\log(1 + 2s_1/s_2)).$$

Consider next an arc γ joining the sides (z_1, z_2) and (z_3, z_4) in Q. For the ρ-length of γ we obtain a minorant if we integrate $1/x$ over a segment with endpoints $s_2/2$ and $s_2/2 + s_1$. Therefore,

$$\int_\gamma \rho(z)|dz| \geq \log(1 + 2s_1/s_2).$$

Setting

$$F(x) = \frac{1 + 2\log(1 + x)}{(\log(1 + x))^2}$$

we thus we have by formula (1.7),

$$1 = M(Q(z_1, z_2, z_3, z_4)) \leq \pi F(2s_1/s_2).$$

From this we obtain, by interchanging the roles of s_1 and s_2,

$$\frac{s_1}{s_2} \geq \frac{2}{F^{-1}(1/\pi)} = \frac{2}{e^{\pi + (\pi^2 + \pi)^{1/2}} - 1} > 10^{-3}. \qquad \square$$

6.7. Characterizations of Quasidiscs

We can now establish the remaining link of our chain.

Theorem 6.6. *A Jordan domain whose boundary satisfies the arc condition is a quasidisc.*

PROOF. Let C be a Jordan curve which satisfies the arc condition and bounds the domains A_1 and A_2. Choose four points z_1, z_2, z_3, z_4 on C such that $A_1(z_1, z_2, z_3, z_4)$ is a quadrilateral with module 1. We shall derive an upper bound for the module of the conjugate quadrilateral $A_2(z_4, z_3, z_2, z_1)$.

Let s_1 denote the distance in A_1 between the sides (z_1, z_2) and (z_3, z_4), and d_1 the same distance measured in the plane. For the remaining sides (z_2, z_3) and (z_4, z_1) these distances are denoted by s_2 and d_2. From Lemma 6.7 it follows that

$$s_1 > 10^{-3} d_2. \tag{6.9}$$

Since C satisfies the arc condition, there is a constant $k \geq 1$ such that one of the sides (z_2, z_3), (z_4, z_1) lies inside a disc of diameter kd_1. From this we conclude that

$$d_1 > \frac{d_2}{10^3 \pi k}. \tag{6.10}$$

For if not, one of the sides (z_2, z_3), (z_4, z_1) lies in a disc of diameter $10^{-3}d_2/\pi$. The other, which is at a distance d_2 from this one, must lie outside this disc. It follows that the sides (z_1, z_2) and (z_3, z_4) can be joined in A_1 by a circular arc of length $\leq 10^{-3}d_2$. This is in contradiction with (6.9), and (6.10) follows.

Since (6.10) is formulated in terms of the distances in the plane, it can be used to estimating the module of A_2. We first conclude as above the existence of a disc $|z - z_0| < kd_2/2$ which contains one of the sides (z_1, z_2), (z_3, z_4). Let

$$r = kd_2/2 + 10^{-3}d_2/\pi k,$$

and define $\rho(z) = 1$ if $|z - z_0| < r$, and $\rho(z) = 0$ elsewhere. By (6.10), the ρ-length of an arc γ joining the first and the third side of A_2 is $\geq 10^{-3}d_2/\pi k$. Hence,

$$M(A_2(z_4, z_3, z_2, z_1)) \leq 10^6 \pi^3 r^2 k^2/d_2^2 = 10^6 \pi (\pi k^2/2 + 10^{-3})^2.$$

By Lemma 6.6, C is a quasicircle, and the theorem is proved. We remark that C is a $c(k)$-quasicircle where $c(k)$ depends only on the constant k in the arc condition. $\qquad\square$

Before summarizing our results we remark that the converse of Theorem 6.6, which we now know to be true, admits a fairly simple direct proof ([LV], p. 101).

Quasicircles passing through ∞ satisfy a particularly simple geometric condition.

Theorem 6.7. *Let C be a K-quasicircle passing through ∞, and z_1, z_2, z_3 finite points of C such that z_2 lies between z_1 and z_3. Then*

$$|z_1 - z_2| + |z_2 - z_3| \leq c(K)|z_1 - z_3|. \tag{6.11}$$

PROOF. Let f be a K-quasiconformal mapping of the plane which maps the real axis onto C such that $f(\infty) = \infty$. Denote $x_i = f^{-1}(z_i)$, $i = 1, 2, 3$, and $C_1 = \{w \mid |w - x_1| = |x_1 - x_2|\}$, $C_2 = \{w \mid |w - x_3| = |x_2 - x_3|\}$. Join z_1 and z_3 by a line segment L, and denote by a_1 and a_2 the first points at which L meets $f(C_1)$ and $f(C_2)$ when one moves along L from z_1 and from z_3. Then

$$|z_1 - a_1| + |z_3 - a_2| \leq |z_1 - z_3|.$$

By Theorem 2.4,

$$|z_1 - z_2| \leq c(K)|z_1 - a_1|, \qquad |z_2 - z_3| \leq c(K)|z_3 - a_2|.$$

These yield the desired inequality (6.11). It follows from the proof and our remark in 2.5 that (6.11) holds for $c(K) = \lambda(K)$. $\qquad\square$

Conversely, assume that a Jordan curve C containing ∞ satisfies (6.11). Then C satisfies the arc condition with the constant $c(K)$, and C is a quasicircle.

The implications in Theorems 6.2–6.6 form an unbroken chain which starts and ends with quasidiscs. This makes it possible to draw the remarkable conclusion that the converse of each of these theorems is true. It follows that quasidiscs can be characterized as domains with the hyperbolic segment property of Theorem 6.2, as uniform domains (Theorem 6.3), as domains which are linearly locally connected (Theorem 6.4) or as domains whose boundary satisfies the arc condition (Theorem 6.5 or Theorem 6.6). Here we assume of course that the domains are simply connected and have more than one boundary point. In addition, Theorem 6.1 tells that quasidiscs are Jordan domains whose boundaries admit quasiconformal reflection.

The special position of infinity necessitates some clarification. In defining uniform domains A we assumed that ∞ is not in A, and Theorem 6.2 is not true if A contains ∞. By contrast, Theorems 6.5 and 6.6 hold without this restriction, which is not utilized in the proofs. It is not hard to prove that the converses of these theorems are also valid for domains in the extended plane. For Theorem 6.6 this follows immediately from the fact that if A is a quasidisc and $\infty \in A$, then ∂A is a bounded quasicircle. Since the converse of Theorem 6.6 holds for quasidiscs of the complex plane, ∂A satisfies the arc condition.

There are also characterizations of quasidiscs in terms of analysis rather than geometry. They sometimes reveal surprising connections between various problems of analysis which on the surface have nothing to do with quasiconformal mappings. In II.4 we shall establish a result of this type, using Schwarzian derivatives (see especially II.4.4). A comprehensive account of the main properties of quasidiscs known in 1982 is given in Gehring's lecture notes [4], which we have utilized to a large extent in this section. These notes also contain an extensive bibliography.

Quasicircles were introduced by Pfluger [2] (1961) and Tienari (1962). In 1963, Ahlfors [4] characterized quasicircles geometrically by proving that the arc condition is necessary and sufficient. In this same paper, he also introduced quasiconformal reflections and used them to prove an important extension theorem for conformal mappings (Theorem II.4.1). Gehring [2] defined the notion of linear local connectivity and proved Theorem 6.5 in 1977. Theorem 6.3 and its converse were established by Martio and Sarvas [1] (1979), and Theorem 6.2 with its converse by Gehring and Osgood [1] (1979).

CHAPTER II

Univalent Functions

Introduction to Chapter II

The theory of univalent analytic functions covers a large part of complex analysis. In this chapter, we deal with certain aspects of the theory which are directly or indirectly connected with Teichmüller theory. The interaction between univalent functions and Teichmüller spaces was already explained briefly in the Introduction to this monograph. A more comprehensive description is provided by Chapters II, III, and V, taken together.

In this chapter a central position is occupied by the Schwarzian derivative of a locally injective meromorphic function. In section 1 we begin with the classical result that the Schwarzian derivative vanishes identically if and only if the function is a Möbius transformation. Some other basic properties of the Schwarzian and its hyperbolic sup-norm are also established.

In section 2 we consider Schwarzian derivatives of conformal mappings of a simply connected domain A. Particular attention is paid to the case in which the image domain is a disc. The norm of the Schwarzian is then intimately related to the geometry of A and provides a measure of the distance of A from a disc.

Sections 3 and 4 deal with a problem which, apart from its intrinsic interest, plays an important part in Teichmüller theory. Let f be quasiconformal in the plane with complex dilatation μ and conformal in a simply connected domain A. The problem is to describe the interrelations existing between μ and the Schwarzian derivative of the restriction $f\,|\,A$.

In section 3 we derive a quantitative estimate which shows that if the sup norm of μ is small, then the norm of the Schwarzian of $f\,|\,A$ is also small. The method of proof can be adapted to the study of many other problems concerning univalent functions with quasiconformal extensions. We establish some results of this type, even though they might veer a little off the main track from the standpoint of applications to the Teichmüller theory.

We return to the main problem in section 4. A basic result says that if a function f, meromorphic in a quasidisc, has a small Schwarzian derivative, then f is univalent and has a quasiconformal extension to the plane with a small μ. It is then proved that the result does not hold in simply connected domains which are not quasidiscs. In this section, the interplay of the concepts complex dilatation, quasidisc and Schwarzian derivative becomes very concrete.

In section 5 we consider meromorphic functions in a disc. Results of section 4 can then be supplemented and expressed in a more explicit form. Again, this is not only of interest in itself but leads to important conclusions in Teichmüller theory.

1. Schwarzian Derivative

1.1. Definition and Transformation Rules

Let us consider a Möbius transformation $z \to f(z) = (az + b)/(cz + d)$. Differentiation yields

$$\frac{f''(z)}{f'(z)} = -\frac{2c}{cz + d}, \qquad \left(\frac{f''}{f'}\right)'(z) = \frac{2c^2}{(cz + d)^2}.$$

Using the notation

$$S_f = \left(\frac{f''}{f'}\right)' - \frac{1}{2}\left(\frac{f''}{f'}\right)^2, \tag{1.1}$$

we conclude that every Möbius transformation satisfies the differential equation

$$S_f = 0. \tag{1.2}$$

Conversely, if we start from the equation (1.2) and set $y = f''/f'$, then $y' = y^2/2$. From this we deduce, by an easy integration, that every solution of (1.2) is a Möbius transformation.

The expression (1.1) is called the *Schwarzian derivative* of the function f. It can of course be defined for a much more general class of functions than Möbius transformations. In order to make clear the notions to be used, we remark that in our terminology, a meromorphic function need not necessarily have poles. A holomorphic function is a meromorphic function without poles. The word "analytic" is a synonym for "meromorphic", but to avoid confusion it is more rarely used. The terms "conformal mapping", "injective meromorphic function", and "univalent analytic (meromorphic) function" all have the same meaning. Sometimes we follow time-honored practice and say "univalent" instead of "univalent analytic".

Let us first assume that f is holomorphic in a domain A in the complex plane and $f'(z) \neq 0$ in A. We then define the Schwarzian derivative S_f of f by means of formula (1.1).

If, in addition, $f(z) \neq 0$, we see from (1.1) that

$$S_f(z) = S_{1/f}(z).$$

We use this formula to define $S_f(z)$ for a meromorphic f at points where f has a first order pole. It follows that if a meromorphic f is locally injective in A, then S_f is defined everywhere in A, and it is holomorphic in A.

Direct computation gives the transformation rule

$$S_{f \circ g} = (S_f \circ g)g'^2 + S_g. \tag{1.3}$$

If g is a Möbius transformation, we have $S_g = 0$, and so

$$S_{f \circ g} = (S_f \circ g)g'^2. \tag{1.4}$$

This formula can be used to define the Schwarzian derivative at infinity. Assume that f is meromorphic and locally injective in a domain which contains ∞, and let φ be defined in a neighborhood of the origin by $\varphi(z) = f(1/z)$. By (1.4), we have $z^4 S_\varphi(z) = S_f(1/z)$. Hence, if we define

$$S_f(\infty) = \lim_{z \to 0} z^4 S_\varphi(z),$$

then S_f is holomorphic at ∞. We see that S_f has a zero of order ≥ 4 at infinity. In conclusion, the Schwarzian derivative can be defined in any domain A for every function f meromorphic and locally injective in A, and it is a holomorphic function in A.

The Schwarzian derivative will play a very central part throughout this chapter on univalent functions and in the presentation of Teichmüller theory in Chapters III and V. It was introduced to complex analysis in 1869 by H. A. Schwarz ([1], p. 78). Schwarz established equation (1.2) and used it to map a simply connected domain bounded by finitely many circular arcs conformally onto a disc.

The special role of Möbius transformations in connection with Schwarzian derivatives appears not only from equation (1.2) but also from the invariance property following from (1.3) and (1.2): If f is a Möbius transformation,

$$S_{f \circ g} = S_g.$$

In studying the local approximation of a meromorphic function by Möbius transformations Martio and Sarvas arrived at the Schwarzian derivative in a way which justifies using the word "derivative" for the operator S_f. Let f be a locally injective meromorphic function in a domain A and z_0 an arbitrary finite point of A. Then there is a unique Möbius transformation h such that

$$\lim_{z \to z_0} \frac{(h \circ f)(z) - z}{(z - z_0)^3}$$

is finite. This limit is equal to $S_f(z_0)/6$.

In the proof we may assume that $f(z_0) \neq \infty$, for if h corresponds to $1/f$, then $z \to h(1/z)$ corresponds to f. For an undetermined Möbius transformation h, we write $(h \circ f)(z) = a_0 + a_1(z - z_0) + a_2(z - z_0)^2 + a_3(z - z_0)^3 + \cdots$. Since h depends on three complex parameters, there is a unique h such that $a_0 = z_0, a_1 = 1, a_2 = 0$. Then $((h \circ f)(z) - z)/(z - z_0)^3$ has the finite limit a_3 as $z \to z_0$. A direct computation, based on $a_1 = 1, a_2 = 0$, and $S_h = 0$, shows that $6a_3 = S_f(z_0)$.

1.2. Existence and Uniqueness

The Schwarzian derivative can be prescribed:

Theorem 1.1. *Let φ be a holomorphic function in a simply connected domain A in the complex plane. Then there is a meromorphic function f in A such that*

$$S_f = \varphi. \tag{1.5}$$

The solution is unique up to an arbitrary Möbius transformation.

PROOF. Through the substitution $y = f''/f'$ equation (1.5) transforms to the Riccati equation $y' - y^2/2 = \varphi$. From this we obtain, by the standard substitution $y = -2w'/w$, the linear second order equation

$$w'' + \tfrac{1}{2}\varphi w = 0. \tag{1.6}$$

It is a well-known result in the classical theory of linear differential equations that given a point $z_0 \in A$, equation (1.6) has a unique holomorphic solution w in a neighborhood of z_0, once we prescribe the values $w(z_0)$ and $w'(z_0)$. It is also easy to verify this directly, with the aid of power series. In fact, we have $\varphi(z) = 2 \sum a_n(z - z_0)^n$ in a disc around z_0. If

$$w(z) = \sum_{n=0}^{\infty} c_n(z - z_0)^n, \tag{1.7}$$

then for w to be a solution we obtain from (1.6)

$$n(n - 1)c_n + \sum_{k=0}^{n-2} a_{n-2-k}c_k = 0, \qquad n = 2, 3, \ldots.$$

The coefficients c_0 and c_1 can be chosen arbitrarily; we take $c_0 = 0, c_1 = 1$. Next we fix an $r, 0 < r < 1$, and a finite number $M > 1$ such that $|a_n| < Mr^{-n}$, $n = 0, 1, \ldots$, and that $Mr^2 < 1$. Then

$$n(n - 1)|c_n| \leq M \sum_{k=0}^{n-2} r^{-n+2+k}|c_k| < r^{-n} \sum_{k=0}^{n-2} r^k|c_k|.$$

The hypothesis $|c_k| \leq r^{-k}$ for $k = 0, 1, \ldots, n - 2$, which is fulfilled for $k = 0$, 1, gives $n|c_n| \leq r^{-n}$. It follows that (1.7) is a holomorphic solution of (1.6) in the disc $|z - z_0| < r$.

Since A is simply connected, we can apply the Monodromy theorem and obtain a global solution of (1.6) by analytic continuation.

Let w_1 and w_2 be two linearly independent holomorphic solutions of (1.6). Since $w_1 w_2'' - w_1'' w_2 = 0$, we have $w_1 w_2' - w_1' w_2 = $ constant. This constant is not zero, because w_1 and w_2 are linearly independent. Set $f = w_1/w_2$; then f is a locally injective meromorphic function in A. Direct computation yields $f''/f' = -2w_2'/w_2$, and so $S_f = -2w_2''/w_2 = \varphi$.

From the invariance of the Schwarzian derivative under Möbius transformations we conclude that, if f is a solution of (1.5) and h an arbitrary Möbius transformation, then $h \circ f$ also is a solution of (1.5).

Assume, conversely, that f and g are solutions of (1.5) in A. Since f is locally injective, we can define $g \circ f^{-1}$ locally. Using (1.3) we deduce from $S_f = S_g$ that $S_{g \circ f^{-1}} = 0$. It follows that locally $g = h \circ f$, where h is a Möbius transformation. But then $g = h \circ f$ with the same Möbius transformation h everywhere in A, and the uniqueness part of the theorem is proved.   □

We supposed in Theorem 1.1 that ∞ is not in A. If $\infty \in A$, it follows from the definition of the Schwarzian derivative at ∞ (see 1.1) that Theorem 1.1 remains valid under the sole restriction that the given function φ must have a zero of order ≥ 4 at infinity. In particular, a function is determined by its Schwarzian derivative up to a Möbius transformation.

1.3. Norm of the Schwarzian Derivative

The Schwarzian derivative S_f measures the deviation of f from a Möbius transformation. In order to make this statement more precise we introduce a norm for S_f.

Let A be a simply connected domain conformally equivalent to a disc and η the Poincaré density of A (cf. I.1.1). For functions φ holomorphic in A we define the norm

$$\| \varphi \|_A = \sup_{z \in A} |\varphi(z)| \eta(z)^{-2}.$$

In particular,

$$\| S_f \|_A = \sup_{z \in A} |S_f(z)| \eta(z)^{-2}. \tag{1.8}$$

There are many reasons to use this "hyperbolic sup-norm" instead of the ordinary norm $\sup|S_f(z)|$. We shall soon see that (1.8) exhibits more invariance with respect to Möbius transformations than $\sup|S_f(z)|$. This is important as such and becomes crucial when we generalize the notion of Schwarzian derivative to Riemann surfaces, because $|S_f|\eta^{-2}$ is a function on a Riemann surface whereas $|S_f|$ is usually not.

It is important to see how the norm (1.8) transforms under conformal mappings. Let f and g be meromorphic functions in a domain A and $h: B \to A$ a conformal mapping. By the transformation rule (1.3),

$$S_{f \circ h} - S_{g \circ h} = (S_f \circ h - S_g \circ h)h'^2.$$

We also know that the hyperbolic metric is conformally invariant: $\eta_B = (\eta_A \circ h)|h'|$. If $w = h(z)$, we thus obtain the basic invariance formula

$$\frac{|S_f(w) - S_g(w)|}{\eta_A(w)^2} = \frac{|S_{f \circ h}(z) - S_{g \circ h}(z)|}{\eta_B(z)^2}. \tag{1.9}$$

Equation (1.9) yields a number of results about the norm. First, it follows immediately from the definition of the norm that we have the invariance

$$\| S_f - S_g \|_A = \| S_{f \circ h} - S_{g \circ h} \|_B.$$

For the special case in which $g = h^{-1}$ is a conformal mapping of A we obtain the formula

$$\| S_f - S_g \|_A = \| S_{f \circ g^{-1}} \|_{g(A)}, \tag{1.10}$$

which will be repeatedly used later. If we choose f here to be the identity mapping, we get the invariance

$$\| S_g \|_A = \| S_{g^{-1}} \|_{g(A)} \tag{1.11}$$

between a conformal mapping and its inverse. Finally, if $g = h^{-1}$ is a Möbius transformation, (1.9) shows the invariance of $|S_f|/\eta^2$ under Möbius transformations, and (1.10) assumes the form

$$\| S_f \|_A = \| S_{f \circ g^{-1}} \|_{g(A)}. \tag{1.12}$$

We see that $\| S_f \|$ is completely invariant with respect to Möbius transformations: If h and g are Möbius transformations, the norms of the Schwarzians are the same for f in A and $h \circ f \circ g$ in $g^{-1}(A)$.

1.4. Convergence of Schwarzian Derivatives

Suppose that the functions f_n, $n = 1, 2, \ldots$, are meromorphic and locally injective in a domain A. If they converge to a locally injective meromorphic function f locally uniformly in A, then the Schwarzians S_{f_n} also tend to S_f locally uniformly in A. However, this does not imply convergence in norm:
From

$$\lim S_{f_n}(z) = S_f(z)$$

locally uniformly in A, it does not necessarily follow that

$$\lim_{n \to \infty} \| S_{f_n} - S_f \|_A = 0. \tag{1.13}$$

A counterexample is obtained as follows. Let r_n, $0 < r_n < 1$, $n = 1, 2, \ldots,$ be numbers tending to 1 and $A = \{z \,|\, |z| > 1\}$. If we set $f_n(z) = z + r_n/z$, $f(z) = z + 1/z$, then f_n and f are meromorphic and even globally injective in A. From $|f_n(z) - f(z)| < 1 - r_n$ we see that $f_n(z) \to f(z)$ uniformly even in the whole domain A. In spite of this, it follows from $\eta_A(z) = (|z|^2 - 1)^{-1}$ and

$$S_{f_n}(z) = -\frac{6r_n}{(z^2 - r_n)^2}, \qquad S_f(z) = -\frac{6}{(z^2 - 1)^2},$$

that

$$\|S_f - S_{f_n}\|_A \geq 6 \lim_{x \to 1} \left(1 - r_n \left(\frac{x^2 - 1}{x^2 - r_n}\right)^2\right) = 6$$

for every n.

The approximating functions f_n map A onto the exteriors of ellipses which collapse to a slit domain $f(A)$. However, if a slit domain is approximated differently, we do get for the Schwarzian derivatives both locally uniform convergence and convergence in norm.

This is seen from the example in which A is the upper half-plane and $f_n(z) = z^{a_n}$, where the numbers a_n are positive and tend increasingly to 2. Then f_n and the limit function $z \to f(z) = z^2$ are univalent in A. From $S_{f_n}(z) = (1 - a_n^2)/(2z^2)$ it follows that

$$|S_{f_n}(z) - S_f(z)| = \frac{4 - a_n^2}{2|z|^2}, \qquad \|S_{f_n} - S_f\|_A = 2(4 - a_n^2).$$

We see that $S_{f_n}(z) \to S_f(z)$ locally uniformly in A and that $S_{f_n} \to S_f$ in norm. In this case $f(A)$ is the plane slit along the non-negative real axis, and the approximating domains $f_n(A) = \{w \,|\, 0 < \arg w < \pi a_n\}$ are infinite sectors.

Figure 2 illustrates the difference between the two cases. In order to get the same limit domain and the same "critical" points, we have replaced f_n in the first example by $2f_n/(1 + r_n)$ and in the second one by $2(f_n + b_n)/(f_n - b_n)$ with $b_n = e^{i\pi a_n/2}$. These changes leave the Schwarzians invariant.

The converse problem is to study whether we can conclude from (1.13) that the mappings f_n converge to f. Since the Schwarzian derivative determines

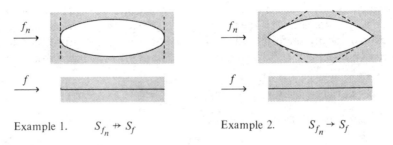

Example 1. $S_{f_n} \nrightarrow S_f$ Example 2. $S_{f_n} \to S_f$

Figure 2

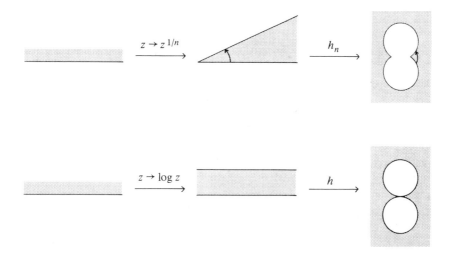

Figure 3

the function only up to a Möbius transformation, some normalization is of course needed for the functions f_n and f.

If no normalization is imposed on the functions f_n in (1.13), they may converge pointwise but the limit can be very different from f. For instance, consider the mappings $z \to f(z) = \log z$ and $z \to f_n(z) = z^{1/n}$ in the upper half-plane. Since $S_f(z) = 1/(2z^2)$, we have

$$\| S_{f_n} - S_f \| = 2/n^2,$$

and so (1.13) is valid. But $f_n(z)$ does not tend to $\log z$ but to the constant 1.

The situation changes if suitable Möbius transformations are applied to f_n and f. In fact, set

$$h_n(w) = \tan \frac{\pi}{4n} \cdot \frac{1 + we^{-(\pi i/2n)}}{1 - we^{-(\pi i/2n)}}, \qquad h(w) = \frac{\pi}{\pi i - 2w}.$$

(For the images of the upper half-plane under $h_n \circ f_n$ and $h \circ f$, see Fig. 3.) We still have of course $S_{h_n \circ f_n} \to S_{h \circ f}$, but now we also have $\lim h_n(z^{1/n}) = h(\log z)$.

In general, the reasoning might go as follows. If we can infer that the functions f_n constitute a normal family, then there is a subsequence f_{n_i} which tends locally uniformly to a limit function f_0. If f_0 is a locally univalent meromorphic function, then $S_{f_{n_i}}(z) \to S_{f_0}(z)$ for every $z \in A$. On the other hand, (1.13) implies that $S_{f_n}(z) \to S_f(z)$ in A. Hence, by the uniqueness part of Theorem 1.1, $f = h \circ f_0$, where h is a Möbius transformation.

Let us give an application. Assume that in (1.13) the functions f_n are conformal mappings of A which have K-quasiconformal extensions to the plane and which keep three fixed boundary points a_1, a_2, a_3 of A invariant. By Theorem I.2.1, the mappings f_n form a normal family. Let f_0 be the limit of a locally uniformly convergent subsequence f_{n_i}. In A the function f_0 is conformal and $S_{f_0} = \lim S_{f_{n_i}}$. Hence, we conclude as above that $f = h \circ f_0$, where h is a Möbius transformation. It follows that f also is a conformal mapping of A with a quasiconformal extension. If f fixes a_1, a_2, a_3, then every convergent subsequence f_{n_i} tends to f. In this case (1.13) implies that the sequence (f_n) itself converges pointwise and $f(z) = \lim f_n(z)$ locally uniformly in A.

We remark that there is a certain analogy in the behavior of Schwarzian derivatives and complex dilatations. In I.4.6 we pointed out that locally uniform convergence of quasiconformal mappings does not imply the convergence in L^∞-norm of their complex dilatations. On the other hand, Theorem I.4.6 shows that the L^∞-convergence of complex dilatations does imply pointwise convergence of suitably normalized mappings.

1.5. Area Theorem

Let us leave Schwarzian derivatives for a moment and establish a classical result on univalent functions, which we shall need in estimating the norm of the Schwarzian derivative of a conformal mapping.

Theorem 1.2 (Area Theorem). *Let f be a univalent meromorphic function in the domain $\{z \mid |z| > 1\}$, with a power series expansion*

$$f(z) = z + \sum_{n=0}^{\infty} b_n z^{-n}. \tag{1.14}$$

Then

$$\sum_{n=1}^{\infty} n|b_n|^2 \leq 1. \tag{1.15}$$

The inequality is sharp.

PROOF. Let C_ρ be the image of the circle $|z| = \rho > 1$ under f. The finite domain bounded by C_ρ has the area

$$m_\rho = \frac{i}{2} \int_{C_\rho} w d\bar{w}.$$

Substituting $w = f(z)$ and considering (1.14) we obtain

$$m_\rho = \pi\rho^2 - \pi \sum_{n=1}^{\infty} n|b_n|^2 \rho^{-2n}.$$

As an area, $m_\rho > 0$ and the result (1.15) follows as $\rho \to 1$.   □

The Area theorem was first proved by Gronwall in 1914 and efficiently used by Bieberbach two years later. It is a historically significant result as it marks the beginning of the systematic theory of univalent functions. We need the following immediate consequence of it:

Under the assumptions of the Area theorem,

$$|b_1| \leq 1. \tag{1.16}$$

Equality holds if and only if $f(z) = z + e^{i\theta}/z$.

For the extremal function of (1.16), the equality sign holds as well in (1.15). For (1.15), there are many other extremals.

The functions satisfying the conditions of the Area theorem with $b_0 = 0$ are said to form the class Σ. Another important class in the theory of univalent functions is S, which consists of functions f univalent and holomorphic in the unit disc with $f(0) = 0$, $f'(0) = 1$. If $f \in S$ and $f(z) = z + \sum_2^\infty a_n z^n$, then $z \to \varphi(z) = 1/f(z^{-2})^{1/2}$ belongs to Σ and $\varphi(z) = z - \frac{1}{2}a_2/z + \cdots$. Hence by (1.16),

$$|a_2| \leq 2, \tag{1.17}$$

with equality only for the Koebe functions $z \to z/(1 + e^{i\theta}z)^2$.

If $f \in S$ and g is an arbitrary conformal self-mapping of the unit disc, then

$$\frac{f \circ g - f(g(0))}{f'(g(0))g'(0)} \tag{1.18}$$

belongs to S. Setting $g(\zeta) = (\zeta + z)/(1 + \bar{z}\zeta)$, we obtain by calculating a_2 for (1.18) and considering (1.17),

$$\left|(1 - |z|^2)\frac{f''(z)}{f'(z)} - 2\bar{z}\right| \leq 4.$$

Integration of this inequality twice leads to the estimate $|f(z)| \geq |z|/(1 + |z|)^2$. As $|z| \to 1$ we get the Koebe one-quarter theorem which we used already in I.1.1.

1.6. Conformal Mappings of a Disc

By Theorem 1.1, any function φ holomorphic in a simply connected domain A of the complex plane is the Schwarzian derivative of a function f which is meromorphic and locally injective in A. It follows that $|S_f(z)|/\eta_A(z)^2$ can grow arbitrarily rapidly as z tends to the boundary of A. In particular, there are many classes of functions f for which the norm of S_f is infinite. Even if f is bounded, it may happen that $\|S_f\|_A = \infty$. An example is the function

$$z \to f(z) = \exp((z + 1)/(z - 1)) \tag{1.19}$$

in the unit disc, for which $S_f(z) = -2(1 - z)^{-4}$.

The behavior of (1.19) differs greatly from that of a Möbius transformation in that (1.19) takes every value belonging to its range infinitely many times. In contrast, a univalent f is analogous to a Möbius transformation in its value distribution, and it will turn out that the norm of its Schwarzian is always finite. We shall first prove this in the case where A is a disc, a disc meaning a domain bounded by a circle or a straight line.

Theorem 1.3. *If f is a conformal mapping of a disc, then*

$$\| S_f \| \leq 6. \tag{1.20}$$

The bound is sharp.

PROOF. By formula (1.12) it does not matter in which disc f is defined. We suppose that f is a conformal mapping of the unit disc D. Let us choose a point $z_0 \in D$ and estimate $|S_f(z_0)| \eta(z_0)^{-2} = (1 - |z_0|^2)^2 |S_f(z_0)|$. By (1.9), this expression is invariant under Möbius transformations. Hence, we may assume that $z_0 = 0$. Also, since f can be replaced by $h \circ f$, where h is an arbitrary Möbius transformation, there is no loss of generality in supposing that $f \in S$. Let a_n denote the nth power series coefficient of f.

The function

$$z \to 1/f(1/z) = z + \sum_{n=0}^{\infty} b_n z^{-n}$$

satisfies the conditions of the Area theorem. From $b_1 = a_2^2 - a_3$ we thus conclude that $|a_2^2 - a_3| \leq 1$. On the other hand, $S_f(0) = 6(a_3 - a_2^2)$. Consequently, $|S_f(0)| \leq 6$, and (1.20) follows.

For the Koebe functions f the coefficient b_1 of $z \to 1/f(1/z)$ is of absolute value 1. Hence, for the Koebe functions $|a_2^2 - a_3| = 1$, and equality holds in (1.20). More generally, in D equality holds in (1.20) for all functions $h \circ f \circ g$, where g is a conformal self-mapping of D, f a Koebe function, and h an arbitrary Möbius transformation. In the upper half-plane, $z \to f(z) = z^2$ is a simple example of a univalent function for which $\| S_f \| = 6$. □

The estimate (1.20) was proved by Kraus [1] in 1932. His paper was forgotten and rediscovered only in the late sixties. Meanwhile, (1.20) was attributed to Nehari ([1]) who proved it in 1949.

2. Distance between Simply Connected Domains

2.1. Distance from a Disc

Let A be an arbitrary simply connected domain which is conformally equivalent to a disc. Even in this general case, the norm of the Schwarzian derivative of a conformal mapping of A is always finite, but the bound may be as high as 12.

In order to study the situation more closely, we introduce the domain constant

$$\delta(A) = \|S_f\|_A,$$

where f is a conformal mapping of A onto a disc. We call $\delta(A)$ *the distance of the domain A from a disc*. In view of the invariance of the norm of the Schwarzian derivative under Möbius transformations, the distance $\delta(A)$ is well defined.

We have $\delta(A) = 0$ if and only if A is a disc. By Theorem 1.3 and formula (1.11),

$$\delta(A) \leq 6$$

for all domains A. The distance $\delta(A)$ measures how much A deviates from a disc, or equivalently, how much a function f mapping A onto a disc deviates from a Möbius transformation.

An illustrative example is provided by the case in which A is the exterior of the ellipse $\{z = e^{i\varphi} + ke^{-i\varphi} | 0 \leq \varphi < 2\pi\}$, $0 \leq k \leq 1$. (For $k = 1$, the ellipse degenerates into the line segment with endpoints ± 2.) The function $z \to f(z) = z + k/z$ maps the disc $E = \{z | |z| > 1\}$ conformally onto A. Because $S_f(z) = -6k(z^2 - k)^{-2}$, it follows that

$$\delta(A) = 6k.$$

We see that $\delta(A)$ changes continuously from 0 to 6 as k increases from 0 to 1. Thus the range of $\delta(A)$ for varying domains A is the closed interval $[0, 6]$.

Another simple example is the angular domain $A = \{z | 0 < \arg z < k\pi\}$, $0 < k \leq 2$. Now $z \to f(z) = z^k$ maps the upper half-plane onto A. From $S_f(z) = (1 - k^2)/(2z^2)$ we obtain

$$\delta(A) = 2|k^2 - 1|. \tag{2.1}$$

Again, $\delta(A)$ covers the whole closed interval from 0 to 6 as k grows from 1 to 2.

It might be expected that domains close to a disc are quasidiscs. This is indeed the case: In section 3 we shall prove that for a K-quasidisc A the distance $\delta(A) \to 0$ as $K \to 1$, and in section 5 that *all* domains with $\delta(A) < 2$ are quasidiscs.

2.2. Distance Function and Coefficient Problems

The problem of estimating the distance function $\delta(A)$ is connected with classical problems regarding the power series coefficients of univalent functions. Theorem 1.3 gave an indication of this. To make this more precise, we first note that in view of formula (1.11), we could define $\delta(A)$ with the aid of conformal mappings of the unit disc D onto A. Let g be a conformal self-mapping of D, such that $g(0) = z_0$. Since $\eta_D(0) = 1$, we conclude from the invariance relation (1.9) that

$$|S_f(z_0)||\eta(z_0)^{-2} = |S_{f \circ g}(0)|.$$

This yields the characterization

$$\delta(A) = \sup\{|S_f(0)| \,|\, f: D \to A \text{ conformal}\} \tag{2.2}$$

for the distance function.

Since composition of f with a Möbius transformation does not change $\delta(A)$, we may further assume that $f(0) = 0$, $f'(0) = 1$ and that ∞ is not in A, i.e., that $f \in S$. Then

$$S_f(0) = 6(a_3 - a_2^2). \tag{2.3}$$

Hence, *determining $\delta(A)$ amounts to maximizing the expression $|a_3 - a_2^2|$.*

2.3. Boundary Rotation

As an example of the use of the characterization (2.2) and the formula (2.3), we shall determine the sharp upper bound for $\delta(A)$ in certain classes of domains characterized geometrically in terms of their boundary. We shall first introduce "boundary rotation".

Let A be a domain of the complex plane whose boundary is a regular Jordan curve. It follows that ∂A is the image of the interval $[0, 2\pi)$ under a continuously differentiable injection γ with a non-zero derivative. Let $\pi\psi(t)$ be the angle between the tangent vector $\gamma'(t)$ to ∂A at $\gamma(t)$ and the vector in the direction of the positive real axis. We assume that $\gamma(t)$ describes ∂A in the positive direction with respect to A as t increases. Then

$$\int_0^{2\pi} d\psi(t) = 2. \tag{2.4}$$

The *boundary rotation* of A is the total variation of $\pi\psi$. If

$$\int_0^{2\pi} |d\psi(t)| = k < \infty, \tag{2.5}$$

the domain A is said to have the boundary rotation $k\pi$.

For an arbitrary domain A in the complex plane conformally equivalent to a disc, boundary rotation is defined as follows. Let A_n, $n = 1, 2, \ldots$, be an exhaustion of A, i.e., $\bar{A}_n \subset A_{n+1} \subset A$ for every n and $\cup A_n = A$. Suppose that $\bar{A}_n \subset B_n$, $\bar{B}_n \subset A$, and that B_n is bounded by a regular Jordan curve. For a fixed n, let α_n be the infimum of the boundary rotations of all such domains B_n. Then the boundary rotation of A is defined to be $\alpha = \lim \alpha_n$. The limit α does not depend on the choice of the exhausting domains A_n, and it agrees with the previously defined boundary rotation for domains whose boundary is a regular Jordan curve.

Boundary rotation can also be defined directly for the above domain A in purely analytic terms. Let f be a conformal mapping of the unit disc D onto A, and let

$$u(z) = 1 + \operatorname{Re} \frac{zf''(z)}{f'(z)}. \tag{2.6}$$

Then the boundary rotation of A is equal to

$$\lim_{r \to 1} \int_0^{2\pi} |u(re^{i\varphi})| \, d\varphi. \tag{2.7}$$

Suppose that A has finite boundary rotation $k\pi$. From the finiteness of (2.7) it follows that the harmonic function u can be represented by means of the Poisson–Stieltjes formula. Integration then yields

$$f'(z) = f'(0)e^{-\int_0^{2\pi} \log(1 - ze^{-i\theta}) \, d\psi(\theta)}. \tag{2.8}$$

Here ψ is a function of bounded variation satisfying (2.4) and (2.5). It agrees with the ψ defined earlier using the tangent if ∂A is a regular Jordan curve. In integrated form (2.8) is a generalization of the classical Schwarz–Christoffel formula for the function mapping a disc conformally onto the interior of a polygon.

Conversely, let ψ be a function of bounded variation satisfying (2.4) and (2.5). Under the additional condition $k \leq 4$, a function f whose derivative is defined by (2.8) is then univalent in D and maps D onto a domain with boundary rotation $k\pi$.

Convex domains of course have finite boundary rotation. More exactly, a domain is convex if and only if its boundary rotation is 2π. Equivalent to this is the assertion that the function u defined by (2.6) is positive in D or that the function ψ is non-decreasing.

For convex domains formula (2.8) was derived by Study in 1913. The notion of boundary rotation was introduced in 1931 by Paatero; his thesis [1] contains detailed proofs for all the results in this subsection.

2.4. Domains of Bounded Boundary Rotation

Using the representation formula (2.8) we can easily estimate $\delta(A)$ for convex domains.

Theorem 2.1. *If A is Möbius equivalent to a convex domain, then*

$$\delta(A) \leq 2. \tag{2.9}$$

Equality holds if A is the image of a parallel strip under a Möbius transformation.

PROOF. We may assume that A itself is convex. Let f be an arbitrary conformal mapping of D onto A. In view of (2.2), inequality (2.9) follows if we prove that $|S_f(0)| \leq 2$. Since we may replace f by the function $z \to cf(ze^{i\varphi})$ for c complex and φ real, there is no loss of generality in assuming that $S_f(0) \geq 0$ and that $f'(0) = 1$.

From (2.8) we obtain by direct computation

$$S_f(0) = \int_0^{2\pi} e^{-2i\theta}\, d\psi(\theta) - \tfrac{1}{2}\left(\int_0^{2\pi} e^{-i\theta}\, d\psi(\theta)\right)^2. \tag{2.10}$$

Since $S_f(0)$ is real and $d\psi(\theta) \geq 0$, it follows that

$$S_f(0) = \int_0^{2\pi} \cos 2\theta\, d\psi(\theta) - \tfrac{1}{2}\left(\int_0^{2\pi} \cos\theta\, d\psi(\theta)\right)^2 + \tfrac{1}{2}\left(\int_0^{2\pi} \sin\theta\, d\psi(\theta)\right)^2$$

$$\leq \int_0^{2\pi} \cos 2\theta\, d\psi(\theta) - \tfrac{1}{2}\left(\int_0^{2\pi} \cos\theta\, d\psi(\theta)\right)^2 + \int_0^{2\pi} \sin^2\theta\, d\psi(\theta) \tag{2.11}$$

$$= \int_0^{2\pi} \cos^2\theta\, d\psi(\theta) - \tfrac{1}{2}\left(\int_0^{2\pi} \cos\theta\, d\psi(\theta)\right)^2 \leq \int_0^{2\pi} \cos^2\theta\, d\psi(\theta) \leq 2.$$

Because $S_f(0) \geq 0$, we have proved (2.9).

Equality holds if

$$\int_0^{2\pi} \cos^2\theta\, d\psi(\theta) = 2, \qquad \int_0^{2\pi} \cos\theta\, d\psi(\theta) = 0.$$

These conditions are fulfilled if ψ has a jump $+1$ at the points 0 and π and is constant on the intervals $(0, \pi)$ and $(\pi, 2\pi)$. Then $S_f(0) = 2$, and it follows from (2.8) that $f'(z) = (1 - z^2)^{-1}$. We conclude that the image of D is a parallel strip. $\qquad\square$

Theorem 2.1 expresses in a quantitative manner the fact that a convex domain is close to a disc: Its distance to a disc is at most 2, while the distance can be as large as 6 in the general case.

Not all domains close to a disc need be Möbius equivalent to a convex domain, as the example $A = \{z \mid 0 < \arg z < k\pi\}$, $k > 1$, shows. Its boundary forms two interior angles $> \pi$, one at 0 and the other at ∞. Since these angles are preserved under Möbius transformations, A is not Möbius equivalent to a convex domain. On the other hand, by formula (2.1) we have $\delta(A) = 2(k^2 - 1)$.

Theorem 2.1 can be generalized.

Theorem 2.2. *Let A be Möbius equivalent to a domain with boundary rotation $\leq k\pi$. If $k \leq 4$, then*

$$\delta(A) \leq \frac{2k + 4}{6 - k}. \tag{2.12}$$

The bound is sharp.

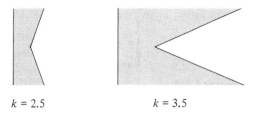

$k = 2.5$ $k = 3.5$

Figure 4. Extremal domains.

The main lines of the proof are the same as those in Theorem 2.1. After similar initial remarks we start from (2.10), assuming this time that $S_f(0) < 0$. In the first line of (2.11) we now ignore the third integral and conclude that

$$|S_f(0)| \leq \tfrac{1}{2} \left(\int_0^{2\pi} \cos \theta \, d\psi(\theta) \right)^2 - \int_0^{2\pi} \cos 2\theta \, d\psi(\theta).$$

With attention paid to (2.4) and (2.5), the estimate (2.12) follows from this after some computation; for the details we refer to Lehto and Tammi [1], p. 255.

From (2.8) we get for the extremal f with $f(0) = 0$ the representation

$$f(z) = \int_0^z \frac{(1 + \zeta)^{k/2 - 1}}{\left(1 + \dfrac{4 - 2k}{6 - k} \zeta + \zeta^2 \right)^{k/4 + 1/2}} \, d\zeta.$$

The corresponding domain $A = f(D)$ is symmetric with respect to the real axis. Its boundary consists of two half-rays in the right half-plane emanating from a point of the real axis and forming the angle $k\pi/2$ in A, and of a vertical line in the left half-plane (Fig. 4). As k grows from 2 to 4, the vertical line moves to the left until it disappears when $k = 4$, i.e., it then reduces to the point ∞.

2.5. Upper Estimate for the Schwarzian Derivative

The use of the distance function δ makes it possible to generalize Theorem 1.3 in a precise form.

Theorem 2.3. *Let A and A' be domains conformally equivalent to a disc and $f: A \to A'$ a conformal mapping. Then*

$$\|S_f\|_A \leq \delta(A) + \delta(A'). \tag{2.13}$$

The estimate is sharp for any given pair of domains A and A'.

PROOF. Let h be a conformal mapping of the unit disc D onto A. From $f = (f \circ h) \circ h^{-1}$ we conclude that $\|S_f\|_A = \|S_{f \circ h} - S_h\|_D$. Since

$$\|S_h\|_D = \delta(A), \qquad \|S_{f \circ h}\|_D = \delta(A'),$$

the triangle inequality yields (2.13).

In order to verify that the estimate (2.13) cannot be improved, we consider conformal mappings $h_1: D \to A$, $h_2: D \to A'$. Given an $\varepsilon > 0$, we choose h_1 and h_2 such that

$$|S_{h_i}(0)| > \|S_{h_i}\|_D - \varepsilon, \qquad i = 1, 2. \tag{2.14}$$

This is possible, because $|S_f(z)|\eta(z)^{-2}$ is invariant under Möbius transformations (cf. the reasoning in 2.2).

Let g be the rotation $z \to e^{i\theta}z$. Then $f = h_2 \circ g \circ h_1^{-1}$ maps A conformally onto A', and

$$\|S_f\|_A = \|S_{h_2 \circ g} - S_{h_1}\|_D.$$

Now

$$\|S_{h_2 \circ g} - S_{h_1}\|_D \geq |S_{h_2 \circ g}(0) - S_{h_1}(0)| = |S_{h_2}(0)e^{2i\theta} - S_{h_1}(0)|.$$

For a suitable θ we obtain from this and (2.14),

$$\|S_f\|_A \geq |S_{h_2}(0)| + |S_{h_1}(0)| > \delta(A') + \delta(A) - 2\varepsilon.$$

Consequently, (2.13) is sharp.   □

From Theorem 2.3 we obtain the new characterization

$$\delta(A) = \tfrac{1}{2}\sup\{\|S_f\|_A | f \text{ conformal self-mapping of } A\}$$

for the distance function.

2.6. Outer Radius of Univalence

Let us introduce another domain constant

$$\sigma_0(A) = \sup\{\|S_f\|_A | f \text{ univalent in } A\}.$$

We call $\sigma_0(A)$ the *outer radius of univalence* of A. In III.5.1 we shall also define the inner radius of univalence of A.

Theorem 2.3 shows that there is a simple connection between the outer radius of univalence and the distance function δ (Lehto [6]).

Theorem 2.4. *Let A be a domain conformally equivalent to a disc. Then*

$$\sigma_0(A) = \delta(A) + 6. \tag{2.15}$$

PROOF. We write the definition of $\sigma_0(A)$ in the form

$$\sigma_0(A) = \sup_{A'}\{\|S_f\|_A | f: A \to A' \text{ conformal}\}.$$

Then it follows from Theorem 2.3 that

$$\sigma_0(A) = \delta(A) + \sup_{A'} \delta(A').$$

Hence, we obtain (2.15) from Theorem 1.3. $\square$

We can thus extend Theorem 1.3 to domains A for which we know $\delta(A)$. For instance, if f is univalent in the sector $A = \{z \,|\, 0 < \arg z < k\pi\}$, $1 \le k \le 2$, then $\| S_f \|_A \le 2k^2 + 4$, or if f is univalent in a convex domain A, then $\| S_f \|_A \le 8$.

Since $\delta(A) \le 6$, we see that in all simply connected domains A and for all functions f univalent in A, we always have

$$\| S_f \|_A \le 12.$$

The maximum 12 can be attained. We must then have $\delta(A) = 6$, and extremals are obtained, for instance, if we consider suitable self-mappings of such domains A.

As one example, let us consider the domain A which is the complement of the line segment $\{x \,|\, -1 \le x \le 1\}$ with respect to the extended plane. In the disc $E = \{w \,|\, |w| > 1\}$ we set $h_\theta(w) = (w + e^{i\theta}/w)/2$, $0 \le \theta \le \pi$, and define

$$f_\theta = e^{-i\theta/2} h_\theta \circ h_0^{-1}.$$

Then every h_θ is a conformal mapping of E, and f_θ a conformal self-mapping of A. We have

$$f_\theta(z) = z \cos\frac{\theta}{2} - i\sqrt{z^2 - 1} \sin\frac{\theta}{2}.$$

From $S_{h_\theta}(w) = -6e^{i\theta}/(w^2 - e^{i\theta})^2$ and $\| S_{f_\theta} \|_A = \| S_{h_\theta} - S_{h_0} \|_E$ it follows that

$$\| S_{f_\theta} \|_A \ge 12 \sin\frac{\theta}{2}.$$

In particular, the maximum value 12 is taken by the function

$$z \to f_\pi(z) = -i\sqrt{z^2 - 1}.$$

There is even a point of A, namely ∞, at which $|S_{f_\pi}(z)|\eta_A(z)^{-2}$ takes the value 12.

2.7. Distance between Arbitrary Domains

Let us consider again two domains A and A' which are conformally equivalent to a disc. As a generalization of the distance to a disc, we introduce the number

$$\delta(A, A') = \inf\{\| S_f \|_A \,|\, f : A \to A' \text{ conformal}\}$$

and call it the distance between the domains A and A'.

The triangle inequality yields the estimates

$$|\delta(A) - \delta(A')| \leq \delta(A, A') \leq \delta(A) + \delta(A').$$

Equality holds in both places if A or A' is a disc. A less trivial example of equality in the lower estimate is provided by the angular domains

$$A = \{z | 0 < \arg z < k\pi\}, \qquad A' = \{z | 0 < \arg z < k'\pi\},$$

$0 \leq k \leq k' \leq 2$. If $f_1(z) = z^k, f_2(z) = z^{k'}$, then $f_2 \circ f_1^{-1}$ is a conformal mapping of A onto A'. Hence

$$\delta(A, A') \leq \|S_{f_2} - S_{f_1}\|_H = 2(k'^2 - k^2).$$

If $k' \leq 1$ or if $k \geq 1$, we have $2(k'^2 - k^2) = 2(1 - k^2) - 2(1 - k'^2) = |\delta(A) - \delta(A')|$, so that in these cases $\delta(A, A') = |\delta(A) - \delta(A')|$.

If $k' > 1, k < 1$, then $2(k'^2 - k^2) = \delta(A) + \delta(A')$, but $\delta(A, A')$ is presumably smaller.

Let Ω be the quotient of the set of domains conformally equivalent to a disc by the group of Möbius transformations. The equivalence class containing all discs can be called the origin of Ω and our previous function $\delta(A)$ the distance to the origin of the equivalence class which contains A.

It is an open problem whether

$$\delta(A, A') = 0$$

implies that A and A' are Möbius equivalent. If the answer is affirmative, then (Ω, δ) is a metric space. Another open problem is to determine the diameter of Ω.

3. Conformal Mappings with Quasiconformal Extensions

3.1. Deviation from Möbius Transformations

Let f be a sense-preserving homeomorphism of the extended plane onto itself. If f is conformal everywhere, then f is a Möbius transformation. If f is a quasiconformal mapping whose complex dilatation μ is small in absolute value, we know that f behaves locally almost like a conformal mapping; we remind the reader of the simple geometric interpretation of $|\mu(z)|$ in I.4.1. Thus the number $\|\mu\|_\infty$ can be regarded as a measure for the deviation of f from a Möbius transformation.

Suppose now that f is quasiconformal in the extended plane with complex dilatation μ and that, furthermore, f is conformal in a simply connected domain A. From the results in the preceding sections we know that there is another measure for the deviation of f from a Möbius transformation, its

Schwarzian derivative: If the norm $\|S_{f|A}\|$ is small, then at least in A the mapping f is close to a Möbius transformation.

It turns out that the norms $\|\mu\|_\infty$ and $\|S_{f|A}\|$ are closely related. If $\|\mu\|_\infty$ is small, then so is $\|S_{f|A}\|$, i.e., the behavior of f in the complement of A, which contains the support of μ, is reflected in the behavior of f in A (Theorem 3.2 in this section). Conversely, if f is conformal in a quasidisc A and $\|S_f\|_A$ is small, then f can be extended to a quasiconformal mapping of the plane with a small $\|\mu\|_\infty$ (Theorem 4.1 in the following section).

The study of the relationships between complex dilatations and Schwarzian derivatives will be one of the leading themes of this monograph. Apart from its intrinsic interest, the possibility of using these two apparently different measures of the deviation from Möbius transformations is important in the theory of Teichmüller spaces.

3.2. Dependence of a Mapping on its Complex Dilatation

In studying the effect of the complex dilatation on the Schwarzian derivative we need a result detailing how a quasiconformal mapping changes when its complex dilatation is multiplied by a complex number. We consider the normalized case of Theorem I.4.3 and make use of the representation formula given therein.

Theorem 3.1. Let μ be a measurable function in the plane with bounded support and $\|\mu\|_\infty < 1$. Let $z \to f(z, w)$ be the quasiconformal mapping of the plane with complex dilatation $w\mu$ and with the property $\lim(f(z, w) - z) = 0$ as $z \to \infty$. Then, for every fixed $z \neq \infty$, the function $w \to f(z, w)$ is holomorphic in the disc $|w| < 1/\|\mu\|_\infty$.

Also, for every fixed z outside the support of μ, the derivatives of the analytic function $z \to f(z, w)$ depend holomorphically on w for $|w| < 1/\|\mu\|_\infty$.

PROOF. By Theorem I.4.3,

$$f(z) = f(z, 1) = z + \sum_{i=1}^{\infty} T\varphi_i(\mu)(z),$$

where we now write $\varphi_i(\mu)$ instead of φ_i to accentuate the dependence of φ_i on μ. From the definition of the functions $\varphi_i(\mu)$ it follows that

$$\varphi_i(w\mu) = w^i \varphi_i(\mu).$$

Hence,

$$f(z, w) = z + \sum_{i=1}^{\infty} T\varphi_i(\mu)(z)w^i. \tag{3.1}$$

From formula (4.15) in I.4.4 we see that $\sum T\varphi_i(\mu)(z)$ converges uniformly whenever $\|\mu\|_\infty < 1$. It follows that the power series (3.1) converges if $|w| \|\mu\|_\infty < 1$. Consequently, $w \to f(z,w)$ is analytic in the disc $|w| < 1/\|\mu\|_\infty$.

Outside the support of μ, the function $z \to f(z,w)$ is a conformal mapping. Also, each function $z \to T\varphi_i(\mu)(z)$ is holomorphic, and $T\varphi_i(\mu)(z)$ is no longer a singular integral. Therefore, we can differentiate in (3.1) with respect to z term by term, without affecting the convergence of the series. If prime denotes differentiation with respect to z, we obtain

$$f'(z,w) = 1 + \sum_{i=1}^\infty (T\varphi_i(\mu))'(z)w^i,$$

and similarly for higher derivatives. It follows that all derivatives of $z \to f(z,w)$ depend holomorphically on w in the disc $|w| < 1/\|\mu\|_\infty$.   □

Theorem 3.1 makes it possible to study the dependence of the power series coefficients of $z \to f(z,w)$ on w. Let

$$f(z,w) = z + \sum_{n=1}^\infty b_n(w)z^{-n}$$

in a neighborhood of infinity. Then *the coefficients $w \to b_n(w)$ are holomorphic in $|w| < 1/\|\mu\|_\infty$*.

To prove this we first note that if f is analytic for $|z| > R$, then $z \to f(Rz)/R$ is in class Σ. Therefore, we may assume without loss of generality that f satisfies the conditions of the Area theorem.

First of all, we have

$$b_1(w) = \lim_{z \to \infty} z(f(z,w) - z).$$

Here the convergence is uniform in w, because Schwarz's inequality and the Area theorem yield the estimate

$$|z(f(z,w) - z) - b_1(w)|^2 \le \sum_{n=2}^\infty |z|^{2-2n}/n \le \frac{1}{|z|^2 - 1}.$$

By Theorem 3.1, the function $w \to z(f(z,w) - z)$ is holomorphic in $|w| < 1/\|\mu\|_\infty$ for every fixed z. Hence $w \to b_1(w)$ is holomorphic, as the uniform limit of holomorphic functions. The analyticity of $w \to b_n(w)$ is deduced similarly from

$$b_n(w) = \lim_{z \to \infty} z^n(f(z,w) - z - \sum_{j=1}^{n-1} b_j(w)z^{-j})$$

by induction.

The special normalization of the mappings is not essential in Theorem 3.1.

Corollary 3.1. *Let μ be a measurable function in the plane which vanishes in the upper half-plane and for which $\|\mu\|_\infty < 1$. Let $f_{w\mu}$ be the quasiconformal mapping of the plane with complex dilatation $w\mu$ which keeps the points 0, 1, ∞*

fixed. Then the function $w \rightarrow f_{w\mu}(z)$ *is holomorphic in* $|w| < 1/\|\mu\|_\infty$ *for every finite z.*

PROOF. Let g be the Möbius transformation which maps the points $0, 1, \infty$ on the points $-1, i, 1$, respectively, and $\tilde{f}_{wv}$ a quasiconformal mapping of the plane whose complex dilatation wv agrees with that of $f_{w\mu} \circ g^{-1}$. Then

$$\mu(z) = v(g(z))\overline{g'(z)}/g'(z).$$

Further, let $\tilde{f}_{wv}$ satisfy the normalization condition $\tilde{f}_{wv}(\zeta) - \zeta \rightarrow 0$ as $\zeta \rightarrow \infty$. By Theorem 3.1, $w \rightarrow \tilde{f}_{wv}(g(z))$ is analytic in $|w| < 1/\|v\|_\infty = 1/\|\mu\|_\infty$.
Set $a_1 = \tilde{f}_{wv}(-1)$, $a_2 = \tilde{f}_{wv}(i)$, $a_3 = \tilde{f}_{wv}(1)$, and

$$h_w(\zeta) = \frac{a_2 - a_3}{a_2 - a_1} \frac{\zeta - a_1}{\zeta - a_3}.$$

Then $h_w \circ \tilde{f}_{wv} \circ g$ has the complex dilatation $w\mu$, and it fixes $0, 1, \infty$. Consequently,

$$f_{w\mu}(z) = h_w(\tilde{f}_{wv}(g(z))).$$

By applying Theorem 3.1 again, we conclude that $h_w(\zeta)$ depends analytically on w. It follows that $w \rightarrow f_{w\mu}(z)$ is holomorphic in the disc claimed. □

From $f_{w\mu} = h_w \circ \tilde{f}_{wv} \circ g$ we also deduce that for every z in the upper half-plane, the derivatives of $z \rightarrow f_{w\mu}(z)$ depend holomorphically on w in the disc $|w| < 1/\|\mu\|_\infty$.

Remark. Theorem 3.1 (and Corollary 3.1) can be generalized: The mapping depends holomorphically on w if its complex dilatation $\mu(\cdot, w)$ is a holomorphic function of w. We shall not need the result in this generality. (A proof, which is still a straightforward application of the representation formula for $f(z, w)$, is in Lehto [3].) For most of our applications, the simple case $\mu(z, w) = w\mu(z)$ is sufficient, but in V.5 we shall also be dealing with complex dilatations of the form $w\mu + v$. In this case, the generalization of Theorem 3.1 is immediate.

More precisely, let us assume that μ and v vanish outside of a disc and that $\|\mu\|_\infty < 1$, $\|v\|_\infty < 1$. Now $T\varphi_i(w\mu + v) = P_i(w)$, $i = 1, 2, \ldots,$ where P_i is a polynomial in w of degree i. We again use the fact that $\sum T\varphi_i(\mu)(z)$ converges uniformly whenever $\|\mu\|_\infty < 1$. It follows that in the representation

$$f(z, w\mu + v) = z + \sum_{n=1}^{\infty} P_n(w)(z),$$

the right-hand series is uniformly convergent in w, provided that $\|w\mu + v\|_\infty < 1$. We conclude that *for every finite z, the function* $w \rightarrow f(z, w\mu + v)$ *is holomorphic in the disc* $|w| < (1 - \|v\|_\infty)/\|\mu\|_\infty$.

By using this result we see that in Corollary 3.1, the complex dilatation $w\mu$

can be replaced by $w\mu + v$. Also, the holomorphic dependence of the derivatives on the parameter w remains in effect if $w\mu$ is replaced by $w\mu + v$.

Ahlfors and Bers realized that in the theory of Teichmüller spaces it is of basic importance to study quasiconformal mappings with varying complex dilatations. In their joint paper [1] they proved the holomorphic dependence of the mapping on its complex dilatation.

3.3. Schwarzian Derivatives and Complex Dilatations

Using Theorem 3.1, we can prove a result which shows that a small complex dilatation forces the Schwarzian derivative to be small.

Theorem 3.2. *Let f be a quasiconformal mapping of the plane which has the complex dilatation μ and which is conformal in a simply connected domain A with at least two boundary points. Then*

$$\|S_{f|A}\|_A \le \sigma_0(A)\|\mu\|_\infty. \tag{3.2}$$

PROOF. If g is a Möbius transformation, we can replace f by $f \circ g$ without changing the norms of either the Schwarzian derivative or the complex dilatation. Also, $\sigma_0(A) = \sigma_0(g^{-1}(A))$. We may therefore assume that $\infty \in A$. Then μ has bounded support.

Let w be a complex number with $|w| < 1$. We consider for a moment the unique quasiconformal mapping $z \to f(z, w/\|\mu\|_\infty)$ of the plane which has the complex dilatation $w\mu/\|\mu\|_\infty$ and the property $f(z, w/\|\mu\|_\infty) - z \to 0$ as $z \to \infty$. (We may assume that $\|\mu\|_\infty > 0$.) By Theorem 3.1, the derivatives of the analytic function $z \to (f|A)(z, w/\|\mu\|_\infty)$ with respect to z depend analytically on w in the unit disc, at every finite point z of A.

Keeping z fixed, we define the function

$$w \to \psi(w) = S_{f(\cdot, w/\|\mu\|_\infty)}(z)\eta_A(z)^{-2}.$$

Since S_f is a rational function of the first three derivatives of $f|A$, we conclude that ψ is analytic in the unit disc $|w| < 1$. Furthermore, the function ψ is bounded: $|\psi(w)| \le \sigma_0(A)$. From the fact that $z \to f(z, 0)$ is the identity mapping it follows that $\psi(0) = 0$. We can therefore apply Schwarz's lemma to ψ and get

$$|\psi(w)| \le \sigma_0(A)|w|.$$

Setting $w = \|\mu\|_\infty$, we get back, modulo a Möbius transformation, the function $z \to f(z)$ we started with, and (3.2) follows. $\qquad\square$

If A is a disc, (3.2) assumes the form

$$\|S_{f|A}\| \le 6\|\mu\|_\infty$$

(Kühnau [1], Lehto [1]). The bound is sharp: For each $\|\mu\|_\infty = k, 0 \le k < 1$, there are mappings f for which $\|S_{f|A}\| = 6k$. For instance, in the case $A = \{z||z| > 1\}$ the function f defined by

$$f(z) = z + k/z \quad \text{if} \quad |z| > 1, \qquad f(z) = z + k\bar{z} \quad \text{if} \quad |z| \le 1, \qquad (3.3)$$

is such an extremal. Other extremals are obtained if this f is composed with Möbius transformations.

Using Theorem 3.2, we can estimate $\delta(A)$ for quasidiscs.

Theorem 3.3. *If A is a K-quasidisc, then*

$$\delta(A) \le 6\frac{K^2 - 1}{K^2 + 1}. \qquad (3.4)$$

PROOF. By Lemma I.6.2, the domain A is the image of the upper half-plane H under a K^2-quasiconformal mapping f of the plane which is conformal in H. By Theorem 3.2,

$$\|S_{f|H}\|_H \le 6\frac{K^2 - 1}{K^2 + 1}.$$

On the other hand, $\|S_{f|H}\|_H = \delta(A)$.   □

It is not known whether the bound in (3.4) is sharp. From the example (3.3) we deduce that the sharp bound is $\ge 6(K - 1)/(K + 1)$. In any event, inequality (3.4) shows that *for a K-quasidisc A, the distance $\delta(A) \to 0$ as $K \to 1$.*

3.4. Asymptotic Estimates

Application of the representation formula (3.1) and reasoning similar to that used in proving Theorem 3.2 make it possible to obtain readily a number of results for conformal mappings with quasiconformal extensions. The rest of this section will be devoted to questions of this type. This entails a brief detour from our main theme, the connection between Schwarzian derivatives and complex dilatations.

Let us consider conformal mappings f which belong to class Σ, i.e., f is univalent in $E = \{z||z| > 1\}$ and has a power series expansion of the form

$$f(z) = z + \sum_{n=1}^{\infty} b_n z^{-n} \qquad (3.5)$$

in E. If f has a quasiconformal extension to the plane with complex dilatation μ satisfying the inequality $\|\mu\|_\infty \le k < 1$, we say that f belongs to the subclass Σ_k of Σ. By abuse of notation, we sometimes use the symbol f both for the conformal mapping of E and for its extension to the plane.

We shall first derive asymptotic estimates for $|f(z) - z|$ and $|b_n|$ in Σ_k as $k \to 0$.

Theorem 3.4. *Let* $f \in \Sigma_k$ *and* $k < k_0 < 1$. *As* $k \to 0$,

$$f(z) = z - \frac{1}{\pi} \int\int_D \frac{\mu(\zeta)}{\zeta - z} d\xi \, d\eta + O(k^2) \tag{3.6}$$

in the whole plane. Here $|O(k^2)| \le ck^2$, *the constant* c *depending only on* k_0.

PROOF. If $p > 2$ and $k_0 \|H\|_p < 1$, we see from formula (4.15) in I.4.4 that

$$\sum_{i=2}^{\infty} |T\varphi_i(z)| \le c_p' \sum_{i=2}^{\infty} (k \|H\|_p)^i \le ck^2.$$

Hence, Theorem 3.4 follows from Theorem I.4.3. □

Corollary 3.2. *The functions* $f \in \Sigma_k$ *satisfy the asymptotic inequality*

$$|f(z) - z| \le \frac{k}{\pi} \int\int_D \frac{d\xi \, d\eta}{|\zeta - z|} + ck^2. \tag{3.7}$$

If

$$\mu(\zeta) = ke^{i\theta} \frac{\overline{\zeta - z}}{|\zeta - z|} \quad a.e., \tag{3.8}$$

then

$$|f(z) - z| = \frac{k}{\pi} \int\int_D \frac{d\xi \, d\eta}{|\zeta - z|} + O(k^2).$$

For $z = 0$, the estimate (3.7) gives $|f(0)| \le 2k + ck^2$. The mapping with extremal μ's (3.8) can be determined:

$$f(z) = \begin{cases} z + k^2 e^{2i\theta}/z & \text{if } |z| > 1, \\ z + 2ke^{i\theta}(|z| - 1) + k^2 e^{2i\theta}\overline{z} & \text{if } |z| \le 1. \end{cases}$$

Hence $|f(0)| = 2k$. These functions f are not only asymptotically extremal but they actually maximize $|f(0)|$ in Σ_k (Kühnau [2], Lehto [3]).

For the coefficients b_n in (3.5), a counterpart of (3.7) can also be easily established.

Theorem 3.5. *In the class* Σ_k,

$$|b_n| \le \frac{2k}{n+1} + ck^2, \qquad n = 1, 2, \ldots, \tag{3.9}$$

with $c \le n^{-1/2}(1 - k)^{-1}$. *If*

$$f_n(z) = \begin{cases} (z^{(n+1)/2} + kz^{-(n+1)/2})^{2/(n+1)} & \text{if } |z| > 1, \\ (z^{(n+1)/2} + k\overline{z}^{(n+1)/2})^{2/(n+1)} & \text{if } |z| \le 1, \end{cases} \tag{3.10}$$

then $f_n \in \Sigma_k$ *and* $b_n = 2k/(n+1)$.

PROOF. We have

$$T\varphi_i(z) = -\frac{1}{\pi}\int\int_D \frac{\varphi_i(\zeta)}{\zeta - z}\,d\xi\,d\eta = \frac{1}{\pi}\sum_{n=1}^{\infty}\left(\int\int_D \varphi_i(\zeta)\zeta^{n-1}\,d\xi\,d\eta\right)z^{-n}$$

for $|z| > 1$. Hence,

$$b_n = \frac{1}{\pi}\sum_{i=1}^{\infty}\int\int_D \varphi_i(\zeta)\zeta^{n-1}\,d\xi\,d\eta. \tag{3.11}$$

Schwarz's inequality and the estimate (4.13) in I.4.4 for $p = 2$ (in which case $\|H\|_p = 1$) yield

$$\left|\int\int_D \varphi_i(\zeta)\zeta^{n-1}\,d\xi\,d\eta\right| \le \pi^{1/2}n^{-1/2}\|\varphi_i\|_2 \le \pi n^{-1/2}k^i.$$

Consequently, we have the asymptotic representation

$$b_n = \frac{1}{\pi}\int\int_D \mu(\zeta)\zeta^{n-1}\,d\xi\,d\eta + O(k^2), \qquad n = 1, 2, \ldots,$$

the remainder term being $\le n^{-1/2}k^2/(1-k)$ in absolute value. From this (3.9) follows.

We conclude by easy computation that $b_n = 2k/(n+1) + O(k^2)$ if

$$\mu(\zeta) = k(\bar{\zeta}/\zeta)^{(n-1)/2} \quad \text{a.e.}$$

Direct verification shows that the function f_n defined by (3.10) has this complex dilatation. We also see that for f_n, the coefficient $b_n = 2k/(n+1)$. □

Inequality (3.9) was established by Kühnau [2] using variational methods. Note that the functions (3.10) are related to each other by

$$f_n(z) = (f_1(z^{(n+1)/2}))^{2/(n+1)}, \qquad n = 1, 2, \ldots.$$

Along with f_n, the functions $z \to e^{-i\theta/(n+1)}f_n(ze^{i\theta/(n+1)})$ are also extremal. Some extremal domains are pictured in Fig. 5.

In 3.6 we shall make a few more remarks on the coefficients b_n in Σ_k.

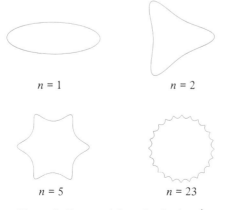

$n = 1$ $n = 2$

$n = 5$ $n = 23$

Figure 5. Extremal domains for $k = \frac{1}{2}$.

3.5. Majorant Principle

By the Area theorem, the coefficients of a function $f \in \Sigma$ satisfy the inequality $|b_n| \leq 1$. It follows that

$$|f(z) - z| \leq \sum_{n=1}^{\infty} |z|^{-n} = \frac{1}{|z| - 1}. \tag{3.12}$$

We conclude from Theorem I.2.1 that Σ is a normal family. The estimate (3.12) also shows that Σ is closed, i.e., it contains the limits of its locally uniformly convergent sequences.

Every subfamily Σ_k is then also normal. Applying Theorem I.2.1 once more we infer that the functions of Σ_k, extended quasiconformally to the plane with $\|\mu\|_\infty \leq k$, constitute a normal family. This implies that every Σ_k is closed also.

Let Φ be a complex-valued functional defined on Σ. We say that Φ is continuous if $\Phi(f_n) \to \Phi(f)$ whenever $f_n(z) \to f(z)$ locally uniformly in E. For a continuous Φ, there are extremal functions which maximize $|\Phi(f)|$ in Σ_k. This follows immediately from the fact that Σ_k is a closed normal family. We set

$$M(k) = \max_{f \in \Sigma_k} |\Phi(f)|.$$

Clearly $M(k)$ is non-decreasing in k.

Let us assume, in particular, that $\Phi(f)$ is a holomorphic function of finitely many of the power series coefficients f and of the values of f and its derivatives f', f'', ..., $f^{(m)}$ at finitely many given points of E. We call such functionals *analytic*. An analytic functional is continuous, because uniform convergence of analytic functions implies uniform convergence of their derivatives. An example of an analytic functional is $\Phi(f) = (|z|^2 - 1)^2 S_f(z)$, considered in Theorem 3.2, which is a rational function of $f'(z)$, $f''(z)$ and $f'''(z)$.

Applying Schwarz's lemma as in the proof of Theorem 3.2, we get the following general result (Lehto [3]).

Theorem 3.6. *Let Φ be an analytic functional on Σ which vanishes for the identity mapping. Then $M(k)/k$ is non-decreasing on the interval $(0, 1)$.*

PROOF. Fix k and k', $0 < k < k' < 1$, and choose an arbitrary mapping $f_0 \in \Sigma_k$. Let μ be the complex dilatation of some extension of f_0, $\|\mu\|_\infty \leq k$. Consider the mappings $f \in \Sigma_{k'}$ which have the complex dilatation $w\mu$ with $|w| < k'/k$. By Theorem 3.1, $\Phi(f)$ depends holomorphically on w in the disc $|w| < k'/k$. If $w = 0$, then f is the identity mapping, so that $\Phi(f)$ vanishes at $w = 0$. Therefore, by Schwarz's lemma

$$|\Phi(f)| \leq \frac{k}{k'} M(k')|w|.$$

For $w = 1$ we have $f = f_0$. Since f_0 is an arbitrary element of Σ_k, we get the desired inequality $M(k) \leq kM(k')/k'$. $\qquad\square$

Corollary 3.3 (Majorant Principle). *If Φ is an analytic functional on Σ which vanishes for the identity mapping, then*

$$\max_{f \in \Sigma_k} |\Phi(f)| \leq k \max_{f \in \Sigma} |\Phi(f)|. \tag{3.13}$$

If equality holds for one value $k \in (0, 1)$, then it holds for all values of k.

PROOF. Inequality (3.13) follows immediately from Theorem 3.6 if we let $k \to 1$.

Suppose that equality holds in (3.13) for some value k, $0 < k < 1$. Let f_k be extremal in this Σ_k and μ the complex dilatation of its extension. For functions f with complex dilatation $w\mu$, $|w| < 1/k$, Schwarz's lemma gives $|\Phi(f)| \leq kM(1)|w|$, where $M(1) = \max|\Phi(f)|$ over Σ. But now equality holds for $w = 1$. It follows that

$$|\Phi(f)| = kM(1)|w| \tag{3.14}$$

in the whole disc $|w| < 1/k$. If $k' \in [0, 1)$ is arbitrarily given, then for $w = k'/k$ the function f is in $\Sigma_{k'}$. Combining (3.13) and (3.14) we deduce that f is extremal in $\Sigma_{k'}$. In other words, if equality holds in (3.13) for one value $k \in (0, 1)$, then it holds for all values of k, and if μ is an extremal complex dilatation, then all dilatations $w\mu$, $|w| < 1/\|\mu\|_\infty$, are extremal. $\qquad\square$

3.6. Coefficient Estimates

Let us illustrate the majorant principle (3.13) with an example. We choose $\Phi(f) = b_n$, for an arbitrary fixed n. This Φ is admissible, because every b_n is zero for the identity mapping. It follows that

$$\max_{\Sigma_k} |b_n| \leq k \max_{\Sigma} |b_n|. \tag{3.15}$$

Assume that for a fixed n, equality holds for some $k > 0$. By what was said about equality in (3.13), equality then holds for all values of k. It follows that the extremal $|b_n|$ for k is k/k_0-times the extremal $|b_n|$ for k_0. Furthermore, if μ_0 is an extremal dilatation for k_0, then $w\mu_0$ is extremal for k with $|w| = k/k_0$. Now consider the formula (3.11). If μ_0 is replaced by $w\mu_0$, then φ_i is multiplied by the power w^i. It follows that all terms in the series (3.11) vanish, except for the first one. But then $|b_n| = 2k/(n + 1)$, and by (3.15),

$$\max_{\Sigma} |b_n| = 2/(n + 1).$$

This is known to be true if $n = 1$, 2. (For classical results on univalent functions we refer to Duren's monograph [1], which also contains an extensive bibliography.) We showed that $b_n = 2k/(n + 1)$ for the extremal (3.10) in Theorem 3.4. Consequently, (3.15) is sharp if $n = 1, 2$.

Thus in Σ_k, $\max|b_1| = k$, $\max|b_2| = 2k/3$. In Σ, $\max|b_3| = 1/2 + e^{-6}$, whereas in Σ_k, $k > 0$, we only know that $\max|b_3|$ is strictly less than $k(1/2 + e^{-6})$. For $n \geq 4$, $\max|b_n|$ is unknown both in Σ and Σ_k. From Theorem 3.5 it follows that

$$\lim_{k \to 0} \max_{\Sigma_k} |b_n|/k = 2/(n + 1)$$

for all values of n.

The Area theorem can also be established for Σ_k. Getting its sharp form requires a small modification of the reasoning which led to (3.13).

Theorem 3.7. *In the class Σ_k,*

$$\sum_{n=1}^{\infty} n|b_n|^2 \leq k^2. \tag{3.16}$$

The estimate is sharp.

PROOF. Given an arbitrary function $f \in \Sigma_k$ with the coefficients b_n, we set $\lambda_n = |b_n|^2/b_n^2$ if $b_n \neq 0$; otherwise $\lambda_n = 1$. Let μ be the complex dilatation of the extended f, and $b_n(w)$ the coefficients of the function $z \to f(z, w/k)$ with Σ-normalization and complex dilatation $w\mu/k$.

For an arbitrary positive integer N, we write

$$\psi(w) = \sum_{n=1}^{N} n\lambda_n b_n(w)^2.$$

By Theorem 3.1, ψ is holomorphic in the unit disc. The Area theorem gives the estimate $|\psi(w)| \leq 1$. Since $b_n(0) = 0$, the function ψ has a zero of order ≥ 2 at the origin. Schwarz's lemma therefore yields the improved estimate

$$|\psi(w)| \leq |w|^2. \tag{3.17}$$

If we set $w = k$, we get back the function f with which we started. Hence

$$\sum_{n=1}^{N} n\lambda_n b_n^2 = \sum_{n=1}^{N} n|b_n|^2 \leq k^2.$$

The desired inequality (3.16) follows as $N \to \infty$.

Equality holds in (3.16) if $f(z) = z + ke^{i\theta}/z$ in $|z| > 1$ and $f(z) = z + ke^{i\theta}\bar{z}$ in $|z| \leq 1$. A relatively simple argument shows that there are no other extremals (Lehto [1]).   □

The result (3.16) is due to Kühnau [2] and Lehto [1].

In V.7.5 we shall again use an estimate of type (3.17) in connection with another problem where we know that the holomorphic function under consideration has at least a double zero at the origin.

The general inequality (3.13) allows many other applications. For instance, the classical Grunsky and Golusin inequalities in Σ can be immediately sharpened for Σ_k. It is also possible to change the original setting and use other normalizations for the mappings. If we consider the class S instead of Σ and define the subclass S_k of S in the same way we defined Σ_k, a difficulty is encountered, because the complex dilatation of the extended mapping does not determine the element of S_k uniquely (S contains many Möbius transformations). Unique correspondence follows, for example, if we require that the extended mappings fix ∞. For this subclass $S_k(\infty)$ we can sharpen a number of results known to be valid in S. (For various applications of the majorant principle, see e.g. Lehto [3], [5].)

Univalent functions with quasiconformal extensions can also be studied by use of variational techniques. Compared to (3.13), such methods often involve much more laborious computations, but are essentially better in many cases where (3.13) fails to give a sharp result. Kühnau is a pioneer in this field. He has had many successors, so that an extensive literature exists today. Delving into these questions would take us too far afield from our main topic, and we content ourselves here with mentioning, besides Kühnau's works [1] and [2], the papers of Schiffer [1] and Schiffer–Schober [1], the lecture notes of Schober [1], and the monographs of Pommerenke [1] and Krushkal–Kühnau [1] among many others.

4. Univalence and Quasiconformal Extensibility of Meromorphic Functions

4.1. Quasiconformal Reflections under Möbius Transformations

Theorem 3.2 establishes a relation between complex dilatations and Schwarzian derivatives in one direction: If a quasiconformal mapping of a plane which is conformal in a simply connected domain A is close to a Möbius transformation in the sense that its complex dilatation is small, then it is close to a Möbius transformation also in the sense that its Schwarzian derivative is small in A.

We shall now establish a result in the opposite direction: If f is meromorphic in a quasidisc A and has a small Schwarzian derivative, then f is univalent and can be extended to a quasiconformal mapping of the plane with a small complex dilatation. We shall then complement this result by proving that the result does not hold for a simply connected domain A unless A is a quasidisc.

The extension of f is carried out by means of smooth quasiconformal reflections. We begin by studying the behavior of quasiconformal reflections under conjugation by Möbius transformations.

Let C be a quasicircle bounding the domains A_1 and A_2 and ψ a K-quasiconformal reflection in C. If h is a Möbius transformation, then $\varphi = h \circ \psi \circ h^{-1}$ is a quasiconformal reflection in $h(C)$. Using the identity

$$h(z_1) - h(z_2) = (h'(z_1)h'(z_2))^{1/2}(z_1 - z_2),$$

valid for all Möbius transformations, we shall establish two formulas, both of which reveal invariance properties of quasiconformal reflections with respect to Möbius transformations.

Writing $\zeta = h(z)$, $\zeta_0 = h(z_0)$, we conclude from $\varphi(\zeta) = h(\psi(z))$ that $\varphi(\zeta) - \zeta_0 = (h'(\psi(z))h'(z_0))^{1/2}(\psi(z) - z_0)$. Hence

$$\frac{\varphi(\zeta) - \zeta_0}{\zeta - \zeta_0} = \left(\frac{h'(\psi(z))}{h'(z)}\right)^{1/2}\frac{\psi(z) - z_0}{z - z_0}.$$

Now fix $z_0 \in C$, such that z_0, $h(z_0) \neq \infty$, and let $z \to z_0$. Then also $\psi(z) \to z_0$, and we obtain our first invariance

$$\limsup_{\zeta \to \zeta_0}\left|\frac{\varphi(\zeta) - \zeta_0}{\zeta - \zeta_0}\right| = \limsup_{z \to z_0}\left|\frac{\psi(z) - z_0}{z - z_0}\right|. \tag{4.1}$$

If C passes through ∞, then by formula (6.2) in I.6.2,

$$\limsup_{z \to z_0}\left|\frac{\psi(z) - z_0}{z - z_0}\right| \leq c_1(K) \tag{4.2}$$

with a finite constant $c_1(K)$ depending only on K. The invariance (4.1) shows that *inequality (4.2) holds for all K-quasiconformal reflections.*

In order to derive the second invariance property, we assume that ψ is continuously differentiable off C. From $h'(\psi(z))\,d\psi(z) = d\varphi(\zeta)$ we then obtain

$$h'(\psi(z))\,d\psi(z)/dz = h'(z)\,d\varphi(\zeta)/d\zeta.$$

Combined with the identity $\varphi(\zeta) - \zeta = (h'(\psi(z))h'(z))^{1/2}(\psi(z) - z)$, this yields

$$(\psi(z) - z)^2\frac{d\psi(z)}{dz}h'(\psi(z))^2 = (\varphi(\zeta) - \zeta)^2\frac{d\varphi(\zeta)}{d\zeta}.$$

Finally, let f be a locally injective meromorphic function in A_2. If $\tilde{f} = f \circ h^{-1}$, then $S_f(\psi(z)) = S_{\tilde{f}}(\varphi(\zeta))h'(\psi(z))^2$, and we arrive at our second invariance

$$(\psi(z) - z)^2\frac{d\psi(z)}{dz}S_f(\psi(z)) = (\varphi(\zeta) - \zeta)^2\frac{d\varphi(\zeta)}{d\zeta}S_{\tilde{f}}(\varphi(\zeta)). \tag{4.3}$$

This invariance gives an estimate which will be needed in what follows:

Let C be a K-quasicircle bounding the domains A_1 and A_2. *There exists a quasiconformal reflection ψ in C such that, for every function f meromorphic and locally injective in A_2,*

$$|\psi(z) - z|^2 |d\psi(z)/dz| \, |S_f(\psi(z))| \le c(K) \|S_f\|_{A_2} \qquad (4.4)$$

at each point $z \in A_1$, $z \ne \infty$, $\psi^{-1}(\infty)$. *Here the finite constant* $c(K)$ *depends only on* K.

PROOF. If C passes through ∞, we take $\psi = \varphi$ as in Lemma I.6.4. Then (4.4) follows directly from the formulas (6.1) and (6.4) in I.6. If C is bounded, we choose the Möbius transformation h so that $\infty \in h(C)$. We then obtain (4.4) from the invariance formula (4.3), because $\|S_f\|_{A_2} = \|S_{\tilde{f}}\|_{h(A_2)}$. $\qquad \square$

4.2. Quasiconformal Extension of Conformal Mappings

We can now prove the main result on the role of the Schwarzian derivative for the univalence and quasiconformal extension of meromorphic functions (Ahlfors [4]).

Theorem 4.1. *Let* A *be a* K-*quasidisc. Then there is a constant* $\varepsilon(K) > 0$, *depending only on* K, *such that every function* f *meromorphic in* A *with the property*

$$\|S_f\|_A < \varepsilon(K) \qquad (4.5)$$

is univalent in A *and can be extended to a quasiconformal mapping of the plane whose complex dilatation satisfies the inequality*

$$\|\mu\|_\infty \le \frac{\|S_f\|_A}{\varepsilon(K)}. \qquad (4.6)$$

PROOF. Let f be meromorphic in A. We may assume that f is locally injective and that ∞ does not lie in $\bar{A}$. Let w_1 and w_2 be solutions of the differential equation $w'' + S_f w/2 = 0$ in A, so normalized that $w_1 w_2' - w_2 w_1' = 1$. In proving Theorem 1.1 we showed that $S_{w_1/w_2} = S_f$. It follows that f is the composition of w_1/w_2 with a Möbius transformation. We may therefore take $f = w_1/w_2$.

The desired extension of f is obtained by an explicit construction. In order to circumvent the difficulty arising from the fact that we have no a priori knowledge about the behavior of f on the boundary of A, we resort to an approximation procedure. We assume first that w_1, w_2 and f are holomorphic on the boundary of A. Having proved the theorem under this additional condition, we obtain the general result by exhausting A with subdomains which are K-quasidiscs and on whose boundaries f has no poles.

Let A_1 denote the complement of the closure of A and ψ a quasiconformal reflection satisfying (4.4); the domain A now plays the role of A_2 in the earlier discussion. We write $z = \psi(\zeta)$ and define a function g in A_1 by

$$g(\zeta) = \frac{w_1(z) + (\zeta - z)w_1'(z)}{w_2(z) + (\zeta - z)w_2'(z)}.$$

At infinity, g is defined as its limit: $g(\infty) = w'_1(\psi(\infty))/w'_2(\psi(\infty))$. We prove that g is a desired continuation of f if (4.5) holds for an appropriate $\varepsilon(K) > 0$.

It is immediate that $g(\zeta) \to f(z_0)$ as ζ approaches a boundary point $z_0 \in \partial A$. Moreover, if ζ is a finite point of A_1 and $g(\zeta) \neq \infty$, then g is continuously differentiable at ζ. In view of the relations $w_1 w'_2 - w_2 w'_1 = 1$ and $w''_i = -S_f w_i/2$, $i = 1, 2$, we obtain by direct computation

$$\partial g(\zeta) = -\frac{1 + S_f(z)(\zeta - z)^2 \partial \psi(\zeta)/2}{(w_2(z) + (\zeta - z)w'_2(z))^2},$$

$$\bar{\partial} g(\zeta) = -\frac{S_f(z)(\zeta - z)^2 \bar{\partial} \psi(\zeta)/2}{(w_2(z) + (\zeta - z)w'_2(z))^2}. \tag{4.7}$$

From this we get an estimate for the complex dilatation $\mu = \bar{\partial} g/\partial g$. With $c(K)$ the constant in (4.4), we choose $\varepsilon(K) = 1/c(K)$. It then follows from (4.4) that

$$|\mu(\zeta)| \leq \frac{c(K)\|S_f\|_A}{2 - c(K)\|S_f\|_A} \leq \frac{\|S_f\|_A}{\varepsilon(K)} < 1. \tag{4.8}$$

We see from (4.7) that with this $\varepsilon(K)$, we have $\partial g(\zeta) \neq 0$. Thus the Jacobian $J_g = |\partial g|^2(1 - |\mu|^2)$ is positive at ζ. Hence g is injective at every finite point ζ at which $g(\zeta) \neq \infty$. By symmetry, $1/g$ is injective at every finite point of A_1 which is not a zero of g. Consequently, g is injective at all finite points of A_1. By considering the function $\zeta \to g(1/\zeta)$ we conclude that g is locally injective at ∞ also, and hence throughout A_1.

Define a function F by $F(z) = f(z)$ in $A \cup \partial A$, and $F(z) = g(z)$ in A_1. Then F is locally injective in A and in A_1. Since $F(z) \neq \infty$ on ∂A, the set $E = \{z | F(z) = \infty\}$ consists of only finitely many points.

Let z_0 be an arbitrary point of ∂A. Considering (4.2) we deduce by direct computation that

$$\lim_{\zeta \to z_0} \frac{g(\zeta) - f(z_0)}{\zeta - z_0} = f'(z_0).$$

Applying (4.4) for $z \to f(z) = \log(z - z_0)$, we conclude that $(\zeta - z)^2 \partial \psi(\zeta) \to 0$ and $(\zeta - z)^2 \bar{\partial} \psi(\zeta) \to 0$ as $\zeta \to z_0$. Consequently, by (4.7), $\partial g(\zeta) \to f'(z_0)$ and $\bar{\partial} g(\zeta) \to 0$ as $\zeta \to z_0$. We conclude that F is continuously differentiable on ∂A, and hence everywhere outside E. On ∂A we have $J_F(z) = |f'(z)|^2$, and so F is locally injective throughout the plane.

By the existence theorem for Beltrami equations (Theorem I.4.4), there is a quasiconformal homeomorphism w of the plane which has the same complex dilatation as F a.e. The function $\varphi = F \circ w^{-1}$ satisfies the Cauchy–Riemann equation $\bar{\partial} \varphi = 0$ a.e. As a quasiconformal mapping w^{-1} has L^2-derivatives. Since the derivatives of F are continuous outside E, we conclude that outside the finite set $w(E)$ the function φ has L^2-derivatives. By the remark made in connection with formula (4.5) in I.4.3, φ is analytic in the complement of $w(E)$. Since φ is continuous everywhere, the points of $w(E)$ cannot be essential singularities, and it follows that φ is meromorphic throughout the extended plane. Hence, φ is rational, and being locally injective, it is a Möbius transformation. We conclude that $F = \varphi \circ w$ is a quasiconformal homeomorphism of the plane. This proves the theorem, provided that f is holomorphic on the boundary of A.

4.3. Exhaustion by Quasidiscs

In order to complete the proof, we shall now show how to relax the restrictive assumption that f is holomorphic on ∂A. Since A is a K-quasidisc, there exists a K-quasiconformal mapping of the plane under which the unit disc maps onto A. It follows that there is an increasing sequence of positive numbers $r_n < 1$, $n = 1, 2, \ldots$, tending to 1, such that the image of each circle $|w| = r_n$ is a K-quasicircle on which $f(z) \neq \infty$. Let A_{1n} and $A_n \subset A$ denote the complementary K-quasidiscs bounded by this quasicircle. As $n \to \infty$, the domains A_n exhaust A.

We proved in I.1.1 that the Poincaré density is smaller in A than in its subdomain A_n. Therefore,

$$\| S_f \|_{A_n} \leq \| S_f \|_A. \tag{4.9}$$

It follows that condition (4.5) can be applied to $f|A_n$. If we choose $\varepsilon(K)$ as in the above proof (note that $\varepsilon(K)$ depends only on K, not on A), we conclude that $f|A_n$ agrees with the restriction to A_n of a quasiconformal mapping F_n of the plane.

Let μ_n be the complex dilatation of F_n. By (4.8) and (4.9)

$$|\mu_n(\zeta)| \leq \frac{c(K)\| S_f \|_{A_n}}{2 - c(K)\| S_f \|_{A_n}} \leq \frac{\| S_f \|_A}{\varepsilon(K)} < 1.$$

We see that the maximal dilatations of the quasiconformal mappings F_n are uniformly bounded. Thus these mappings constitute a normal family. The limit of a locally uniformly convergent subsequence is a quasiconformal mapping of the plane whose restriction to A is f. Its complex dilatation also satisfies (4.8), i.e., (4.6) is true and the theorem is proved. $\qquad\square$

4.4. Definition of Schwarzian Domains

Theorem 4.1 leads to the question whether the condition that A be a quasidisc is necessary for the conclusion that a meromorphic function with a small Schwarzian derivative is univalent in A and has a quasiconformal extension. Or we may restrict ourselves to classical complex analysis and pose the simpler question: In which domains does a small Schwarzian derivative imply univalence?

Let us introduce the following definition: A simply connected domain A with more than one boundary point is called a *Schwarzian domain* if there is a positive constant ε such that every meromorphic function f with $\|S_f\|_A \le \varepsilon$ is univalent in A. More precisely, we say that A is then an ε-Schwarzian domain. This notion is due to Gehring [2].

Theorem 4.1 tells us that every quasidisc is a Schwarzian domain. It was for a long time an open problem whether or not this sufficient condition was also necessary, and few conjectures were expressed favoring either direction. Finally, in 1977, Gehring [2] solved the problem by proving that every Schwarzian domain is a quasidisc. Thus a small Schwarzian derivative forces a function to be univalent in A if and only if A is a quasidisc. This is one of the unexpected characterizations of quasidiscs in terms of analysis we were talking about in I.6.7. Unexpected, because it shows that quasiconformal mappings are intrinsic to the problem of relating the injectivity of meromorphic functions to their Schwarzian derivative, a problem which on the surface has nothing to do with quasiconformality.

We shall now present Gehring's proof, which is based on the fact that a linearly locally connected domain conformally equivalent to a disc is a quasidisc (cf. I.6.4 and I.6.7).

4.5. Domains Not Linearly Locally Connected

Topological properties of plane domains can often be expressed in analytic terms with the aid of the complex logarithm. We shall derive a result of this type for domains which fail to be linearly locally connected with a given constant. (For the definition of linear local connectivity, see I.6.4.)

Lemma 4.1. *Let A be a simply connected domain which is not linearly locally connected for a constant $c > 1$. Then there are two points z_1, z_2 of A and two finite points w_1, w_2 outside A, such that the function $z \to h(z) = \log((z - w_1)/(z - w_2))$ satisfies the inequality*

$$|h(z_1) - h(z_2) - 2\pi i| \le \frac{4}{c - 1}. \tag{4.10}$$

PROOF. It follows from the assumption and from the definition of linear local connectivity that there is a disc $D(z_0, r)$ with the following property: Either there are two points p_1 and p_2 in $A \cap \bar{D}(z_0, r)$ which cannot be joined in $A \cap \bar{D}(z_0, cr)$, or else there are two points in $A \backslash D(z_0, r)$ which cannot be joined in $A \backslash D(z_0, r/c)$. Suppose initially that the former alternative is true.

Consider the line segment with endpoints p_1, p_2 and a simple arc in A from p_1 to p_2 which meets the line segment at finitely many points only. Among these points of intersection there are two adjacent, z_1 and z_2, which cannot be joined in $A \cap \bar{D}(z_0, cr)$. Let α be the line segment with endpoints z_1, z_2, and β the subarc from z_2 to z_1 of the arc joining p_1 and p_2. Then $\alpha \cup \beta$ bounds two Jordan domains A_1 and A_2; we assume that infinity lies in A_2.

For brevity we write $D(z_0, cr) = U$. Suppose that there are points $w_i \in \partial U \cap A_i$, $i = 1, 2$, in the complement CA of A. Then

$$h(z_1) - h(z_2) = \int_\beta ((z - w_1)^{-1} - (z - w_2)^{-1}) \, dz = 2\pi i (n(w_1) - n(w_2))$$

$$- \int_\alpha ((z - w_1)^{-1} - (z - w_2)^{-1}) \, dz,$$

where $n(w_i)$ is the winding number of $\alpha \cup \beta$ with respect to w_i. Since $w_1 \in A_1$, $w_2 \notin A_1$, we have $n(w_1) = n = \pm 1$, $n(w_2) = 0$. It follows that

$$|h(z_1) - h(z_2) - 2\pi n i| \leq \int_\alpha \left(\frac{1}{|z - w_1|} + \frac{1}{|z - w_2|} \right) |dz|. \qquad (4.11)$$

Because $\alpha \subset \bar{D}(z_0, r)$ and $w_1, w_2 \in \partial D(z_0, cr)$, we see that $|z - w_i| \geq (c - 1)r$ for every $z \in \alpha$. Therefore

$$\int_\alpha \left(\frac{1}{|z - w_1|} + \frac{1}{|z - w_2|} \right) |dz| \leq \frac{2}{(c - 1)r} \int_\alpha |dz| \leq \frac{4}{c - 1}. \qquad (4.12)$$

If $n = 1$, we thus get (4.10) from (4.11) and (4.12). If $n = -1$, we obtain (4.10) by interchanging the roles of w_1 and w_2.

We still have to prove that it is possible to find points $w_i \in \partial U \cap A_i$ which are not in A, i.e., that

$$CA \cap \partial U \cap A_i \neq \emptyset, \qquad i = 1, 2. \qquad (4.13)$$

This requires topological arguments.

We first observe that, because $\alpha \subset U$ and β has points outside the closure of U, the curve $\alpha \cup \beta$ meets ∂U in at least two points. By Kerékjártó's theorem (Newman [1], p. 168), each component of the complement of $\alpha \cup \beta \cup \partial U$ is a Jordan domain. In particular, each component of $A_1 \cap U$ is a Jordan domain. Since a Jordan domain is locally connected at every boundary point, there is a neighborhood V of z_1 such that each pair of points in $A_1 \cap V$ can be joined in $A_1 \cap U$. Let $z \in A_1 \cap V$ and let A_1' be the z-component of $A_1 \cap U$. Every point p of $A_1 \cap V$ can be joined with z in $A_1 \cap U$, and the joining arc lies in A_1'. It follows that $p \in A_1'$, and so $A_1 \cap V \subset A_1' \cap V$, i.e., A_1' is the only component of $A_1 \cap U$ whose boundary contains z_1.

If $z' \in \alpha$, we can join z' and z_1 by an arc whose inner points lie in $A_1 \cap U$. Then this arc is in the closure of A_1'. Hence $z' \in \partial A_1'$, and we conclude that $\alpha \subset \partial A_1'$. If γ is the complement of α with respect to $\partial A_1'$, then $\gamma \subset A \cup (\partial U \cap A_1)$. On the other hand, γ joins z_1 and z_2 in the closure of U. Thus γ is not contained in A, and (4.13) follows for $i = 1$. Similar reasoning yields (4.13) for $i = 2$.

Finally, we have to consider the case in which there are two points in $A \backslash D(z_0, r)$ which cannot be joined in $A \backslash D(z_0, r/c)$. Let $f(z) = z_0 + 1/(z - z_0)$. By what we just proved, there are points ζ_1, ζ_2 in $f(A)$ and points t_1, t_2 outside $f(A)$, such that the function $z \to g(z) = \log((z - t_1)/(z - t_2))$ satisfies the inequality

$$|g(\zeta_1) - g(\zeta_2) - 2\pi i| \le \frac{4}{c-1}. \tag{4.14}$$

If $z_i = f^{-1}(\zeta_i)$, $w_i = f^{-1}(t_i)$, $i = 1, 2$, then direct calculation yields

$$h(z) = g(f(z)) + \log\frac{z_0 - w_1}{z_0 - w_2}.$$

Hence, (4.10) follows from (4.14).   □

4.6. Schwarzian Domains and Quasidiscs

The result we set out to prove can now be established without difficulty.

Theorem 4.2. *A Schwarzian domain is a quasidisc.*

PROOF. Let A be an a-Schwarzian domain. Then A is trivially a'-Schwarzian for $a' \le a$. We may suppose, therefore, that $a \le 2$. (In III.5 we shall show that, in fact, no domain A is a-Schwarzian for $a > 2$, but here this result is not needed.) Nor is there any loss of generality in assuming that ∞ does not lie in A.

We shall show that A is linearly locally connected with constant

$$c = 1 + 16/a. \tag{4.15}$$

The theorem then follows from Theorems I.6.5 and I.6.6. More precisely, we conclude that an a-Schwarzian domain is a $K(a)$-quasidisc, where the constant $K(a)$ depends only on a.

The proof is indirect. Suppose that A is not linearly locally connected with the constant $c = 1 + 16/a$. By Lemma 4.1, there are points z_1, z_2 in A and w_1, w_2 outside A, such that (4.10) holds. Clearly $h(z_1) \ne h(z_2)$.

Define

$$f(z) = e^{bh(z)}, \qquad b = \frac{2\pi i}{h(z_1) - h(z_2)}.$$

Then $f(z_1)/f(z_2) = 1$, so that f is not univalent.

From $S_f = -b^2 h'^2/2 + S_h$ we get by an easy computation

$$S_f(z) = \frac{1-b^2}{2}\left(\frac{w_1 - w_2}{(z - w_1)(z - w_2)}\right)^2.$$

If η denotes the Poincaré density of A, then formula (1.5) in I.1.1 yields the estimate

$$|S_f(z)|\eta(z)^{-2} \le 8|b^2 - 1|. \tag{4.16}$$

Since $a \le 2$, we see from (4.15) that $c - 1 \ge 8$. Then $|h(z_1) - h(z_2)| \ge 2\pi - 1/2$, and so

$$|b - 1| \le \frac{4}{(c-1)(2\pi - 1/2)} < \frac{4}{5(c-1)} \le \frac{1}{10}.$$

It follows that

$$|b^2 - 1| \le \frac{21}{10} \cdot \frac{4}{5(c-1)} < \frac{a}{8}.$$

We conclude from (4.16) that $\| S_f \|_A < a$. Since A is an a-Schwarzian domain, f is univalent. This is a contradiction, and so A is linearly locally connected with the constant $c = 1 + 16/a$. ☐

The effect of the Schwarzian derivative on the univalence and quasiconformal extension of meromorphic functions will be studied in more detail in sections III.4 and III.5.

5. Functions Univalent in a Disc

5.1. Quasiconformal Extension to the Complement of a Disc

Certain special curves admit simple quasiconformal reflections. In such cases Theorem 4.1 can be expressed in a more explicit form. Particularly important is the case where the domain is a disc. From the expression for the complex dilatation of the extended mapping in this special setting we shall draw in Chapter V important conclusions regarding the theory of Teichmüller spaces.

Theorem 5.1. *Let f be meromorphic in a disc D. If*

$$\| S_f \|_D < 2,$$

then f is univalent and can be extended to a quasiconformal mapping of the plane. The constant 2 is best possible.

If D is the unit disc, there is an extension with complex dilatation

$$\mu(1/\bar{z}) = -\tfrac{1}{2}(z/\bar{z})^2 (1 - |z|^2)^2 S_f(z) \tag{5.1}$$

for z in D, while if D is the upper half-plane, an extension exists with

$$\mu(\bar{z}) = -2y^2 S_f(z) \tag{5.2}$$

for z in D.

PROOF. Suppose first that D is the unit disc. In this case we have the simple quasiconformal reflection $\zeta \to 1/\bar{\zeta}$. More precisely, in proving Theorem 4.1 we use in the approximation stage of the argument the reflection $\psi(\zeta) = r_n^2/\bar{\zeta}$, $r_n < 1$. By letting $r_n \to 1$, we then obtain from (4.7) the expression (5.1) for the complex dilatation $\mu = \bar{\partial}g/\partial g$. Since $\|\mu\|_\infty = \|S_f\|/2$, it follows that we can take $\varepsilon(K) = 2$.

After this it is clear that $\varepsilon(K) = 2 \, (= \varepsilon(1))$ will do for any disc, because all discs are Möbius equivalent.

If the domain is the upper half-plane, we have the reflection $\zeta \to \bar{\zeta}$. In this case (4.7) yields the expression (5.2) for μ.

In order to show that the bound 2 cannot be replaced by any larger, we consider the function $z \to f(z) = \log z$ in the upper half-plane H. This function f is univalent, and from $S_f(z) = (2z^2)^{-1}$ it follows that $\|S_f\| = \sup 2y^2/|z|^2 = 2$. But the image of H is the parallel strip $\{w|0 < \operatorname{Im} w < \pi\}$, which is not a Jordan domain. Consequently, f does not even possess a homeomorphic extension to the plane.   □

Here we obtained Theorem 5.1 as a corollary of the general Theorem 4.1, but actually Theorem 5.1, proved by Ahlfors and Weill [1] in 1962, was discovered before Theorem 4.1. Thanks to the simple reflections $z \to 1/\bar{z}$ for the unit disc and $z \to \bar{z}$ for the half-plane, the proof of Theorem 5.1 is considerably shorter than that of Theorem 4.1. We can bypass the considerations which guarantee the existence of the special reflection ψ needed in the case of an arbitrary quasidisc.

Theorem 5.1 gives the result we mentioned in 2.1:

Let A be a simply connected domain with more than one boundary point. If the distance $\delta(A)$ from a disc is <2, then A is a quasidisc.

Let us return to (5.2) and set $\bar{\varphi}(z) = -S_f(\bar{z})/2$. We then have in the lower half-plane

$$\mu = \bar{\varphi}/\eta^2,$$

where φ is a holomorphic function and η the Poincaré density.

Supplementing Theorem 4.1, Bers proved that the same result holds even in the general case: *A function f which is meromorphic in a quasidisc A and has a sufficiently small $\|S_f\|$ is univalent and has a quasiconformal extension with a complex dilatation $\bar{\varphi}/\eta^2$, where φ is a holomorphic function and η the Poincaré density in the complement of the closure of A.*

The complex dilatations $\bar{\varphi}/\eta^2$ turn out to be important in Chapter V when we consider quasiconformal mappings which are lifts to the universal covering surface of quasiconformal mappings between Riemann surfaces. In our applications we shall get along with the case of the half-plane and content ourselves therefore to indicating only briefly how the reasoning goes in the general case. For the complete proof we refer to Bers [7] or [9].

Let M be the set of complex dilatations which are zero in A and of the form $\bar{\varphi}/\eta^2$ outside the closure of A, and Q the space of holomorphic functions of A with a finite hyperbolic sup-norm. If f is a quasiconformal mapping with complex dilatation μ in M, then $S_{f|A}$ is uniquely determined by μ. By methods of functional analysis, Bers proves that the mapping $\mu \rightarrow S_{f|A}$ of M into Q is invertible in a neighborhood of the origin of Q. The proof makes use of the representation formula for f, of the reproducing property of the Bergman kernel function and, just as in Theorem 4.1, of a Lipschitz-continuous quasi-conformal reflection.

5.2. Real Analytic Solutions of the Boundary Value Problem

Interrupting briefly the study of functions meromorphic in a disc, we show how Theorem 5.1 can be utilized for constructing real analytic solutions of the boundary value problem discussed in I.5.

Let a k-quasisymmetric function h be given (cf. I.5.2). We assume first that $k < \sqrt{2}$. Let $\tilde{h}$ be a quasiconformal self-mapping of the upper half-plane H with boundary values h. We construct $\tilde{h}$ by using the Beurling–Ahlfors method so that the maximal dilatation of $\tilde{h}$ does not exceed k^2 (cf. I.5.3). Because $k^2 < 2$, the complex dilatation μ of $\tilde{h}$ satisfies the condition

$$\|\mu\|_\infty < 1/3. \tag{5.3}$$

By Theorem I.4.4, there exists a quasiconformal mapping f of the plane which has the complex dilatation μ in H and which is conformal in the lower half-plane H'. By Theorem 3.2 and formula (5.3),

$$\|S_{f|H'}\|_{H'} \le 6\|\mu\|_\infty < 2.$$

Thus we can apply Theorem 5.1 and obtain a new quasiconformal extension $\tilde{f}$ for $f|H'$. Then $f^* = \tilde{h} \circ (f|H)^{-1} \circ \tilde{f}$ is a quasiconformal self-mapping of H with boundary values h. From the expression

$$\tilde{f}(\bar{z}) = \frac{w_1(z) + (\bar{z} - z)w_1'(z)}{w_2(z) + (\bar{z} - z)w_2'(z)},$$

$z \in H'$, we see that $\tilde{f}$ is real analytic. By the Uniqueness theorem I.4.2, the mapping $\tilde{h} \circ (f|H)^{-1}$ is conformal. Hence f^* is real analytic.

In the case of an arbitrary quasisymmetric function h we write $h = h_n \circ \cdots \circ h_2 \circ h_1$, where each h_i is k-quasisymmetric for $k < \sqrt{2}$. Using formula (4.17) in I.4.7 we remark that this is possible, because the boundary function of a K-quasiconformal self-mapping of H fixing ∞ is $\lambda(K)$-quasisymmetric and $\lambda(K) \rightarrow 1$ as $K \rightarrow 1$. If f_i^* is a real analytic solution corresponding to h_i, then $f_n^* \circ \cdots \circ f_1^*$ is a real analytic solution for h. We have thus given a proof for Theorem I.5.3.

5.3. Criterion for Univalence

Theorem 1.3 says that if f is univalent in a disc, then $\|S_f\| \le 6$. By use of Theorem 5.1, we obtain a converse to this theorem.

Theorem 5.2. *Let f be meromorphic in a disc. If*

$$\|S_f\| \leq 2,$$

then f is univalent. The bound 2 is best possible.

PROOF. Consider functions f_n, $n = 1, 2, \ldots$, which are meromorphic in the given disc, fix three points of the disc, and have Schwarzians $(1 - 1/n)S_f$; by Theorem 1.1 such functions exist. Since $\|S_{f_n}\| < 2$, every f_n is univalent owing to Theorem 5.1. They form a normal family, and the limit of a locally uniformly convergent subsequence is a univalent function. Since the limit function shares the same Schwarzian derivative with f, we conclude that f is univalent.

In order to prove that the bound 2 cannot be replaced by a larger number, consider the analytic function $z \to f(z) = z^{i\varepsilon}$, $\varepsilon > 0$, in the upper half-plane. Then $S_f(z) = (1 + \varepsilon^2)(2z^2)^{-1}$, and so $\|S_f\| = 2(1 + \varepsilon^2)$. On the other hand, f is not univalent for any $\varepsilon > 0$: For instance, f takes the same value at the points i and $i \exp(2\pi/\varepsilon)$ of the upper half-plane.   □

Here we derived Theorem 5.2 as an easy corollary of Theorem 5.1. However, as might be expected, quasiconformal mappings are not needed for the proof of Theorem 5.2, which is in fact a much older result than Theorem 5.1. It was proved in 1949 by Nehari [1], and in the same year Hille showed that the bound 2 is sharp.

5.4. Parallel Strips

We shall now show that the condition $\|S_f\| \leq 2$ in a disc allows for conclusions about the function f beyond its univalence. The results in the remaining part of this section are due to Gehring and Pommerenke [1].

Let f be a meromorphic function in a disc D, h a conformal mapping of a domain A onto D, and $g = f \circ h$. Since $\|S_f\|_D = \|S_g - S_h\|_A$, we can transfer Theorem 5.2 to A. If we can construct h explicitly, the modified theorem can be of interest. In particular, if A is a parallel strip, certain technical advantages are gained which will be utilized in the following.

Let us normalize the domains and assume that D is the unit disc and $A = \{z \mid |\mathrm{Im}\, z| < \pi/2\}$. Then

$$z \to h(z) = \tanh(z/2)$$

is a conformal map of A onto D. It maps the real axis $\mathbb{R}$ onto the real diameter of D, and it has the Schwarzian $S_h(z) = -1/2$. The Poincaré density of A at $z = x + iy$ has the value $(2\cos y)^{-1}$.

We list two immediate consequences. First, Theorem 5.2 assumes the following form: Let g be meromorphic in $A = \{z \mid |\mathrm{Im}\, z| < \pi/2\}$. If

$$\left| S_g(z) + \frac{1}{2} \right| \le \frac{1}{2\cos^2 y},$$

then g is univalent in A. The bound is best possible.

Second, if x is a point of the real axis and $w = h(x)$, then by the basic invariance formula (1.9) in 1.3,

$$4|S_g(x) + \tfrac{1}{2}| = (1 - |w|^2)^2 |S_f(w)|. \tag{5.4}$$

This equation remains valid if h is replaced by an arbitrary conformal mapping of A onto D. The image of the real axis is always a geodesic line in the Poincaré metric of the unit disc, i.e., it is a circular arc which intersects the unit circle orthogonally.

If f has no poles in D, then g is holomorphic in A. Direct computation shows that on the real axis $\mathbb{R}$, the function $x \to v(x) = |g'(x)|^{-1/2}$ then satisfies the differential equation

$$v'' = \varphi v \tag{5.5}$$

with

$$\varphi = -\frac{1}{2}\operatorname{Re} S_g + \left(\frac{1}{2}\operatorname{Im} \frac{g''}{g'} \right)^2. \tag{5.6}$$

We remark that equation (5.5) with φ determined by (5.6) is an identity which holds on $\mathbb{R}$ for every g holomorphic in A. Note that by (5.4), condition $\|S_f\|_D \le 2$ implies $\operatorname{Re} S_g(x) \le 0$ and hence $\varphi(x) \ge 0$.

5.5. Continuous Extension

We shall prove that if $\|S_f\|_D \le 2$, then f always has a continuous extension to the boundary of D. Using (5.5) and (5.6), we first establish as a preparatory result an estimate on $|f'|$ when f satisfies certain additional conditions.

Lemma 5.1. *Let f be a function meromorphic in the unit disc D satisfying $f''(0) = 0$, $\|S_f\|_D \le 2$, and $(1 - |w|^2)^2 |S_f(w)| \le 1$ in a neighborhood $|w| \le a < 1$ of the origin. Then f is holomorphic in D and*

$$|f'(w)| \le \frac{M|f'(0)|}{1 - |w|^2} \left(\log \frac{1 + |w|}{1 - |w|} \right)^{-2}, \tag{5.7}$$

where M is a constant depending only on a.

PROOF. By symmetry, it is sufficient to consider the case $w > 0$. Assume first that f is holomorphic in D.

We make use of the conformal mapping $h: A \to D$ defined in 5.4, the composition $g = f \circ h$, and the function $x \to v(x) = |g'(x)|^{-1/2}$. Because $h''(0) = 0$, we conclude from the assumption $f''(0) = 0$ that $g''(0) = 0$. Hence

$$v'(0) = -\frac{1}{2}v(0)\operatorname{Re}\frac{g''(0)}{g'(0)} = 0.$$

The hypothesis $\|S_f\|_D \leq 2$ implies, as we mentioned at the end of 5.4, that $\varphi(x) \geq 0$ on $\mathbb{R}$. By (5.5) we then have $v''(x) \geq 0$, so that v' is increasing in x. From $v'(0) = 0$ it follows that $v'(x) \geq 0$ for $x \geq 0$. Consequently, $v(x) \geq v(0)$, and by (5.5), $v''(x) \geq \varphi(x)v(0)$.

From $(1 - |w|^2)^2|S_f(w)| \leq 1$ in $|w| \leq a$ we infer, in view of (5.4) and the relation $z = h^{-1}(w) = \log((1 + w)/(1 - w))$, that $\operatorname{Re}S_g(x) \leq -1/4$ if $0 \leq x \leq b = \log((1 + a)/(1 - a))$. Hence, by (5.6), $\varphi(x) \geq 1/8$, and so $v''(x) \geq v(0)/8$ for $0 \leq x \leq b$. This yields $v'(x) \geq xv(0)/8$ for $0 \leq x \leq b$. We conclude that

$$v(x) \geq v(0)(1 + b(x - b)/8)$$

for $x \geq b$. If $c = \min(1/b, b/8)$, it follows that $v(x) \geq cv(0)x$, i.e.,

$$|g'(x)| \leq \frac{|g'(0)|}{c^2x^2},$$

for $x \geq 0$. Because $2|g'(x)| = 2|h'(x)f'(w)| = (1 - |w|^2)|f'(w)|$, $2g'(0) = f'(0)$, and $x = \log((1 + w)/(1 - w))$, this is (5.7) with $M = c^{-2}$.

Now drop the assumption that f is holomorphic in D and set $r = \inf\{|w_0| \,|\, w_0 \text{ a pole of } f\}$. Repetition of the above reasoning shows that (5.7) then holds in $|w| < r$. We see that if $r < 1$ and $|w_0| = r$, then $|f'(w)|$ remains bounded as $w \to w_0$. This is a contradiction, and so f has no poles in D. $\quad\square$

Inequality (5.7) yields the following preliminary result: A function f satisfying the conditions of Lemma 5.1 has a continuous extension to the boundary of D. This follows from the fact that

$$\psi(r) = \frac{M|f'(0)|}{1 - r^2}\left(\log\frac{1 + r}{1 - r}\right)^{-2}$$

is integrable over the interval $[a, 1]$. Hence, we see from

$$|f(r_2e^{i\theta}) - f(r_1e^{i\theta})| \leq \int_{r_1}^{r_2}\psi(r)\,dr, \qquad r_2 > r_1,$$

that the limit of $f(re^{i\theta})$ as $r \to 1$ exists uniformly in θ. It defines, therefore, a continuous extension of f.

5.6. Image of Discs

We can now prove the main theorem about the image of a disc under a conformal mapping f for which $\|S_f\| \leq 2$. The result (Gehring and Pommerenke [1]) complements Theorem 5.2.

Theorem 5.3. *Let f be meromorphic in a disc D and $\|S_f\| \leq 2$. Then $f(D)$ is either a Jordan domain or the image of a parallel strip under a Möbius transformation.*

PROOF. We may assume that D is the unit disc. We first note that

$$m = \inf_{w \in D} (1 - |w|^2)^2 |S_f(w)| = 0. \tag{5.8}$$

For if $m > 0$, the function $1/S_f$ is holomorphic in D. From

$$\frac{1}{|S_f(w)|} \leq \frac{(1 - |w|^2)^2}{m}$$

and the maximum principle we then arrive at the impossible conclusion that $1/S_f$ vanishes identically in D.

From (5.8) we conclude the existence of a point $w_0 \in D$ at which

$$|S_f(w_0)| < (1 - |w_0|^2)^{-2}.$$

We may assume that $f(w_0) = 0$. If $g_1(w) = (w + w_0)/(1 + \bar{w}_0 w)$ and $f_1 = f \circ g_1$, then $\|S_{f_1}\| \leq 2$ and

$$|S_{f_1}(0)| = |S_f(w_0)|(1 - |w_0|^2)^2 < 1.$$

Thus there is an $a > 0$ such that

$$|S_{f_1}(w)| < (1 - |w|^2)^{-2}$$

for $|w| < a$. Next set $c = f_1''(0)/2f_1'(0)^2$, $g_2(w) = w/(cw + 1)$, and $f_2 = g_2 \circ f_1$. Then $S_{f_2} = S_{f_1}$ and $f_2''(0) = 0$, so that f_2 fulfills all conditions of Lemma 5.1. It follows that f_2 has a continuous extension to ∂D, and hence $f = g_2^{-1} \circ f_2 \circ g_1^{-1}$ also has this property.

If the extended f is injective in the closure of D, then it is a homeomorphism of the closure of D onto its image. In this case $f(D)$ is a Jordan domain.

Suppose then that the extended f is not injective. Let f take the same value at the points w_1 and w_2 in the closure of D. Since the condition $\|S_f\| \leq 2$ implies that f is injective in D, it follows that w_1 and w_2 lie on ∂D. Let γ be the noneuclidean line of D with endpoints w_1 and w_2, and w_0 a fixed interior point of γ. We assume first that the value the extended f takes at w_1 and w_2 is ∞. Then the images of the components of $\gamma \backslash \{w_0\}$ under f both have infinite length.

Let h now be a conformal mapping of the parallel strip A onto D which maps the real axis onto γ, and $g = f \circ h$. Then the g-images of the components of $\mathbb{R} \backslash \{x_0\}$, $x_0 = h^{-1}(w_0)$, are the same as the f-images of $\gamma \backslash \{w_0\}$. Hence,

$$\int_{-\infty}^{x_0} |g'(x)| \, dx = \int_{x_0}^{\infty} |g'(x)| \, dx = \infty. \tag{5.9}$$

On the other hand, we get upper estimates for these integrals by studying the function $v = |g'|^{-1/2}$. As before, we deduce from (5.5) that v' is increasing in x. If $v'(x_0) > 0$, it follows that $v(x) \geq v(x_0) + v'(x_0)(x - x_0)$ for $x \geq x_0$. Hence,

$$\int_{x_0}^{\infty} |g'(x)| \, dx \leq \frac{1}{v(x_0)v'(x_0)}.$$

Similarly, if $v'(x_0) < 0$,

$$\int_{-\infty}^{x_0} |g'(x)| \, dx \leq -\frac{1}{v(x_0)v'(x_0)}.$$

In conjunction with (5.9) these estimates show that $v'(x_0) = 0$, i.e., that v' vanishes on the real axis. It follows that $\operatorname{Re}(g''/g') = -2v'/v = 0$ on $\mathbb{R}$. By (5.6) and (5.5),

$$(\tfrac{1}{2}\operatorname{Im}(g''/g'))^2 \leq \varphi = 0.$$

We conclude that g'' vanishes on $\mathbb{R}$ and hence in A. Thus g is a similarity transformation, and $f(D) = g(A)$ is a parallel strip.

If the extended f takes a finite value c at w_1 and w_2, we can apply the above reasoning to the function $f_1 = 1/(f - c)$. We infer that $f_1(D)$ is a parallel strip, and the proof of the theorem is completed.   □

5.7. Homeomorphic Extension

The following result (Gehring–Pommerenke [1]), an easy consequence of Theorem 5.3, fits in the narrow gap between Theorems 5.1 and 5.2.

Theorem 5.4. *Let f be meromorphic and satisfy*

$$|S_f(z)| < \frac{2}{(1 - |z|^2)^2}$$

in the unit disc. Then f is univalent and has a homeomorphic extension to the plane.

PROOF. By Theorem 5.2, f is univalent. The image $f(D)$ is a Jordan domain if and only if f has a homeomorphic extension to the plane. Hence, if a homeomorphic extension does not exist, then by Theorem 5.3, $f(D)$ is the image of the parallel strip A under a Möbius transformation. If h again denotes the conformal mapping $z \to \tanh(z/2)$ of A onto D, then $g = f \circ h$ is a Möbius transformation. It follows from (5.4) that

$$(1 - |z|^2)^2 |S_f(z)| = 2$$

at every point of $h(\mathbb{R})$. This is in contradiction with the hypothesis.   □

Summarizing, we obtain the following precise classification for functions f meromorphic in the unit disc: Condition

$$(1 - |z|^2)^2 |S_f(z)| \leq 2$$

implies that f is univalent, condition

$$(1 - |z|^2)^2 |S_f(z)| < 2$$

that f is univalent and has a homeomorphic extension to the plane, and condition

$$\sup_z (1 - |z|^2)^2 |S_f(z)| < 2$$

that f is univalent and has a quasiconformal extension to the plane.

CHAPTER III

Universal Teichmüller Space

Introduction to Chapter III

The notion of universal Teichmüller space was crystallized in connection with the problem of imbedding the Teichmüller space of a Riemann surface into a space of Schwarzian derivatives. In the general case, the Schwarzians in question are holomorphic quadratic differentials for a group of Möbius transformations (see V.4). The universal Teichmüller space corresponds to the situation in which the group is trivial. The Schwarzians are then just holomorphic functions, and the machinery developed in Chapter II can be applied directly. It follows that Chapter III, devoted to the study of the universal Teichmüller space, provides a bridge between univalent functions and Teichmüller spaces.

The other end of the bridge will not be visible until Teichmüller spaces of Riemann surfaces are introduced in Chapter V. The drawback is that in section 1 of this chapter we are unable to motivate the definition of the universal Teichmüller space. In fact, plenty of explanation is required before the role of the universal space in Teichmüller theory becomes clear; a reader who wishes to get an early idea of this role may want to consult V.3.

In section 1, various models of the universal Teichmüller space T are introduced, and the group structure of T is discussed.

Following Teichmüller's classical example, we define in section 2 a distance function which makes T a metric space. This space is shown to be pathwise connected and complete.

In section 3, we study the model of T provided by the family of normalized quasisymmetric functions. This characterization offers certain technical advantages. Using it, we prove that T is contractible and that T is not a topological group.

Section 4 deals with the problem of mapping T into the Banach space of Schwarzian derivatives with finite norm. By appealing to the results of Chapter II, we prove that the mapping is a homeomorphism of T onto its image and that the image agrees with the interior of the set consisting of the Schwarzians of univalent functions. The mapping could also be used to introduce a natural complex analytic structure into the universal Teichmüller space, but we shall not take up this question until V.5, in connection with an arbitrary Teichmüller space.

In section 5 we introduce the inner radius of univalence of a simply connected domain. Its use makes it possible to analyze further the image of the universal Teichmüller space in the space of Schwarzian derivatives.

1. Models of the Universal Teichmüller Space

1.1. Equivalent Quasiconformal Mappings

Let us consider the family of all quasiconformal mappings of a fixed domain in the plane. In this section we assume that this domain is the upper half-plane H. We wish to introduce additional structure to this family and begin by regarding two mappings as equivalent if they differ by a conformal mapping. In view of the Riemann mapping theorem, we may then restrict ourselves to self-mappings of H and require that they are normalized so as to keep fixed the three boundary points 0, 1 and ∞. We denote by F the family of such normalized mappings. (Recall: every element of F can be extended to a homeomorphic self-mapping of the closure of H. It is actually the extended mappings to which the normalization requirements apply.)

By the existence and uniqueness theorems for Beltrami equations (Theorems I.4.4 and I.4.2), there is a one–one correspondence between F and the open unit ball B of the Banach space which consists of all L^∞-functions on H.

A more interesting space is obtained if we introduce a weaker equivalence relation.

Definition. Two mappings of the family F are equivalent if they agree on the real axis. The complex dilatations of equivalent mappings are also said to be equivalent. The set of the equivalence classes is the universal Teichmüller space T.

We thus have two models for T: Its points are classes of equivalent mappings in the family F or of equivalent functions on the ball B.

A third model is obtained in terms of quasisymmetric functions. We recall that a quasisymmetric function is said to be normalized if it fixes the points 0 and 1. Let X denote the class of all normalized quasisymmetric functions. If

$[f]$ is the point of T represented by the mapping $f \in F$, then

$$[f] \to f | \mathbb{R} \tag{1.1}$$

is a bijective mapping of T onto X. For it is clear from the definition of T that (1.1) is well defined and injective. By Theorem I.5.1, it maps T into X, and by Theorem I.5.2, it is surjective. It follows that we can rephrase the definition of the universal Teichmüller space: T *is the set of all normalized quasisymmetric functions.*

This observation allows an important conclusion:

Theorem 1.1. *Every point of the universal Teichmüller space can be represented by a real analytic quasiconformal mapping $f \in F$ or by a real analytic complex dilatation $\mu \in B$.*

PROOF. The result follows immediately from Theorem I.5.3. (For a complete proof, see II.5.2.)   □

We shall see in V.3.2 that the universal Teichmüller space contains as a subset the Teichmüller space of any Riemann surface which allows a half-plane as its universal covering surface. It was Bers [7, 8] who recognized the importance of this largest and, in many ways, simplest Teichmüller space and gave it the name universal.

1.2. Group Structures

If f belongs to F, then so does its inverse f^{-1}; along with f and g in F, the composition $f \circ g$ is also in F. The family F can thus be regarded as a group. From the definition of the universal Teichmüller space it follows that T inherits this group structure: T *is the quotient of the group F of all normalized quasiconformal self-mappings of the upper half-plane by the normal subgroup of mappings equivalent to the identity.*

If $f, g \in F$, the rule

$$[f] \circ [g] = [f \circ g] \tag{1.2}$$

defines the group operation in T. The neutral element, i.e., the point of T determined by the identity mapping (or by the complex dilatation which is identically zero) is called the origin of T and denoted by 0.

Normalized quasisymmetric functions also form a group under composition. This follows from Theorems I.5.1 and I.5.2 or from an easy elementary computation. We see that the mapping (1.1) is an isomorphism between the groups T and X.

Let us consider, for a moment, quasiconformal self-mappings of H which are not necessarily normalized. If f_1 and f_2 are two such mappings, we still say that f_1 is equivalent to f_2 if $(f_2 \circ f_1^{-1}) | \mathbb{R}$ is the identity. Let $f \in F$ be a normalized and g an arbitrary quasiconformal self-mapping of H. We choose a Möbius transformation h, mapping H onto itself, such that $h \circ f \circ g^{-1} \in F$, and define $\omega = \omega_{[g]} \colon T \to T$ by the formula

$$\omega([f]) = [h \circ f \circ g^{-1}].$$

Then ω, which depends only on the equivalence class $[g]$, is a well defined bijection of T onto itself. Further, $[g] = 0$ implies that ω is the identity and

$$\omega_{[g_1]} \circ \omega_{[g_2]}([f]) = [h_1 \circ h_2 \circ f \circ g_2^{-1} \circ g_1^{-1}] = \omega_{[g_1] \circ [g_2]}([f]).$$

We conclude that when g runs through all quasiconformal self-mappings of H, the transformations $\omega_{[g]} \colon T \to T$ form a group. It is called the *universal modular group* M.

We now return to normalized quasiconformal self-mappings of H and consider the subgroup M_t of the universal modular group consisting of transformations $\omega_{[g]}$ with $g \in F$. Then

$$\omega_{[g]}([f]) = [f \circ g^{-1}],$$

i.e., elements of M_t are right translations of the group T.

In 2.1 we shall introduce a metric into T and prove that every $\omega_{[g]}$ is an isometry with respect to this metric.

The group M_t of right translations is transitive: If $p_1 = [f_1]$ and $p_2 = [f_2]$ are given points of T, there is an $\omega_{[g]} \in M_t$ such that $p_2 = \omega_{[g]}(p_1)$. This is clearly the case if $g = f_2^{-1} \circ f_1$.

1.3. Normalized Conformal Mappings

The mappings belonging to F can be continued quasiconformally to the plane by reflection in the real axis. However, such an extension does not give new insight into the properties of T. It was a fundamental observation of Bers [4] that one should extend, not the mappings of F but rather their complex dilatations, in such a way that the corresponding extended mappings are conformal in the lower half-plane. The machinery developed in Chapter II can then be applied to the study of T.

Let $\mu \in B$ and f^μ be the mapping of F with complex dilatation μ. We extend μ to the lower half-plane H' by giving it there the value 0. Let f_μ be the quasiconformal mapping of the plane which fixes 0, 1, ∞ and whose complex dilatation agrees with the extended μ. Then $f_\mu | H'$ is conformal.

Theorem 1.2. *The complex dilatations μ and v are equivalent if and only if the conformal mappings $f_\mu | H'$ and $f_v | H'$ coincide.*

PROOF. Suppose first that $f_\mu | H' = f_v | H'$. The mappings $f_\mu \circ (f^\mu)^{-1}$ and $f_v \circ (f^v)^{-1}$ are both conformal in the upper half-plane H, which they map onto the same quasidisc. Because they fix 0, 1, ∞, it follows that they agree in H, and hence also on the real axis $\mathbb{R}$. Since $f_\mu = f_\nu$ on $\mathbb{R}$, we conclude that $f^\mu = f^\nu$ on $\mathbb{R}$, i.e., μ and ν are equivalent.

Assume, conversely, that $f^\mu = f^\nu$ on $\mathbb{R}$. We define a mapping w of the plane by the requirements $w = f_\mu \circ f_\nu^{-1}$ in $f_\nu(H' \cup \mathbb{R})$, and $w = f_\mu \circ (f^\mu)^{-1} \circ f^\nu \circ f_\nu^{-1}$ in $f_\nu(H)$. From the hypothesis $f^\mu = f^\nu$ on $\mathbb{R}$ it follows that w is a homeomorphism of the plane. In addition, $w|f_\nu(H')$ is conformal. But so is also $w|f_\nu(H)$, because $f_\mu \circ (f^\mu)^{-1}$ and $f^\nu \circ f_\nu^{-1}$ are conformal. Since $f_\nu(\mathbb{R})$ is a quasicircle, we infer from Lemma I.6.1 that w is a Möbius transformation. Owing to the normalization, w is the identity mapping, and so $f_\mu = f_\nu$ in H'. $\qquad\qquad\Box$

Let F^* be the family of all quasiconformal mappings of the plane which fix the points 0, 1, ∞ and are conformal in the lower half-plane. Two mappings f_μ and f_ν of F^* are said to be equivalent if they agree in the lower half-plane.

Theorem 1.1 says that every equivalence class $[f^\mu]$ contains real analytic mappings. It follows that each class $[f_\mu]$ has representatives which are real analytic in the upper half-plane H. For in H,

$$f_\mu = f_\mu \circ (f^\mu)^{-1} \circ f^\mu,$$

where $f_\mu \circ (f^\mu)^{-1}$ is conformal. Therefore, $f_\mu|H$ is real analytic whenever f^μ is.

In particular, there are mappings f_μ which are conformal in H' and real analytic in H but which, nonetheless, are very irregular on the real axis $\mathbb{R}$. We recall that the Hausdorff dimension of the image curve $f_\mu(\mathbb{R})$ can be arbitrarily close to 2 (cf. I.6.1).

By Theorem 1.2, the space T can be regarded as the set of the equivalence classes $[f_\mu]$. Or more explicitly: *The universal Teichmüller space is the set of the normalized conformal mappings $f_\mu|H'$.*

1.4. Sewing Problem

The characterization of T by means of the conformal mappings $f_\mu|H'$ leads to far-reaching conclusions. Section 4, in particular, will be devoted to considerations emanating from this model of T.

Anticipating a need in section 1.5, we give a solution to the following *sewing problem*: Let h be a strictly increasing continuous function on the real axis, growing from $-\infty$ to $+\infty$. Find conformal mappings f_1 and f_2 of the upper and lower half-plane, respectively, onto complementary Jordan domains such that

$$f_1^{-1} \circ f_2 = h$$

on the real axis. We call the pair (f_1, f_2) normalized if f_1 and f_2 both fix 0, 1 and ∞.

Depending on h, the sewing problem, to which many questions in complex analysis seem to lead, need not have any solution or it may have infinitely many pairs of solutions. However, if h is a normalized quasisymmetric function, the existence of a unique normalized solution can be easily established by aid of the quasiconformal mappings f^μ and f_μ. It is in this form that the result will be used for studying the universal Teichmüller space.

Lemma 1.1. *Let h be a normalized quasisymmetric function. Then the sewing problem has a unique normalized pair of solutions.*

PROOF. Given a function $h \in X$, there is a mapping $f^\mu \in F$ such that $f^\mu|\mathbb{R} = h$. Then

$$f_1 = (f_\mu|H) \circ (f^\mu)^{-1}, \qquad f_2 = f_\mu|H',$$

is a solution of the sewing problem. This can be verified immediately.

Suppose that the pair (g_1, g_2) is also a normalized solution. Then $g_2|\mathbb{R} = g_1 \circ h = g_1 \circ f^\mu|\mathbb{R}$. Hence, the mapping w which agrees with $g_1 \circ f^\mu$ in $H \cup \mathbb{R}$ and with g_2 in H' is a homeomorphism of the plane. Off the real axis it is quasiconformal. By Lemma I.6.1, w is quasiconformal everywhere. Since w has the same complex dilatation as f_μ and both mappings fix 0, 1 and ∞, it follows from the uniqueness theorem (Theorem I.4.2) that $w = f_\mu$. Comparison of the definitions of w, f_1 and f_2 then shows that $g_1 = f_1, g_2 = f_2$. $\quad\square$

Note that f_1 and f_2 map the half-planes onto quasidiscs. Lemma 1.1 is due to Pfluger [2]; in [LV], p. 92, it was proved without the use of the existence theorem for Beltrami equations.

1.5. Normalized Quasidiscs

We shall now express in geometric terms the fact that points of the universal Teichmüller space can be represented by the conformal mappings $f_\mu|H'$. We call a quasidisc normalized if its boundary passes through the points 0, 1, ∞ and is so oriented that the direction from 0 to 1 to ∞ to 0 is negative with respect to the domain. Let Δ denote the class of all normalized quasidiscs.

If $f \in F^*$, then

$$[f] \to f(H') \tag{1.3}$$

is a bijective mapping of T onto Δ. We first conclude from Theorem 1.2 that (1.3) is well defined. If $f(H') = g(H')$, then $g^{-1} \circ f$ is a conformal self-mapping of H' fixing 0, 1, ∞. Hence $f|H' = g|H'$, i.e., $[f] = [g]$, and it follows that (1.3) is injective. Finally, by Lemma I.6.2, every quasidisc is the image of H' under a quasiconformal mapping of the plane which is conformal in H'. Since the required normalization is achieved by use of a suitable Möbius transformation, we conclude that (1.3) is surjective.

The bijection (1.3) provides one more model for the universal Teichmüller space: T is the collection of all normalized quasidiscs.

Using Lemma 1.1 we obtain a connection between normalized quasidiscs and the group structure of T (Gardiner [1]).

Theorem 1.3. *Two points* $[f^\mu]$, $[f^\nu] \in T$ *are inverse elements of the group* T *if and only if the quasidiscs* $f_\mu(H)$ *and* $f_\nu(H')$ *are mirror images with respect to the real axis.*

PROOF. Assume first that $[f^\mu]$ and $[f^\nu]$ are inverse; we can then take $f^\nu = (f^\mu)^{-1}$. Let $f_{\mu*}$ be the quasiconformal mapping of the plane which fixes the points 0, 1, ∞ and whose complex dilatation μ^* vanishes in H and equals $\bar\mu(\bar z)$ at almost all points $z \in H'$. We write $g_1 = f_{\mu*}|H$ and denote by g_2 the unique conformal mapping of H' onto $f_{\mu*}(H')$ which keeps 0, 1, ∞ fixed. Then g_1 and g_2 are normalized conformal mappings of the upper and lower half-planes, respectively, onto complementary quasidiscs.

In order to study $g_1^{-1} \circ g_2$ on the real axis $\mathbb{R}$, we continue f^μ by reflection in $\mathbb{R}$ and use the same notation f^μ for the extended mapping. Then

$$f^\mu = g_2^{-1} \circ f_{\mu*}$$

in H', because both sides are normalized quasiconformal self-mappings of H' with the same complex dilatation. Hence, on $\mathbb{R}$

$$g_1^{-1} \circ g_2 = (f^\mu)^{-1} = f^\nu.$$

Now set

$$f_1 = f_\nu \circ (f^\nu|H)^{-1}, \qquad f_2 = f_\nu|H'.$$

Then f_1 and f_2 are also normalized conformal mappings of the upper and lower half-planes onto complementary quasidiscs. On the real axis, $f_1^{-1} \circ f_2 = f^\nu = g_1^{-1} \circ g_2$. We conclude from Lemma 1.1 that $g_1 = f_1$, $g_2 = f_2$.

From the definition of $f_{\mu*}$ it follows that $f_{\mu*}(\bar z) = \bar f_\mu(z)$; one way to verify this is to compute the partial derivatives. Since $f_\nu(H) = f_1(H) = g_1(H) = f_{\mu*}(H)$, we obtain

$$f_\nu(H') = f_{\mu*}(H') = \overline{f_\mu(H)},$$

and the first part of the theorem has been proved.

After this the converse is easily established. Suppose that $\bar f_\mu(H) = f_\nu(H')$. By what was just proved, there is a quasiconformal mapping f_λ, where λ is determined by $f^\lambda = (f^\mu)^{-1}$, such that $\bar f_\mu(H) = f_\lambda(H')$. Since the mapping (1.3) is injective, we conclude that λ is equivalent to ν. It follows that $[f^\mu]$ and $[f^\nu]$ are inverse elements of T. $\qquad\square$

In II.2.1 we defined the distance $\delta(f(H')) = \|S_{f|H'}\|_{H'}$ between the domains $f(H')$ and H'. This notion was generalized in II.2.7 to apply to two arbitrary domains conformally equivalent to discs. When the domains are normalized quasidiscs, we can give a new definition: If $f, g \in F^*$, then the distance between $f(H')$ and $g(H')$ is defined by

$$q(f(H'), g(H')) = \| S_f - S_g \|_{H'}.$$

It is easy to check that q, so defined, is a metric on Δ. By use of the bijection (1.3), the metric can be transferred to T. This q-metric will be studied in detail in section 4. Before that in section 2, a different metric will be defined for T in a more direct manner.

2. Metric of the Universal Teichmüller Space

2.1. Definition of the Teichmüller Distance

In addition to the group structure, the universal Teichmüller space has a natural metric. We obtain this metric by measuring the distance between quasiconformal mappings in terms of their maximal dilatations. When representing points of T by mappings f it does not matter whether we assume that $f \in F$ (normalized self-mapping of H) or that $f \in F^*$ (normalized mapping of the plane, conformal in H'). This follows from the fact that f^μ and f_μ have the same maximal dilatation.

The distance between the points p and q of T is defined by

$$\tau(p, q) = \tfrac{1}{2} \inf \{ \log K_{g \circ f^{-1}} | f \in p, g \in q \}, \tag{2.1}$$

where K denotes the maximal dilatation and $f, g \in F$ or $f, g \in F^*$. This is called the *Teichmüller distance* between p and q. Teichmüller [1] used this idea for defining distance in his studies on compact Riemann surfaces (cf. V.2.2).

Before proving that the Teichmüller distance makes T into a metric space, we show that τ admits various other formulations. In order to fix the ideas, we assume for a moment that the quasiconformal mappings representing points of T are in the class F^*.

Fix a representative $f_0 \in p$ and set

$$\tau_1(p, q) = \tfrac{1}{2} \inf \{ \log K_{g \circ f_0^{-1}} | g \in q \}. \tag{2.2}$$

Alternatively, we can fix both $f_0 \in p$ and $g_0 \in q$, consider the class W of all quasiconformal mappings of the plane which agree in $f_0(H')$ with $g_0 \circ f_0^{-1}$, and set

$$\tau_2(p, q) = \tfrac{1}{2} \inf \{ \log K_w | w \in W \}. \tag{2.3}$$

Lemma 2.1. *The functions τ, τ_1 and τ_2 are the same.*

PROOF. Clearly, $\tau \leq \tau_1$. If $w \in W$, then $g = w \circ f_0 \in q$, so that $\tau_1 \leq \tau_2$. Finally, if $f \in p, g \in q$, then $g \circ f^{-1} \in W$, and so $\tau_2 \leq \tau$. ☐

By our previous remark, Lemma 2.1 holds also if the points of T are represented by mappings from F. The class W must of course be modified, being now the family of all quasiconformal self-mappings of H which agree on $\mathbb{R}$ with $g_0 \circ f_0^{-1}$. If we set $h = g_0 \circ f_0^{-1}\mathbb{R}$, we thus reencounter the family we considered in I.5.7, there under the name F_h. We proved that F_h contains an extremal mapping which has the smallest maximal dilatation in F_h. If this is denoted by K_h, then by Lemma 2.1,

$$\tau(p,q) = \tfrac{1}{2}\log K_h. \tag{2.4}$$

One important consequence of the existence of an extremal mapping in F_h is that in expressions (2.1), (2.2), and (2.3), we can replace inf by min. After this observation, it is readily seen that (2.1) (or (2.2), (2.3), or (2.4)) defines a metric in T. Clearly τ is non-negative and symmetric, and $\tau(p,p) = 0$. Suppose that $\tau(p,q) = 0$. Then it follows from (2.4) that F_h contains a conformal mapping, and because of the normalization, this mapping is the identity. Hence $f_0|\mathbb{R} = g_0|\mathbb{R}$, which implies $p = q$. The triangle inequality follows from the property $K_{g \circ f} \leq K_f K_g$ of the maximal dilatation.

The Teichmüller distance is invariant under the universal modular group. For the subgroup M_t, the trivial observation

$$g \circ f^{-1} = (g \circ f_0^{-1}) \circ (f \circ f_0^{-1})^{-1}$$

yields immediately the invariance

$$\tau([f],[g]) = \tau([f \circ f_0^{-1}],[g \circ f_0^{-1}]).$$

In particular,

$$\tau([f],[g]) = \tau(0,[g \circ f^{-1}]),$$

so that all distances can be measured from the origin. The general result (which we shall not make use of) follows from the fact that conformal mappings do not change maximal dilatation.

2.2. Teichmüller Distance and Complex Dilatation

If f and g have the complex dilatations μ and v, the norm of the complex dilatation of $g \circ f^{-1}$ is equal to $\|(\mu - v)/(1 - \bar{\mu}v)\|_\infty$. Therefore, in terms of complex dilatations, the Teichmüller distance assumes the form

$$\tau(p,q) = \frac{1}{2}\min\{\log\frac{1 + \|(\mu - v)/(1 - \bar{\mu}v)\|_\infty}{1 - \|(\mu - v)/(1 - \bar{\mu}v)\|_\infty}\,|\,\mu \in p, v \in q\}.$$

The right-hand expression displays a striking similarity to the hyperbolic distance h in the unit disc D given by formula (1.1) in I.1.1. We introduce the number

$$\|h(\mu,v)\|_\infty = \operatorname*{ess\,sup}_{z \in H} h(\mu(z), v(z))$$

and call it the hyperbolic distance between the complex dilatations μ and v.

From (1.1) in I.1.1 we then obtain the characterization

$$\tau(p,q) = \min\{\|h(\mu,v)\|_\infty \,|\, \mu \in p, v \in q\}$$

of the Teichmüller distance as the minimal hyperbolic distance.

Consider the bounded function

$$\beta(p,q) = \inf\left\{\left\|\frac{\mu - v}{1 - \bar{\mu}v}\right\|_\infty \,\Big|\, \mu \in p, v \in q\right\}.$$

From $\beta = \tanh \tau$ we see that in the definition of β, inf can be replaced by min, that β also makes T into a metric space, that β is invariant under the universal modular group and that the metrics defined by τ and β are topologically equivalent. In other words, in studying topological properties of T we may use the metric provided by β instead of the Teichmüller metric. Let us give a simple application.

Theorem 2.1. *The universal Teichmüller space is pathwise connected.*

PROOF. Consider the origin of T, i.e., the point represented by the function of B which is identically zero, and an arbitrary point $p \in T$ represented by μ. For $0 \le t \le 1$, let p_t be the point represented by the function $t\mu$ of B. Then

$$\beta(p_{t_1}, p_{t_2}) \le \left\|\frac{t_1\mu - t_2\mu}{1 - t_1 t_2 |\mu|^2}\right\|_\infty \le \frac{|t_1 - t_2|}{1 - \|\mu\|_\infty^2}.$$

We see that the mapping $t \to p_t$ is continuous, i.e., it is a path in T joining the origin to p. $\qquad\square$

In 3.3 we shall prove that the universal Teichmüller space is not only pathwise connected but even contractible. This means that T can be deformed continuously to a point. The result is less trivial than it would seem at first glance, because for equivalent μ and v, the complex dilatations $t\mu$ and tv need not be equivalent for $0 < t < 1$ (Gehring [1]).

2.3. Geodesics for the Teichmüller Metric

The length of an arc $\gamma: [0,1] \to (T, \tau)$ is the supremum of $\Sigma\, \tau(\gamma(t_{j-1}), \gamma(t_j))$ for all subdivisions $0 = t_0 < t_1 < \cdots < t_n = 1$ of the unit interval. An arc γ is a geodesic if the length of every subarc α of γ is equal to the distance between the endpoints of α.

Geodesics of T can be described explicitly with the help of extremal complex dilatations. We say that $\mu \in p$ is extremal if $\|\mu\|_\infty = \min\{\|v\|_\infty \,|\, v \in p\}$.

Theorem 2.2. *If μ is an extremal complex dilatation for the point $p \in T$, then*

$$\mu_t = \frac{(1+|\mu|)^t - (1-|\mu|)^t}{(1+|\mu|)^t + (1-|\mu|)^t}\frac{\mu}{|\mu|}, \qquad 0 \le t \le 1, \tag{2.5}$$

is extremal for the point $p_t = [\mu_t]$. The arc $t \to p_t$ is a geodesic from 0 to p, and

$$\tau(p_t, 0) = t\tau(p, 0). \tag{2.6}$$

PROOF. From (2.5) we see that $\mu_t(z)$ is the point which divides the hyperbolic length (in the unit disc) of the line segment from 0 to $\mu(z)$ in the ratio $t: (1 - t)$ (cf. formula (4.16) in I.4.7).

If f^μ has maximal dilatation K, then by Theorem I.4.7, the mapping f^{μ_t} has maximal dilatation K^t and $f^\mu \circ (f^{\mu_t})^{-1}$ has maximal dilatation K^{1-t}. Suppose that $w \in p_t$. Then $\varphi = f^\mu \circ (f^{\mu_t})^{-1} \circ w \in p$, and so $K \leq K_\varphi \leq K^{1-t} K_w$. Consequently, $K_w \geq K^t$. We conclude that μ_t is extremal for the point p_t. This reasoning also shows that $\tau(p_t, 0) = \frac{1}{2} \log K^t = t\tau(p, 0)$, i.e., the validity of (2.6).

Since $f^\mu \circ (f^{\mu_t})^{-1}$ has maximal dilatation K^{1-t}, we conclude that $\tau(p_t, p) \leq (1 - t)\tau(p, 0)$. Consequently, $\tau(0, p_t) + \tau(p_t, p) = \tau(0, p)$ for every t. Finally, if we repeat the above argument for an arbitrary subarc of $t \to p_t$, we see that $t \to p_t$ is a geodesic. $\qquad\square$

Since the extremal μ need not be unique (cf. I.5.7), we cannot conclude that the geodesic $t \to [\mu_t]$ is unique.

2.4. Completeness of the Universal Teichmüller Space

We shall now prove that the space (T, τ) is complete. We first describe explicitly the construction on which our proof of the completeness is based and list the pertinent facts associated with that construction.

Lemma 2.2. *Every Cauchy sequence in (T, τ) contains a subsequence whose points are represented by complex dilatations μ_n with the following properties:*

1° $\lim \mu_n(z) = \mu(z)$ *exists almost everywhere;*
2° $[f_{\mu_n}] \to [f_\mu]$ *in the Teichmüller metric;*
3° $f_{\mu_n}(z) \to f_\mu(z)$ *uniformly in the spherical metric;*
4° $f^{\mu_n}(z) \to f^\mu(z)$ *locally uniformly in the upper half-plane.*

PROOF. In order to simplify the notation, we renumber functions each time that we pass from a sequence to its subsequence. Also, we write $f_n = f_{\mu_n}$.

Let $([f_n])$ be a Cauchy sequence in (T, τ). We shall construct inductively a subsequence with the properties 1°–4° using suitably chosen mappings f_n.

First, fix a mapping f_i so that

$$\inf \log K_{f_{i+p} \circ f_i^{-1}} < \tfrac{1}{2}, \qquad p = 1, 2, \ldots,$$

where for each p, the infimum is taken over all mappings of $[f_{i+p}]$. Since $([f_n])$ is a Cauchy sequence, such a mapping f_i exists, as can be seen from formula (2.2) and Lemma 2.1. We renumber the sequence by setting $f_i = f_1$.

After this, we choose for every $n > 1$ the mapping f_n from its equivalence class so that

$$\log K_{f_n \circ f_1^{-1}} < \tfrac{1}{2}.$$

From this new sequence (f_n) we choose a mapping f_k so that

$$\inf \log K_{f_{k+p} \circ f_k^{-1}} < \tfrac{1}{4},$$

where again for each p the infimum is taken over all mappings of the class $[f_{k+p}]$. We set $f_k = f_2$, and for $n > 2$, choose a representative of $[f_n]$ so that

$$\log K_{f_n \circ f_2^{-1}} < \tfrac{1}{4}.$$

Continuing this procedure we obtain a sequence (f_n), such that $([f_n])$ is a subsequence of the given Cauchy sequence and such that, for any two consecutive indexes, the maximal dilatations satisfy the inequality

$$\log K_{f_{n+1} \circ f_n^{-1}} < 2^{-n}, \qquad n = 1, 2, \ldots.$$

It follows that

$$\log K_{f_{n+p} \circ f_n^{-1}} \leq \sum_{j=1}^{p} 2^{-(n+j-1)} < 2^{-n+1}, \tag{2.7}$$

for $n, p = 1, 2, \ldots$.

Considering the connection between the maximal dilatation and the norm of the complex dilatation, we deduce from (2.7) that the complex dilatations μ_n of f_n satisfy the inequality

$$\| \mu_{n+p} - \mu_n \|_\infty \leq 2 \left\| \frac{\mu_{n+p} - \mu_n}{1 - \bar{\mu}_n \mu_{n+p}} \right\|_\infty < 2 \tanh 2^{-n}.$$

Thus (μ_n) is a Cauchy sequence in L^∞. Since L^∞ is complete, the limit $\mu = \lim \mu_n$ exists in L^∞. Thus the validity of condition $1°$ follows. From (2.7) we conclude that the mappings f_n (and hence also f^{μ_n}) are K-quasiconformal for a fixed K. It follows that $\|\mu\|_\infty < 1$. Therefore, $[\mu] = \lim[\mu_n]$ in (T, β) and hence also in (T, τ). This means that the statement $2°$ is true. Since the mappings f_n and f^{μ_n} keep the three points $0, 1, \infty$ fixed, the families $\{f_n\}$ and $\{f^{\mu_n}\}$ are normal, by Theorem I.2.1. Hence $3°$ and $4°$ follow, after possible passage to further subsequences.   □

Theorem 2.3. *The universal Teichmüller space is complete.*

PROOF. In view of statement $2°$ in Lemma 2.2, it is enough to observe that if a Cauchy sequence contains a convergent subsequence, then the sequence itself is convergent.   □

Having proved Theorem 2.3 we see that in Lemma 2.2, we need not pick a subsequence: *If $p_n \to p$ in (T, τ), there are complex dilatations $\mu_n \in p_n$, $\mu \in p$, such that the conditions $1°$, $3°$, and $4°$ of Lemma 2.2 hold* (cf. Theorem I.4.6).

In view of the simple relation between the distances τ and β, it is easy to show that the metric space (T, β) is also complete. In contrast, using Schwarzian derivatives we shall obtain in section 4 one more model for T which is a metric space homeomorphic to (T, τ), but is not complete. (It is, in fact, the metric space (Δ, q) discussed at the end of 1.5.)

3. Space of Quasisymmetric Functions

3.1. Distance between Quasisymmetric Functions

Let us consider again the space X of normalized quasisymmetric functions discussed in 1.1. For $h \in X$ we defined in I.5.1 the maximal dilatation K_h^*. Imitating the method we used in defining the Teichmüller distance, we set

$$\rho(h_1, h_2) = \tfrac{1}{2} \log K_{h_2 \circ h_1^{-1}}^*$$

for $h_1, h_2 \in X$. It is easy to verify that ρ defines a metric on X.

Theorem 3.1. *The group isomorphism*

$$[f] \to f | \mathbb{R} \tag{3.1}$$

is a homeomorphism of (T, τ) onto (X, ρ).

PROOF. We proved in 1.1 that (3.1) is a bijection of T onto X. From (2.4) and the left-hand inequality (5.10) in I.5.7 it follows that

$$\rho(f_1 | \mathbb{R}, f_2 | \mathbb{R}) \leq \tau([f_1], [f_2]). \tag{3.2}$$

Hence (3.1) is continuous. From Lemma I.5.5 (or from the right-hand inequality (5.10) in I.5.7) we conclude that the inverse of (3.1) is continuous. □

From the double inequality (5.10) in I.5.7 we can draw another conclusion: *The space (X, ρ) is complete.* For we conclude from the right-hand inequality (5.10) that the preimage of a Cauchy sequence in (X, ρ) is a Cauchy sequence in (T, τ). The inequality (3.2) then shows that (3.1) maps a convergent sequence of (T, τ) onto a convergent sequence of (X, ρ).

Suppose that $h, h_n \in X$, $n = 1, 2, \ldots$, and that $\lim \rho(h_n, h) = 0$. Then $h_n \to h$ locally uniformly in the euclidean metric. For by Lemma I.5.1, $\{h_n\}$ is a normal family. If $\tilde{h}$ is the limit of a convergent subsequence (h_{n_i}) of (h_n), we have $K_{\tilde{h} \circ h^{-1}}^* \leq \lim K_{h_{n_i} \circ h^{-1}}^* = 1$. Hence $\tilde{h} = h$. Since every convergent subsequence of (h_n) has the limit h, the sequence itself tends to h.

The converse is not true. A counterexample is obtained if we set $h_n(x) = x$ for $x \leq n$, and $h_n(x) = 2x - n$ for $x > n$. Then $h_n \in X$ and $\lim h_n(x) = h(x) = x$, uniformly on every bounded interval. But the quasisymmetry constant of h_n is 2, so that by the remark in I.5.2, $\liminf \rho(h_n, h) \geq (\log \lambda^{-1}(2))/2 > 0$.

3.2. Existence of a Section

Using the space (X, ρ), we can prove that the universal Teichmüller space is *contractible*, i.e., there exists a continuous mapping of $T \times \{t|0 \leq t \leq 1\}$ into T which is the identity mapping for $t = 0$ and a constant mapping for $t = 1$.

As a preparation for the proof, we modify the mapping (3.1) by changing its domain of definition. Let L^∞ be the space of functions bounded and measurable in the upper half-plane, and $B = \{\mu \in L^\infty | \|\mu\|_\infty < 1\}$ its open unit ball. For $\mu \in B$, we now consider the mapping ψ of B onto X, defined by

$$\psi(\mu) = f^\mu | \mathbb{R}.$$

In view of the definition of the Teichmüller distance, inequality (3.2) can be written in the form

$$\rho(f^\mu | \mathbb{R}, f^\nu | \mathbb{R}) \leq \frac{1}{2} \log \frac{1 + \|(\mu - \nu)/(1 - \bar{\mu}\nu)\|_\infty}{1 - \|(\mu - \nu)/(1 - \bar{\mu}\nu)\|_\infty}.$$

It follows that ψ is continuous.

Of course, the mapping $\psi \colon B \to X$ is not invertible. However, there exists a *section* $s \colon X \to B$, i.e., a continuous mapping of X into B such that $\psi \circ s$ is the identity mapping of X.

A section s can be constructed with the aid of the Beurling–Ahlfors extension of a quasisymmetric function. Given a function $h \in X$, we set as in I.5.3,

$$f(x + iy) = \frac{1}{2} \int_0^1 (h(x + ty) + h(x - ty)) \, dt + i \int_0^1 (h(x + ty) - h(x - ty)) \, dt.$$

Let μ be the complex dilatation of f. We shall prove that the mapping $s \colon X \to B$, defined by $s(h) = \mu$, is a section.

For $h_i \in X$, $i = 1, 2$, we denote their Beurling–Ahlfors extensions by f_i, and set $s(h_i) = \mu_i$. Let K be the maximal dilatation and k the quasisymmetry constant of $h_2 \circ h_1^{-1}$. We showed in I.5.2 that $k \leq \lambda(K)$. Hence,

$$\rho(h_1, h_2) = \frac{1}{2} \log K \geq \frac{1}{2} \log \lambda^{-1}(k).$$

Now suppose that $h_2 \to h_1$ in (X, ρ). It follows from the above inequality that $k \to 1$. By Lemma I.5.3, the maximal dilatation of $f_2 \circ f_1^{-1}$ then tends to 1. This is equivalent to μ_2 converging to μ_1, and we have proved that s is continuous.

Trivially, $(\psi \circ s)(h) = f^{s(h)} | \mathbb{R} = h$, so that $\psi \circ s$ is the identity mapping of X. Consequently, $s \colon X \to B$ is a section.

3.3. Contractibility of the Universal Teichmüller Space

After these preparations, the desired result can be easily established (Earle and Eells [1]).

Theorem 3.2. *The universal Teichmüller space is contractible.*

PROOF. Every point of T is an equivalence class $[s(h)]$, $h \in X$. We show that

$$([s(h)], t) \to [(1 - t)s(h)] \tag{3.3}$$

deforms T continuously to the point 0 as t increases from 0 to 1.

In proving this, we make use of Theorem 3.1 which says that $[\mu] \to f^\mu | \mathbb{R}$ is a homeomorphism of (T, τ) onto (X, ρ). It means that instead of (3.3), we can consider the induced mapping

$$(h, t) \to f^{(1-t)s(h)} | \mathbb{R} \tag{3.4}$$

of $X \times [0, 1]$ into X. Clearly, $(h, 0) \to h$ and $(h, 1) \to$ identity. The theorem follows if we prove that (3.4) is continuous.

The mapping (3.4) is the composition of the three mappings

$$(h, t) \to (s(h), t), \qquad (s(h), t) \to (1 - t)s(h), \qquad (1 - t)s(h) \to f^{(1-t)s(h)} | \mathbb{R}.$$

The first one is continuous, because we just proved that $h \to s(h)$ is a continuous map of X into B. The second maps $B \times [0, 1]$ continuously into B, since $\|(1 - t_1)s(h_1) - (1 - t_2)s(h_2)\|_\infty \le \|s(h_1) - s(h_2)\|_\infty + |t_1 - t_2|$. Finally, the third mapping is continuous, because we showed that $\mu \to f^\mu | \mathbb{R}$ maps B continuously into X. Hence, (3.4) is a continuous contraction of X to a point, and (3.3) has the same property with respect to T. $\qquad\square$

3.4. Incompatibility of the Group Structure with the Metric

We showed at the end of 3.1 that pointwise convergence of functions of X does not imply ρ-convergence. A slightly more complicated counterexample leads to the conclusion that the topological structure and the group structure of X are not compatible. We express the result in terms of T.

Theorem 3.3. *The universal Teichmüller space is not a topological group.*

PROOF. The theorem follows if we find an $[f] \in T$ and a sequence of points $[g_n] \in T$, such that $[g_n]$ tends to $[g]$ but $[f \circ g_n]$ does not tend to $[f \circ g]$. Because the mapping (3.1) is a group isomorphism and a homeomorphism, the counterexample can be constructed in X. We follow a suggestion of P. Tukia.

In order to simplify notation we write f instead of $f | \mathbb{R}$. We define f as follows: $f(x) = x$ if $x \ge 0$, $f(x) = x/2$ if $-2 \le x < 0$, and $f(x) = x + 1$ if $x < -2$. Then f is a 2-quasisymmetric function of X. Set $\iota_n(x) = x$ if $x \ge 0$ and $\iota_n(x) = (1 + 1/n)x$ if $x < 0$, $n = 1, 2, \dots$. Then $\iota_n \in X$ is $(1 + 1/n)$-quasisymmetric, and therefore $(1 + 1/n)^2$-quasiconformal (cf. I.5.3). If ι denotes the identity mapping of $\mathbb{R}$ onto itself, we thus have

$$\rho(\iota_n, \iota) \le \log(1 + 1/n).$$

Let us define $g_n = \iota_n \circ f^{-1}$. Because $\rho(g_n, f^{-1}) = \rho(\iota_n, \iota)$, we deduce that

$$\lim_{n \to \infty} \rho(g_n, f^{-1}) = 0. \qquad (3.5)$$

We prove that $f \circ g_n = f \circ \iota_n \circ f^{-1}$ (which converges to the identity ι pointwise) does not tend to $f \circ (\lim g_n) = \iota$ in the ρ-metric.

Direct calculation yields

$$f(g_n(x)) = \begin{cases} (1 + 1/n)x & \text{if } -n/(n+1) \le x < 0, \\ 2(1 + 1/n)x + 1 & \text{if } -1 \le x < -n/(n+1). \end{cases}$$

It follows that $f \circ g_n$ has a quasisymmetry constant ≥ 2 for every n. Consequently,

$$\rho(f \circ g_n, \iota) \ge \tfrac{1}{2} \log \lambda^{-1}(2).$$

In conjunction with (3.5), this shows that (X, ρ), and hence (T, τ), is not a topological group. □

We see that, unlike the right translation, the left translation $[f] \to [f_0 \circ f]$, f_0 fixed, need not be continuous in T.

4. Space of Schwarzian Derivatives

4.1. Mapping into the Space of Schwarzian Derivatives

The universal Teichmüller space was defined by means of quasiconformal self-mappings of the upper half-plane. In this section and in section 5, we change the roles of the upper and lower half-planes. Now f^μ is a self-mapping of the lower half-plane H' and f_μ is conformal in the upper half-plane H. This change, which does not affect any of the results in sections 1–3, simplifies notation here, because we are now dealing primarily with the conformal part of the mappings f_μ.

It follows from what we proved in 1.3 that each point of the universal Teichmüller space T can be represented by a normalized conformal mapping $f_\mu | H$. It was Bers [6] who noticed the importance of forming the Schwarzian derivative and defining the mapping

$$[\mu] \to S_{f_\mu | H}$$

of T. The image points, as Schwarzian derivatives, are holomorphic functions in H, for which the norm defined in II.1.3 is pertinent.

This leads us to introduce the space Q of all functions φ holomorphic in H for which the hyperbolic sup norm

$$\| \varphi \| = \sup_{z \in H} 4y^2 |\varphi(z)|,$$

$z = x + iy$, is finite. The space Q has a natural linear structure over the complex numbers.

Furthermore, Q is complete. For if (φ_n) is a Cauchy sequence in Q, then $\varphi_n \eta^{-2}$, $\eta(z) = (2y)^{-1}$, is a Cauchy sequence in L^∞. Since L^∞ is complete, there is a $\psi \in L^\infty$ such that $\varphi_n \eta^{-2}$ converges to ψ in L^∞. Here ψ can be taken to be continuous. Then φ_n converges locally uniformly to $\varphi = \psi \eta^2$, because η is locally bounded. It follows that φ is holomorphic, $\|\varphi_n - \varphi\| \to 0$, and $\varphi \in Q$. We conclude that Q is a Banach space. Its points are Schwarzian derivatives: By Theorem II.1.1 every function $\varphi \in Q$ is the Schwarzian derivative of a function f meromorphic in H.

Going back to the functions f_μ, we write $s_\mu = S_{f_\mu | H}$. By Theorem II.1.3, $\|s_\mu\| \le 6$. Therefore, $[\mu] \to s_\mu$ maps T into Q. The mapping is well defined, for if ν is equivalent to μ, we have $f_\nu | H = f_\mu | H$ and hence $s_\nu = s_\mu$.

4.2. Comparison of Distances

We shall prove that the mapping $[\mu] \to s_\mu$ is a homeomorphism of T onto its image in Q. To this end we shall compare the β-distance of two given points $[\mu]$ and $[\nu]$ of T to the distance of their images s_μ and s_ν in Q. We write $q(\varphi_1, \varphi_2) = \|\varphi_1 - \varphi_2\|$ for points of Q.

In the special case $\nu = 0$, estimates in both directions can be obtained directly from our previous results. If $\nu = 0$, then also $s_\nu = 0$, and $q(s_\mu, 0) = \|s_\mu\|$. Moreover, $\beta([\mu], 0) = \inf \|\mu\|_\infty$. By Theorem II.3.2, $\|s_\mu\| \le 6 \|\mu\|_\infty$. This holds no matter how μ is chosen from the equivalence class. Consequently,

$$q(s_\mu, 0) \le 6\beta([\mu], 0). \tag{4.1}$$

We remark that $q(s_\mu, 0)$ is equal to the distance $\delta(f_\mu(H))$ of $f_\mu(H)$ from H (cf. the remark at the end of 1.5).

In order to get an inequality in the opposite direction, we choose an arbitrary $\varphi \in Q$ such that $\|\varphi\| < 2$. By Theorems II.1.1 and II.5.1, there is a normalized quasiconformal mapping f of the plane which is conformal in H, for which $S_{f|H} = \varphi$, and whose complex dilatation μ in the lower half-plane is obtained from the formula $\mu(\bar{z}) = -2y^2 \varphi(z)$, $z \in H$. For this mapping $f = f_\mu$ we have $\|\mu\|_\infty = \|s_\mu\|/2$. Hence,

$$q(s_\mu, 0) \ge 2\beta([\mu], 0). \tag{4.2}$$

We assumed that $q(s_\mu, 0) < 2$. But since all β-distances are <1, (4.2) holds trivially if $q(s_\mu, 0) \ge 2$.

We shall now generalize (4.1) and (4.2) for arbitrary points $[\mu]$ and $[\nu]$. We start with the transformation rule

$$\|s_\mu - s_\nu\|_H = \|S_{f_\mu \circ f_\nu^{-1}}\|_{f_\nu(H)} \tag{4.3}$$

for the Schwarzian derivatives (formula (1.10) in II.1.3). We write $A_\nu = f_\nu(H)$ and apply Theorem II.3.2 to the conformal mapping $w = f_\mu \circ (f_\nu^{-1} | A_\nu)$ in the quasidisc A_v. It follows that

$$\|S_w\|_{A_v} \leq \sigma_0(A_v) \left\| \frac{\mu - v}{1 - \bar{\mu}v} \right\|_\infty ;$$

here $\sigma_0(A_v) = 6 + \delta(A_v)$ is the outer radius of univalence of A_v. In view of (4.3), this yields the estimate

$$q(s_\mu, s_v) \leq \sigma_0(A_v)\beta([\mu], [v]), \tag{4.4}$$

which is the desired generalization of (4.1). Since $\sigma_0(A_v) \leq 12$, we could use as the coefficient on the right-hand side the absolute constant 12. On the other hand, $\sigma_0(A_v)$ can be replaced by $\min(\sigma_0(A_\mu), \sigma_0(A_v))$, where $A_\mu = f_\mu(H)$.

It is more difficult to generalize the inequality (4.2). We choose v from its equivalence class so that f_v has the smallest possible maximal dilatation K_v. After this, we consider Schwarzian derivatives s_μ which are so close to s_v that $q(s_\mu, s_v) < \varepsilon(K_v)$, where ε is the constant of Theorem II.4.1. Then, by formula (4.3),

$$\|S_w\|_{A_v} < \varepsilon(K_v).$$

We know that w has a quasiconformal extension, namely, $f_\mu \circ f_v^{-1}$. However, we prefer to extend w by utilizing Theorem II.4.1, which makes it possible to estimate the complex dilatation. By that theorem, w has a quasiconformal extension to the plane such that the complex dilatation κ of the extended mapping satisfies the inequality

$$\|\kappa\|_\infty \leq \frac{\|S_w\|_{A_v}}{\varepsilon(K_v)} = \frac{q(s_\mu, s_v)}{\varepsilon(K_v)}. \tag{4.5}$$

If the extended w is also denoted by w, then $f_\lambda = w \circ f_v$ is a quasiconformal extension of $f_\mu | H$ to the lower half-plane. Thus λ is equivalent to μ. Because $w = f_\lambda \circ f_v^{-1}$, we have, therefore,

$$\|\kappa\|_\infty = \left\| \frac{\lambda - v}{1 - \bar{\lambda}v} \right\|_\infty \geq \beta([\mu], [v]).$$

Combining this with (4.5) we finally arrive at the inequality

$$q(s_\mu, s_v) \geq \varepsilon(K_v)\beta([\mu], [v]), \tag{4.6}$$

valid also if $q(s_\mu, s_v) \geq \varepsilon(K_v)$. This contains (4.2) as a special case: If $v = 0$, then $K_v = 1$ and $\varepsilon(K_v) = 2$. Since the roles of μ and v can be interchanged, we can replace $\varepsilon(K_v)$ in (4.6) by $\max(\varepsilon(K_\mu), \varepsilon(K_v))$.

4.3. Imbedding of the Universal Teichmüller Space

The estimates (4.4) and (4.6) show that the β- and q-metrics are topologically equivalent. Thus a new important model is obtained for the topological space (T, τ).

Theorem 4.1. *The mapping*

$$[\mu] \to S_{f_\mu|H}, \tag{4.7}$$

is a homeomorphism of the universal Teichmüller space onto its image in Q.

PROOF. We noted already in 4.1 that (4.7) is well defined in T. If $[\mu]$ and $[\nu]$ have the same image, it follows from the normalization that $f_\mu|H = f_\nu|H$, i.e., μ and ν are equivalent. Hence (4.7) is injective. Inequality (4.4) shows that (4.7) is continuous, and (4.6) that its inverse is continuous. □

In 4.5 we shall show that the image of T under the homeomorphism (4.7) is open in Q. (We have almost proved this in establishing (4.6).) The mapping (4.7), which will later be considered in connection with an arbitrary Teichmüller space, is called the *Bers imbedding* of Teichmüller space.

By Theorem 4.1, the convergence $S_{f_{\mu_n}} \to S_{f_\mu}$ in Q implies that $[\mu_n] \to [\mu]$ in T. Hence, by Lemma 2.2 and the remark following Theorem 2.3, $f_{\mu_n}(z) \to f_\mu(z)$ uniformly in H.

Anticipating developments in Chapter V, we denote the image of T under (4.7) by $T(1)$. When there is no fear of confusion, we often identify $T(1)$ with the universal Teichmüller space. Like X, the space $T(1)$ is simpler than T in that its points are functions and not equivalence classes of functions.

We can also define $T(1) = \{S_f | f$ is conformal in H and has a quasiconformal extension to the plane$\}$. For such an f is equal to a normalized mapping $f_\mu|H$ modulo a Möbius transformation, which does not change the Schwarzian derivative.

By Theorem II.5.1, the set $T(1)$ contains the open ball $B(0,2) = \{\varphi \in Q | \|\varphi\| < 2\}$. In this ball, the inverse of the mapping (4.7) can be described explicitly:

$$\varphi \to [\mu], \qquad \mu(\bar{z}) = -2y^2 \varphi(z).$$

The space $(T(1), q)$ *is not complete*, even though it is homeomorphic to the complete spaces (T, τ), (T, β) and (X, ρ). In order to prove this, it is sufficient to find an $S_f \in Q \setminus T(1)$ and functions $S_{f_n} \in T(1)$, $n = 1, 2, \ldots$, such that $S_{f_n} \to S_f$ in Q. Then (S_{f_n}) is a Cauchy sequence in $T(1)$ but its limit is not in $T(1)$. An example is provided by the functions $z \to f(z) = \log z$, $z \to f_n(z) = z^{1/n}$ in H, which we considered, for another purpose, in II.1.4. Since

$$S_{f_n}(z) = \left(1 - \frac{1}{n^2}\right) \frac{1}{2z^2},$$

we have $\|S_{f_n}\|_H = 2(1 - 1/n^2) < 2$. By Theorem II.5.1, the mapping f_n has a quasiconformal extension to the plane. Hence $S_{f_n} \in T(1)$. In II.1.4 we saw already that

$$\|S_{f_n} - S_f\| = 2/n^2 \to 0.$$

But since $z \to \log z$ does not even have a homeomorphic extension, S_f is not in $T(1)$.

4.4. Schwarzian Derivatives of Univalent Functions

Let us define the set

$$U = \{S_f | f \text{ univalent in } H\}.$$

Then trivially $T(1) \subset U$, and by Theorem II.1.3, $U \subset Q$. More precisely, we conclude from Theorem II.1.3 that U is contained in the closure of the ball $B(0, 6) = \{\varphi \in Q | \|\varphi\| < 6\}$. On the other hand, it follows from Theorem II.5.2 that U contains the closure of $B(0, 2)$.

Let $f_1(w) = w + 1/w, f_2(w) = w - 1/w$. If h is a conformal mapping of the upper half-plane onto $\{w | |w| > 1\}$, then $S_{f_1 \circ h}, S_{f_2 \circ h} \in U$. From the calculations in II.2.6 it follows that $\|S_{f_1 \circ h} - S_{f_2 \circ h}\| = 12$. We conclude that the diameter of U is 12.

The set U is closed in Q. For suppose that $S_{f_n} \in U$, $n = 1, 2, \ldots$, and that S_{f_n} converges to S_f in Q. We show that f is univalent.

We are free to compose the functions f_n with arbitrary Möbius transformations. There is no loss of generality, therefore, in assuming that every f_n fixes the same three points a_1, a_2, a_3 in H. By Theorem I.2.1, the family $\{f_n\}$ is then normal. Consequently, (f_n) contains a subsequence which is locally uniformly convergent in H. By renumbering the functions we may assume that (f_n) itself has this property. The limit $g = \lim f_n$ fixes a_1, a_2 and a_3, and is therefore univalent in H. At every point $z \in H$ we have $\lim S_{f_n}(z) = S_g(z)$ and also $\lim S_{f_n}(z) = S_f(z)$. Hence, f differs from g by a Möbius transformation, and so f is univalent.

Since U is closed and $T(1) \subset U$, the closure of $T(1)$ is contained in U. If $S_f \in U$, we can always find functions f_n with $S_{f_n} \in T(1)$ such that

$$f(z) = \lim f_n(z) \qquad \text{locally uniformly in } H. \qquad (4.8)$$

An approximating sequence f_n with $S_{f_n} \in T(1)$ is obtained as follows. Set

$$g_n(z) = \frac{z + i/n}{1 - iz/n}, \qquad n = 2, 3, \ldots.$$

Then g_n maps H onto the disc $D_n = \{w | |w - i| < ((n - 1)/(n + 1)) |w + i|\}$, whose closure lies in H. As $n \to \infty$, the discs D_n exhaust H, and $g_n(z) \to z$, locally uniformly. Hence, by setting $f_n = f \circ g_n$, we obtain a sequence of functions for which (4.8) is true. The property $S_{f_n} \in T(1)$ follows from the fact that $f_n(\mathbb{R}) = f(\partial D_n)$ is a quasicircle (cf. the remark in I.6.1).

If (4.8) holds, the derivatives of f_n converge to the derivatives of f. Hence,

$$S_f(z) = \lim_{n \to \infty} S_{f_n}(z)$$

locally uniformly in H. However, as we showed in II.1.4, it does not necessarily follow that $S_{f_n} \to S_f$ in Q. We cannot conclude, therefore, that the closure of $T(1)$ coincides with U. A counterexample such as the one in II.1.4 does not disprove this either, but actually the closure of $T(1)$ is not the whole of U. This will be explained in 4.6.

4.5. Univalent Functions and the Universal Teichmüller Space

From Theorems II.4.1 and II.4.2 we obtain a remarkable connection between the sets $T(1)$ and U (Gehring [2]).

Theorem 4.2. *The set $T(1)$ is the interior of U.*

PROOF. We prove first that $T(1)$ is an open subset of Q. Fix an arbitrary point S_f of $T(1)$. For $S_h \in Q$ we write $g = h \circ f^{-1}$, and conclude that g is meromorphic in the quasidisc $f(H)$. By Theorem II.4.1, there exists a positive constant ε such that if $\|S_g\|_{f(H)} < \varepsilon$, then g is univalent in $f(H)$ and has a quasiconformal extension to the plane. Now choose $S_h \in Q$ such that $\|S_h - S_f\|_H < \varepsilon$. Then

$$\|S_g\|_{f(H)} = \|S_h - S_f\|_H < \varepsilon.$$

Because $h = g \circ f$, we conclude that $S_h \in T(1)$. It follows that $T(1)$ is open.

Since $T(1) \subset U$, the proof will be complete if we show that int $U \subset T(1)$. Choose a point $S_f \in$ int U. We then have an $\varepsilon > 0$ such that

$$V = \{\varphi \in Q \mid \|\varphi - S_f\| \leq \varepsilon\} \subset U.$$

Let g be an arbitrary meromorphic function in the domain $f(H)$, with the property $\|S_g\|_{f(H)} \leq \varepsilon$. If $h = g \circ f$, then

$$\|S_h - S_f\|_H = \|S_g\|_{f(H)} \leq \varepsilon.$$

It follows that $S_h \in V \subset U$, i.e., h is univalent in H. But then $g = h \circ f^{-1}$ is univalent in $f(H)$. What we have proved is that $f(H)$ is an ε-Schwarzian domain. Hence, by Theorem II.4.2, the domain $f(H)$ is a quasidisc. We conclude that $S_f \in T(1)$ (cf. Lemma I.6.2, statement 3°) as we wished to show. $\qquad\square$

The result that $T(1)$ is open in Q was first proved by Ahlfors [4].

4.6. Closure of the Universal Teichmüller Space

For a long time it was a famous open problem, raised by Bers, whether the closure of $T(1)$, which is contained in U, actually agrees with U. In 1978, Gehring [3] showed that the answer to this question is in the negative. He constructed a counterexample with the help of the simply connected domain G which is the complement of the curve

$$\gamma = \{z = \pm e^{(-a+i)t} \mid 0 \leq t < \infty\} \cup \{0\},$$

where $a > 0$ is small (Fig. 6). This G is not a Jordan domain, but more than that, at the origin its boundary is so rigid that G possesses the following

Figure 6. S_h not in the closure of $T(1)$.

property: There is a positive constant ε such that if f is a conformal mapping of G and $\|S_f\|_G \leq \varepsilon$, then $f(G)$ is not a Jordan domain.

For the proof we refer to Gehring [3]. With the aid of this result the negative answer to the question of Bers is readily established.

Theorem 4.3. *The closure of $T(1)$ is a proper subset of U.*

PROOF. Let G be the domain defined above and $\varepsilon > 0$ the associated constant. If h is a conformal mapping of the upper half-plane onto G, we prove that S_h does not lie in the closure of $T(1)$.

Consider an arbitrary point S_w of the neighborhood $\{\varphi \in Q \mid \|\varphi - S_h\|_H < \varepsilon\}$. For $f = w \circ h^{-1}$ we then have $\|S_f\|_G = \|S_w - S_h\|_H < \varepsilon$. Therefore, either f is not univalent or f is univalent but $f(G) = w(H)$ is not a Jordan domain. It follows that S_w is not in $T(1)$.   □

Recently, Theorem 4.3 has been strengthened: There exists a conformal mapping $h: H \to G$, where G is now a Jordan domain, such that $S_h \notin \overline{T(1)}$ (Flinn [1]).

Theorem 4.3 gives rise to the study of the boundary of $T(1)$. If $T(1)$ is visualized as the collection of quasidiscs $f_\mu(H)$, then Flinn's result means that there are Jordan domains which do not belong to the boundary of $T(1)$. More information about the boundary of $T(1)$ is provided by some recent results in the joint paper of Astala and Gehring [1].

We showed in 4.4 that the diameter of U is 12. Since the closure of $T(1)$ does not coincide with U, we cannot conclude immediately that the diameter of $T(1)$ is also 12. But this can be proved if we modify slightly the example which we used for U. For every positive $r < 1$, the functions $w \to f_1(w) = w + r/w$, $w \to f_2(w) = w - r/w$ are not only univalent in $E = \{w \mid |w| > 1\}$ but have the quasiconformal extensions $w \to w + r\bar{w}$ and $w \to w - r\bar{w}$, respectively. Moreover, an easy calculation shows that

$$\lim_{w \to \infty} (|w|^2 - 1)^2 |S_{f_1}(w) - S_{f_2}(w)| \geq 12r.$$

If $h: H \to E$ is a conformal mapping, $S_{f_1 \circ h}$, $S_{f_2 \circ h} \in T(1)$, and by the invariance formula (1.9) in II.1.3, $\|S_{f_1 \circ h} - S_{f_2 \circ h}\| \geq 12r$. It follows that *the set $T(1)$ has diameter 12*.

Theorem 2.1 says that the universal Teichmüller space is pathwise connected. Therefore, $T(1)$ and its closure are connected, and by Theorem 3.2, the set $T(1)$ is even contractible.

Since U is not the closure of $T(1)$, the connectedness of $T(1)$ does not imply that U is connected. In fact, the author just learned of a striking result of Thurston [1] which asserts that U possesses one-point components. (Thurston's result contains Theorem 4.3 as a corollary.)

5. Inner Radius of Univalence

5.1. Definition of the Inner Radius of Univalence

Let A be a simply connected domain of the extended plane whose boundary consists of more than one point. In II.2.6 we defined the outer radius of univalence

$$\sigma_0(A) = \sup\{\|S_f\|_A \,|\, f \text{ univalent in } A\}$$

of the domain A. We proved that $\sigma_0(A)$ is directly connected with the distance $\delta(A)$ of A from a disc (defined in II.2.1): $\sigma_0(A) = \delta(A) + 6$ for all domains A.

Let us now define the *inner radius of univalence*

$$\sigma_I(A) = \sup\{a \,|\, \|S_f\|_A \leq a \Rightarrow f \text{ univalent in } A\}$$

of A. Note that $a = 0$ is always an admissible number, because $\|S_f\| = 0$ implies that f is a Möbius transformation and hence univalent. Like the distance δ and the radius σ_0, the inner radius σ_I is also invariant under Möbius transformations, i.e., two Möbius equivalent domains have the same σ_I.

The set $\{S_f \,|\, f \text{ univalent in } A\}$ is closed in the family of functions holomorphic in A, when the topology is defined by the hyperbolic sup norm. We proved this in 4.4 in the case where A was the upper half-plane, and the same proof applies to an arbitrary A. It follows that we can replace sup by max in the definition of $\sigma_I(A)$. In other words, if $\|S_f\|_A = \sigma_I(A)$, then f is univalent.

Theorems II.4.1 and II.4.2 imply that $\sigma_I(A) > 0$ if and only if A is a quasidisc. As we remarked before, this is an interesting result because quasiconformal mappings do not appear in the definition of the inner radius of univalence.

For a disc,

$$\sigma_I(A) = 2.$$

This follows directly from Theorem II.5.2. We shall see in 5.7 that for all other domains A, the inner radius of univalence is smaller. Before that, we shall show how to get information about σ_I from the results derived in the previous section. For this purpose, these results must be slightly generalized so that the special position of the half-plane in the definition of the universal Teichmüller space is removed.

5.2. Isomorphic Teichmüller Spaces

Throughout this section the roles of the upper and lower half-planes in connection with the universal Teichmüller space are as in section 4, i.e., f^μ is a self-mapping of the lower half-plane and f_μ is conformal in the upper half-plane.

Let A be a quasidisc whose boundary contains the points $0, 1, \infty$ and for which the direction from 0 to 1 to ∞ orients ∂A positively with respect to A. The universal Teichmüller space T_A is defined by means of quasiconformal self-mappings of the complement of A which fix $0, 1,$ and ∞, exactly as we did it in 1.1 in the case of a half-plane. Theorem 1.2 can be proved for A word for word as in the case of a half-plane. It follows that the points of T_A can be represented by conformal mappings of A which are restrictions to A of quasiconformal mappings of the plane and which fix the points $0, 1,$ and ∞.

The given quasidisc A can be regarded as a point of the universal Teichmüller space $T = T_H$. By this we mean that there is a unique point $p \in T_H$ such that $f_{\mu_0}(H) = A$ whenever $\mu_0 \in p$ (cf. 1.5).

Take a fixed mapping f_{μ_0} with the property $f_{\mu_0}(H) = A$, and consider the transformation

$$f_\mu | H \to f_\mu \circ (f_{\mu_0}^{-1} | A). \tag{5.1}$$

This is an isomorphism between T_H and T_A in the sense that it is a bijective isometry. Obviously, (5.1) is well defined in T_H and a bijection of T_H onto T_A. If $w_i = f_i \circ f_{\mu_0}^{-1}, i = 1, 2$, then $w_2 \circ w_1^{-1} = f_2 \circ f_1^{-1}$, and so (5.1) preserves Teichmüller distances.

Note that under (5.1) the origin is shifted: the point $[\mu_0] \in T_H$ maps to the origin of T_A. More generally, $[\mu]$ maps to the point represented by the complex dilatation of $f_\mu \circ f_{\mu_0}^{-1}$.

Suppose $[\mu] \in T_A$. In order to study the mapping $[\mu] \to S_{f_\mu | A}$, we define Q_A as the space of functions φ holomorphic in A for which the norm $\sup |\varphi(z)| |\eta_A(z)|^{-2}$ is finite. As in the case $A = H$, we set $U_A = \{\varphi = S_f | f$ univalent in $A\}$; by Theorem II.2.3, this is a subset of Q_A. Finally, $T_A(1) = \{S_f \in U_A | f$ has a quasiconformal extension to the plane$\}$. Our previous space Q will now be denoted by Q_H.

The function $w = f \circ f_{\mu_0}^{-1} | A$ is meromorphic in A if and only if f is meromorphic in H. From $\|S_w\|_A = \|S_f - S_{f_{\mu_0}}\|_H$ we get

$$\|S_f\|_H - \|S_{f_{\mu_0}}\|_H \leq \|S_w\|_A \leq \|S_f\|_H + \|S_{f_{\mu_0}}\|_H$$

and conclude that $\|S_w\|_A$ is finite if and only if $\|S_f\|_H$ is finite. It follows that the mapping

$$S_f \to S_{f \circ f_{\mu_0}^{-1} | A} \tag{5.2}$$

is a bijection of Q_H onto Q_A. Clearly, (5.2) maps U_H onto U_A and $T_H(1)$ onto $T_A(1)$. If $w_i = f_i \circ f_{\mu_0}^{-1} | A, i = 1, 2$, then by formula (1.9) in II.1.3,

$$\|S_{w_1} - S_{w_2}\|_A = \|S_{f_1} - S_{f_2}\|_H.$$

We see that the mapping (5.2) of Q_H onto Q_A is also an isometry.

Let ψ_1 be the mapping $[\mu] \to S_{f_\mu|H}$ of T_H onto $T_H(1)$ and ψ_2 the mapping $[\mu] \to S_{f_\mu|A}$ of T_A onto $T_A(1)$. Then (5.1) followed by ψ_2 is the same as ψ_1 followed by (5.2)

5.3. Inner Radius and Quasiconformal Extensions

Let A be a quasidisc. We define the spaces Q_A, U_A, and $T_A(1)$ as in 5.2. Let $h: H \to A$ be a conformal mapping. Generalizing (5.2) we can define a bijective isometry of Q_H onto Q_A by

$$S_f \to S_{f \circ h^{-1}}. \tag{5.3}$$

We conclude that for varying quasidiscs A, all spaces U_A and $T_A(1)$ are isomorphic. In particular, Theorem 4.2 can be generalized:

For all quasidiscs A,

$$T_A(1) = \text{int } U_A. \tag{5.4}$$

This relation leads to a new characterization of the inner radius of univalence. It follows from the definition of σ_I and from the subsequent remark about sup and max in the definition that the closed ball $\{\varphi \in Q_A | \, \|\varphi\|_A \leq \sigma_I(A)\}$ is contained in U_A. By (5.4), the interior of this ball lies in $T_A(1)$. In other words, *if*

$$\|S_f\|_A < \sigma_I(A),$$

then f is not only injective in A but has a quasiconformal extension to the plane.

This result generalizes the statements of Theorems II.5.1 and II.5.2 which are concerned with a disc. It sheds new light on the inner radius of univalence and explains why quasiconformal mappings play a role. We conclude that the definition of the inner radius of univalence can be expressed in the form

$$\sigma_I(A) = \inf\{\|S_f\|_A | f \text{ univalent in } A, f(A) \text{ not a quasidisc}\}. \tag{5.5}$$

This characterization yields upper bounds for $\sigma_I(A)$.

Another way of expressing (5.5) is that $B(0, \sigma_I(A)) = \{\varphi \in Q_A | \, \|\varphi\|_A < \sigma_I(A)\}$ is the largest open ball in Q_A centered at the origin which is contained in $T_A(1)$. The inverse of the isomorphism (5.3) takes this ball onto the ball $B(S_h, \sigma_I(A))$ of Q_H. This gives a characterization for the inner radius of univalence in geometric terms: *If h is a conformal mapping of H onto a quasidisc A, then $\sigma_I(A)$ is the distance from the point S_h to the boundary of $T_H(1)$.*

This result gives additional information about the mapping $[\mu] \to S_{f_\mu|H}$ of T_H onto $T_H(1)$. We proved in section 4 that $T_H(1)$ is open in Q_H. Now we can express this result in a more precise form:

Theorem 5.1. *The largest open ball of Q_H which is centered at the point $S_{f_\mu|H}$ and lies in $T_H(1)$ has the radius $\sigma_I(f_\mu(H))$.*

We recall that the domain constant δ also has a similar geometric interpretation: $\delta(A)$ is the distance from the point S_h to the origin of $T_H(1)$. The geometric interpretations of σ_I and δ can actually be exploited. Before doing it we shall estimate the inner radius for certain specific domains. We first derive a lower estimate for $\sigma_I(A)$ with the aid of quasiconformal reflections.

5.4. Inner Radius and Quasiconformal Reflections

The outer radius of univalence can often be readily determined thanks to the relation $\sigma_0 = \delta + 6$. In contrast, there seems to be no easy way to find the exact value of the inner radius of univalence of a quasidisc. A modification of the proof of Theorem II.4.1 yields a lower bound for $\sigma_I(A)$ which turns out to be sharp in certain cases.

Lemma 5.1. *Let A be a bounded quasidisc which can be exhausted by the domains $A_r = \{rz | z \in A\}$, $0 < r < 1$, and λ a quasiconformal reflection in ∂A which is continuously differentiable off ∂A. Let $A' = A \backslash \{\lambda^{-1}(\infty)\}$. Then*

$$\sigma_I(A) \geq 2 \inf_{z \in A'} \frac{|\bar{\partial}\lambda(z)| - |\partial\lambda(z)|}{|\lambda(z) - z|^2 \eta_A(z)^2}. \tag{5.6}$$

PROOF. Let f be a meromorphic function in A. Suppose first that f is holomorphic on ∂A. As in the proof of Theorem II.4.1, we set $f = w_1/w_2$ and write

$$\varphi(z) = \frac{w_1(z) + (\lambda(z) - z)w_1'(z)}{w_2(z) + (\lambda(z) - z)w_2'(z)}.$$

Let $g = \varphi \circ \lambda^{-1}$ and $\mu_g = \bar{\partial}g/\partial g$ be the complex dilatation of g. The reasoning in the proof of Theorem II.4.1 shows that if $\|\mu_g\|_\infty < 1$, then f is univalent and g is a quasiconformal extension of f. (In proving Theorem II.4.1 we resorted to inequality (4.4) to determine the boundary values of ∂g and $\bar{\partial}g$. Here we can proceed more directly, since we may assume that the right-hand expression in (5.6) is positive.)

Now

$$|\mu_g \circ \lambda| = \left| \frac{\mu_\varphi - \mu_\lambda}{1 - \bar{\mu}_\lambda \mu_\varphi} \right|.$$

Since λ is a sense-reversing quasiconformal mapping, $\mu_\lambda = \partial\lambda/\bar{\partial}\lambda$ is bounded away from 1 in absolute value. It follows that $\|\mu_g\|_\infty < 1$ if and only if $\|\mu_\varphi\|_\infty < 1$.

Direct computation gives

$$\mu_\varphi(z) = \frac{\partial\varphi(z)}{\bar{\partial}\varphi(z)} = \frac{\partial\lambda(z) + (\lambda(z) - z)^2 S_f(z)/2}{\bar{\partial}\lambda(z)}.$$

We conclude that $\|\mu_\varphi\|_\infty < 1$ if

$$|\partial\lambda(z) + (\lambda(z) - z)^2 S_f(z)/2| \le t|\bar{\partial}\lambda(z)|$$

for some $t < 1$. A fortiori, this is the case if

$$|S_f(z)| \le 2\frac{t|\bar{\partial}\lambda(z)| - |\partial\lambda(z)|}{|\lambda(z) - z|^2}.$$

This in turn holds whenever

$$\|S_f\|_A < 2 \inf_{z \in A'} \frac{|\bar{\partial}\lambda(z)| - |\partial\lambda(z)|}{|\lambda(z) - z|^2 \eta_A(z)^2}. \tag{5.7}$$

Under this condition S_f is a point of the ball $B(0, \sigma_I(A))$ of Q_A.

If f is not holomorphic on ∂A, we apply the above reasoning to $f|A_r$ for suitable values of r tending to 1. The function $z \to r\lambda(z/r)$ is a quasiconformal reflection in ∂A_r, and $r\eta_{A_r}(z) = \eta_A(z/r)$. It follows that whenever (5.7) holds, the mappings $f|A_r$ are univalent and have quasiconformal extensions with uniformly bounded maximal dilatations. As in II.4.3, a normal family argument then shows that under (5.7), $S_f \in T_A(1)$, and (5.6) follows. ☐

Remark. Let h be a Möbius transformation, $\varphi = h \circ \lambda \circ h^{-1}$, and $\zeta = h(z)$. Then

$$\frac{|\bar{\partial}\lambda(z)| - |\partial\lambda(z)|}{|\lambda(z) - z|^2 \eta_A(z)^2} = \frac{|\bar{\partial}\varphi(\zeta)| - |\partial\varphi(\zeta)|}{|\varphi(\zeta) - \zeta|^2 \eta_{h(A)}(\zeta)^2}.$$

This can be verified by direct computation (cf. II.4.1). It follows that Lemma 5.1 holds for quasidiscs which are Möbius equivalent to a quasidisc A fulfilling the conditions of the lemma.

Let us test the accuracy of (5.6). If A is the upper half-plane and $\lambda(z) = \bar{z}$, then

$$\frac{|\bar{\partial}\lambda(z)| - |\partial\lambda(z)|}{|\lambda(z) - z|^2 \eta_A(z)^2} = \frac{4y^2}{|\bar{z} - z|^2} = 1.$$

Hence (5.6) gives $\sigma_I(A) \ge 2$. We get the same result if A is the unit disc and $\lambda(z) = 1/\bar{z}$. Consequently, in these two cases the lower bound in (5.6) is equal to $\sigma_I(A)$.

If A is the exterior of the ellipse $\{z = e^{i\varphi} + ke^{-i\varphi}|0 \le \varphi < 2\pi\}$, $0 \le k < 1$, we have the reflection $z \to 1/\bar{w} + k/w$ with $2w = z + (z^2 - 4k)^{1/2}$. Now (5.6) gives the simple estimate $\sigma_I(A) \ge 2(1 - k)^2$. This lower bound is asymptotically correct as $k \to 0$ or $k \to 1$, but is not sharp, as we shall see in 5.6.

5.5. Inner Radius of Sectors

Lemma 5.1, combined with the characterization (5.5), makes it possible to determine $\sigma_I(A)$ for sectors (Lehto [7], Lehtinen [2]).

Theorem 5.2. *Let A be the sectoral region $\{z | 0 < \arg z < k\pi\}, 0 < k < 2$. Then*

$$\sigma_I(A) = \begin{cases} 2k^2 & \text{if } 0 < k \le 1, \\ 4k - 2k^2 & \text{if } 1 < k < 2. \end{cases} \tag{5.8}$$

PROOF. A continuously differentiable quasiconformal reflection λ in ∂A can be defined by the formula

$$\lambda(z) = z^{1-1/k} \bar{z}^{1/k}, \qquad z \in A.$$

It follows that

$$\frac{|\bar\partial \lambda(z)| - |\partial \lambda(z)|}{|\lambda(z) - z|^2} = \frac{1/k - |1 - 1/k|}{|z^{1-1/k} \bar{z}^{1/k} - z|^2}.$$

Setting $z = re^{i\theta}$ we obtain

$$z^{1-1/k} \bar{z}^{1/k} - z = -2rie^{i(1-1/k)\theta} \sin(\theta/k).$$

In addition,

$$\frac{1}{\eta(z)} = \frac{2k \operatorname{Im} z^{1/k}}{|z|^{1/k-1}} = 2kr \sin(\theta/k).$$

Therefore,

$$|z^{1-1/k} \bar{z}^{1/k} - z| = \frac{1}{k\eta(z)}.$$

Thus (5.6) gives, in view of the Remark in 5.4,

$$\sigma_I(A) \ge 2k(1 - |k - 1|), \tag{5.9}$$

which is (5.8) with $\ge$ instead of equality.

We still have to show that equality holds in (5.9). Suppose first that $0 < k \le 1$. In this case it is easy to prove directly, without making use of Lemma 5.1, that $\sigma_I(A) = 2k^2$. The proof is based on the fact that $\sigma_I(A)$ is equal to the distance from S_h to $\partial T_H(1)$, where $h: H \to A$ is a conformal mapping. Now $z \to h(z) = z^k$ is such a mapping. Hence, $S_h(z) = (1 - k^2)/(2z^2)$ and $\|S_h\|_H = 2(1 - k^2)$. For the function $z \to g(z) = \log z$ we have $S_g(z) = 1/(2z^2)$, and we know that S_g is not in $T_H(1)$, because $g(H)$ is not a Jordan domain. It follows that

$$\sigma_I(A) \le \|S_g - S_h\|_H = \sup 4y^2 k^2/(2|z|^2) = 2k^2.$$

In conjunction with (5.9), this yields the first result (5.8).

As we said, in this case the inequality (5.9) is readily obtained directly. In fact, if φ is a boundary point of $T_H(1)$ nearest to S_h, then

$$\sigma_I(A) = \|\varphi - S_h\| \ge \|\varphi\| - \|S_h\|.$$

Since φ is not in $T_H(1)$, we have $\|\varphi\| \ge 2$, and so

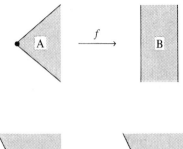

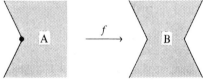

Figure 7. $S_f \in \partial T_A(1)$ closest to the origin.

$$\sigma_I(A) \geq 2 - 2(1 - k^2) = 2k^2.$$

(If no nearest point exists, an obvious ε-reasoning yields this inequality.)

In order to complete the proof, we still have to show that $\sigma_I(A) \leq 4k - 2k^2$ if $1 < k < 2$. In view of (5.5), this follows if we find a conformal mapping $f: A \to B$ such that $\|S_f\|_A = 4k - 2k^2$ and B is not a Jordan domain.

Set

$$B = \{z \,|\, |\arg z| < k\pi/2\} \cap \{z \,|\, |\arg(1 - z)| < k\pi/2\}.$$

This is not a Jordan domain because of the behavior of ∂B at ∞ (Fig. 7). The Schwarzian derivative of the conformal mapping $w_1: H \to H \cap B$, with $w_1(0) = 1$, $w_1(\infty) = 0$, $w_1(1) = \infty$, can be computed (see Nehari [2], p. 203). It follows that

$$S_{w_1}(z) = \frac{4 - k^2}{8z^2} + \frac{2k - k^2}{2(z - 1)^2} + \frac{k^2 - 2k}{2z(z - 1)}.$$

A conformal mapping $w: H \to B$ is obtained if the map $z \to w_1(z^2)$, defined in the first quadrant of the plane, is reflected in the imaginary axis. The composition rule for Schwarzian derivatives yields

$$S_w(z) = \frac{1 - k^2}{2z^2} + \frac{4k - 2k^2}{(z^2 - 1)^2}.$$

With $h(z) = z^k$, the function $f = w \circ h^{-1}$ maps A conformally onto B. Since $\|S_f\|_A = \|S_w - S_h\|_H$, we finally arrive at the desired result

$$\|S_f\|_A = \sup_{y>0} \frac{4y^2(4k - 2k^2)}{(x^2 - y^2 - 1)^2 + 4x^2y^2} = 4k - 2k^2.$$

This completes the proof of (5.8). The example of the conformal mapping $f: A \to B$ is due to Lehtinen [2]. $\square$

5.6. Inner Radius of Ellipses and Polygons

In many cases Theorem 5.2, in conjunction with the fact that the hyperbolic metric depends monotonically on the domain, provides an efficient means for deriving estimates of the inner radius. The underlying ideas, which are due to Lehtinen ([5]; cf. also [4]), are in the following two lemmas.

Lemma 5.2. *Let A be a quasidisc which is contained in a domain B_k Möbius equivalent to the sector $A_k = \{z\,|\,|\arg z| < k\pi/2\}$. If $0 < k \leq 1$, assume that a vertex v of B_k lies on ∂A. Then*

$$\sigma_I(A) \leq 2k^2.$$

If $1 < k < 2$, assume that there are points z_1 and z_2 in ∂A such that for a Möbius transformation g mapping B_k onto A_k, $g(z_1) = e^{ik\pi/2}$, $g(z_2) = e^{-ik\pi/2}$. Then

$$\sigma_I(A) \leq 4k - 2k^2.$$

PROOF. Suppose first that $0 < k \leq 1$. Let g be a Möbius transformation mapping B_k onto A_k with $g(v) = 0$. Set $f(z) = \log g(z)$. Then $f(A)$ is not a quasidisc, and so by (5.5), $\sigma_I(A) \leq \|S_f\|_A$. By the monotonicity of the hyperbolic metric (formula (1.2) in I.1.1), $\|S_f\|_A \leq \|S_f\|_{B_k} = 2k^2$.

Suppose next that $1 < k < 2$. From the proof of Theorem 5.2 we deduce the existence of a conformal mapping f of A_k such that $\|S_f\|_{A_k} = 4k - k^2$ and that $f(e^{ik\pi/2}) = f(e^{-ik\pi/2}) = \infty$. Then $f(g(A))$ is not a Jordan domain, and by reasoning as in the case $0 < k \leq 1$, we arrive at the desired estimate. ☐

The second lemma, which does not rest on Theorem 5.2, is more general.

Lemma 5.3. *Let A be a quasidisc. If every two-point subset of A is contained in the closure of a quasidisc $B \subset A$ for which $\sigma_I(B) \geq m$, then*

$$\sigma_I(A) \geq m.$$

PROOF. Let an $\varepsilon > 0$ be given. There exists a meromorphic function f in A for which $\|S_f\|_A < \sigma_I(A) + \varepsilon$ but which is not univalent. Let z_1 and z_2 be two different points of A such that $f(z_1) = f(z_2)$, and $B \subset A$ a quasidisc such that $\{z_1, z_2\} \subset \bar{B}$ and $\sigma_I(B) \geq m$. Since either f is not univalent in B or else $f(B)$ is not a quasidisc, $\|S_f\|_B \geq \sigma_I(B)$. By the monotonicity of the hyperbolic metric, $\|S_f\|_A \geq \|S_f\|_B$. Hence $\sigma_I(A) > m - \varepsilon$, and the lemma follows. ☐

As a first application, let us reconsider the case in which A is the exterior of the ellipse $\{z = e^{i\varphi} + ke^{-i\varphi}\,|\,0 \leq \varphi < 2\pi\}$, $0 \leq k < 1$. The domain A is contained in the infinite domain B whose boundary consists of two circular arcs of which one passes through the points $1 + k$, $i(1 - k)$, $-(1 + k)$ and the other through $1 + k$, $-i(1 - k)$, $-(1 + k)$. Since B is Möbius equivalent to the angle A_l with $l = 1 + (4/\pi)\arctan k$, Lemma 5.2 yields the upper estimate

$$\sigma_I(A) \leq 2 - (32/\pi^2)(\arctan k)^2.$$

On the other hand, simple geometry shows that each pair of points in A lies in a domain $B' \subset A$ Möbius equivalent to the angle $A_{l'}$ with $l' = 1 + (2/\pi)$ $\arctan(4k/((1 - k)(k^2 + 6k + 1)^{1/2}))$. By Lemma 5.3,

$$\sigma_I(A) \geq 2 - (8/\pi^2)(\arctan(4k/((1 - k)(k^2 + 6k + 1)^{1/2})))^2.$$

This is sharper than the simple estimate $\sigma_I(A) \geq 2(1 - k)^2$ we obtained from Lemma 5.1.

In the second application of Lemmas 5.2 and 5.3, we suppose that A is a finite polygonal domain with interior angles $k_i\pi$, $i = 1, 2, \ldots, n$. Then Lemma 5.2 yields immediately the upper estimate

$$\sigma_I(A) \leq \min_i \{2k_i(1 - |k_i - 1|)\}. \tag{5.10}$$

In certain cases, we can refine this result with the aid of Lemma 5.3 so as to obtain the exact value of the inner radius. *If A is a triangle or a regular n-sided polygon, then* (5.10) *holds as an equality.*

It follows that for a triangle A,

$$\sigma_I(A) = 2\left(\frac{\alpha}{\pi}\right)^2, \tag{5.11}$$

where α is the smallest angle of A, and for a regular n-sided polygon

$$\sigma_I(A) = 2\left(\frac{n - 2}{n}\right)^2. \tag{5.12}$$

For the proofs we refer to Lehtinen [5]; the results (5.11) and (5.12) are also due to Calvis [1].

5.7. General Estimates for the Inner Radius

Let $h: H \to A$ be a conformal mapping. Then $\delta(A)$ is the distance from S_h to the origin of Q_H, and $\sigma_I(A)$ the distance from S_h to $U_H \backslash T_H(1)$. The set $T_H(1)$ contains the ball $B(0, 2)$, and U_H is contained in the closure of $B(0, 6)$. It follows that *the double inequality*

$$2 \leq \delta(A) + \sigma_I(A) \leq 6$$

holds for all domains A conformally equivalent to a disc.

The left-hand inequality $\sigma_I(A) \geq 2 - \delta(A)$ is an equality for all sectors $A = \{z | 0 < \arg z < k\pi\}$, $0 < k \leq 1$, because by Theorem 5.2, $\sigma_I(A) = 2k^2$, and we have earlier computed $\delta(A) = 2(1 - k^2)$. Other extremals are obtained as follows. Consider a domain for which $\|S_h\| = 2$ and which is not a quasidisc. If $0 < r \leq 1$ and $S_f = rS_h$, then for $A = f(H)$, we have $\delta(A) = 2r$ and $\sigma_I(A) = 2(1 - r)$.

The upper estimate $\sigma_I(A) \leq 6 - \delta(A)$ is of interest only if $\delta(A)$ is close to 6. For $\delta(A) < 4$, a better estimate can be derived.

Theorem 5.3. *For all domains A conformally equivalent to a disc,*

$$\sigma_I(A) \leq 2. \tag{5.13}$$

Equality holds if and only if A is a disc.

PROOF. Let A be an arbitrary quasidisc. Every Jordan domain is Möbius equivalent to a subdomain of H having 0 and ∞ as boundary points. We may assume, therefore, that A itself is such a domain.

In A, we consider the function $z \to f(z) = \log z$, for which $S_f(z) = 1/(2z^2)$. From the monotonicity of the hyperbolic metric it follows that $\|S_f\|_A \leq 2$. Because f maps both 0 and ∞ to infinity, $f(A)$ is not a Jordan domain. Hence, we obtain (5.13) from the characterization (5.5) of the inner radius.

The same idea, in a refined form, can be used to prove that $\sigma_I(A) = 2$ only if A is a disc (Lehtinen [2]). We now assume that $A \subset H$ has two finite boundary points on the real axis. If A is not H, there are two finite points in $\partial A \cap \mathbb{R}$ such that the open interval on $\mathbb{R}$ between these points lies in the complement of ∂A. A simple geometric argument shows that A then lies in a non-convex sector both of whose sides contain a point of ∂A at an equal distance from the vertex (for the details, see Lehtinen [2]). Therefore, we may assume that A lies in an angle $A_k = \{z | 0 < \arg z < k\pi\}$, $1 < k < 2$, such that the points 1 and $e^{k\pi i}$ are on the boundary of A.

Instead of the logarithm, we now consider the extremal mapping f of the sectoral domain A_k exhibited in the proof of Theorem 5.2. Since $f(1) = f(e^{k\pi i})$, the image of A under $f|A$ is not a Jordan domain. It follows from (5.5), the monotonicity of the hyperbolic metric, and Theorem 5.2, that

$$\sigma_I(A) \leq \|S_{f|A}\|_A \leq \|S_f\|_{A_k} = 4k - 2k^2 < 2. \qquad \square$$

By Theorem 5.3, every point $\neq 0$ of $T_H(1)$ has a distance < 2 from the boundary of $T_H(1)$.

It is an open question what values the inner radius of univalence can assume for K-quasidiscs. The sectoral domains show that

$$\inf\{\sigma_I(A) | A \text{ } K\text{-quasidisc}\} \leq \frac{8}{(K+1)^2}.$$

We may also ask whether Theorem II.4.1 holds if $\varepsilon(K)$ is replaced by $\sigma_I(A)$.

CHAPTER IV

Riemann Surfaces

Introduction to Chapter IV

A number of textbooks have been written on the subject of Riemann surfaces. In spite of this, we found it advisable to include in our presentation a chapter in which we have collected the material on Riemann surfaces that will come into play in Chapter V. A brief survey of the general theory of Riemann surfaces is given in sections 1–3 and of groups of Möbius transformations in section 4. We have occasionally lingered on some topics slightly longer than would be strictly necessary for later needs, in order to provide the reader with a broader background.

In section 1 standard definitions of manifolds and Riemann surfaces and of functions and differentials are given. We have also treated in some detail the classical problem of Gauss to map a portion of a surface imbedded in euclidean three-space conformally into the plane. This problem inaugurated the theory of quasiconformal mappings around 1825. It also gives a first hint of the intrinsic role of quasiconformal mappings in the theory of Riemann surfaces.

Section 2 deals with covering surfaces and their topology. The main results are formulated but proofs are usually only sketched. References for complete proofs are to the monograph Ahlfors and Sario [1].

In section 3, results of section 2 are applied to Riemann surfaces. In conjunction with the general uniformization theorem, they yield the fundamental result that, modulo conformal equivalence, every Riemann surface is the quotient of a disc or the finite plane or the extended plane by a discontinuous group of Möbius transformations. The section concludes with a study of how mappings between Riemann surfaces induce mappings between the covering surfaces and the covering groups.

Section 4 provides a survey of some main features of the theory of groups of Möbius transformations. For detailed proofs, reference is usually made to the monograph of Lehner [1]. In section 5 we have collected various results on compact Riemann surfaces, using Springer [1] as a reference.

Quadratic differentials play a remarkable role in the theory of Teichmüller spaces. Therefore, the geometry and the metric induced by a quadratic differential are studied quite extensively in sections 6 and 7. Here we have largely utilized the recent monograph of Strebel [6], to which we also refer for more details.

1. Manifolds and Their Structures

1.1. Real Manifolds

A real *n-dimensional manifold* M is a Hausdorff space with a countable base for topology which is locally homeomorphic to the euclidean space $\mathbb{R}^n$. This means that to every point $p \in M$ there is an open neighborhood $U \subset M$ of p and a homeomorphism h of U onto an open set in $\mathbb{R}^n$. Such a mapping h is called a *local parameter* on M. A set of local parameters is said to be an *atlas* of M if the union of their domains covers M.

Let h_1, h_2 be local parameters on M such that their domains U_1 and U_2 have a non-void intersection. Restricting h_1 and h_2 to $U_1 \cap U_2$, we obtain a homeomorphic mapping $h_2 \circ h_1^{-1}$ between the open sets $h_1(U_1 \cap U_2)$ and $h_2(U_1 \cap U_2)$ in $\mathbb{R}^n$. With the help of this induced mapping, which we call a *parameter transformation*, properties definable in $\mathbb{R}^n$ can be transported to the manifold M.

An atlas is called differentiable if for all pairs of local parameters, the parameter transformations are differentiable where defined. A maximal differentiable atlas is called a differentiable structure on the manifold, and a manifold with a differentiable structure is a *differentiable manifold*. When speaking in the following of the local parameters on a differentiable manifold, we always assume that the parameters belong to its differentiable structure.

With obvious modifications, the method used for defining differentiable manifolds leads to manifolds with other structures. For instance, if every parameter transformation is of class C^k (has continuous partial derivatives up to order k), we speak of C^k-manifolds. Or if the word "differentiable" is replaced by "real-analytic", we obtain real-analytic manifolds.

A continuous map f of a differentiable manifold M into a differentiable manifold N is said to be differentiable at a point $p \in M$ if there are local parameters h and k, defined in neighborhoods of p and $f(p)$, respectively, such that the composition $k \circ f \circ h^{-1}$ is differentiable at $h(p)$. If f is differentiable at p, then $k \circ f \circ h^{-1}$ is differentiable at $h(p)$ for all local parameters h and

k near p and $f(p)$. The map $f: M \to N$ is said to be differentiable if it is differentiable at every point of M.

A path on a manifold M is a continuous mapping of an interval into M. If M is differentiable, we can speak of a differentiable path.

Let f be a mapping of an n-dimensional C^1-manifold M into a manifold N. The function f is said to be (Lebesgue) measurable on M if for every local parameter h on M defined in an open set U, the function $f \circ h^{-1}$ is measurable with respect to n-dimensional Lebesgue measure, i.e., if for every open set $G \subset N$, the inverse image $(f \circ h^{-1})^{-1}(G) = h(U \cap f^{-1}(G))$ is a measurable subset of $h(U)$. If the preimage $h(U \cap f^{-1}(G))$ is always a Borel set, f is Borel measurable on M. This latter definition can be used for all manifolds, because Borel sets are preserved under homeomorphisms.

A *surface* is a connected two-dimensional manifold. A surface is always pathwise connected. This follows by standard reasoning from the facts that a surface is connected and, clearly, locally pathwise connected. A surface may or may not be compact. In the classical literature a compact surface is called closed, a non-compact surface open.

1.2. Complex Analytic Manifolds

If we replace in the definition of an n-dimensional manifold in 1.1 the euclidean space $\mathbb{R}^n$ by the space $\mathbb{C}^n$ of n-tuples of complex numbers, we get a complex n-dimensional manifold. The euclidean space $\mathbb{R}^{2n}$ becomes identified with the space $\mathbb{C}^n$ if we associate $(x_1, \ldots, x_n, y_1, \ldots, y_n) \in \mathbb{R}^{2n}$ with $(x_1 + iy_1, \ldots, x_n + iy_n) \in \mathbb{C}^n$. Therefore, a complex n-dimensional manifold can be regarded as a real $2n$-dimensional manifold.

Let M be a complex n-manifold. Suppose M has an atlas in which all parameter transformations are biholomorphic, i.e., along with their inverses they are holomorphic functions of n complex variables. Maximal atlases with this property are called complex analytic structures on M. We say that a manifold equipped with such a structure is a *complex analytic manifold*.

Definition. A one-dimensional connected complex analytic manifold is a Riemann surface.

In the case $n = 1$ we also call the complex analytic structure conformal. Thus a Riemann surface is a surface with a conformal structure.

Besides playing a central role in complex analysis, the notion of a Riemann surface has initiated or influenced a multitude of other mathematical disciplines. The basic observation that the natural habitat of an analytic function is not a subdomain of the complex plane but a surface which is locally conformally equivalent to a plane domain appears already in Riemann's thesis [1] from 1851. A rigorous definition of Riemann surface in modern terms was given as early as 1913 by Weyl [1]. Weyl's monograph also contains the first precise definition for a surface.

A domain in the complex plane is to be regarded as the Riemann surface with its natural conformal structure induced by the identity mapping, unless otherwise stated. The Riemann sphere is another example of a Riemann surface with a natural conformal structure.

An open subset of a Riemann surface S is assumed to have the conformal structure induced by the conformal structure of S. When speaking of the local parameters on a Riemann surface, we always assume that the parameters belong to its conformal structure.

Analytic functions can be defined on Riemann surfaces. A continuous map f of a Riemann surface S into a Riemann surface W is analytic at a point $p \in S$ if there are local parameters h and k, defined in a neighborhood of p and $f(p)$, such that $k \circ f \circ h^{-1}$ is holomorphic at $h(p)$. This is an invariant definition, independent of the choice of the local parameters near p and $f(p)$. If f is analytic at each point, f is an analytic mapping. In case W is the complex plane, an analytic mapping f is said to be holomorphic, while if W is the extended plane, f is termed meromorphic.

An injective analytic function on S is called a conformal mapping of S. It follows from the definitions that local parameters of S are conformal mappings of open subsets of S into the complex plane.

1.3. Border of a Surface

A *bordered surface* is a connected Hausdorff space with a countable base for topology in which there exists an open covering by sets homeomorphic with sets open in the closed half-plane $\bar{H} = \{x + iy | y \geq 0\}$. The concepts of a local parameter and an atlas can be introduced in an obvious manner for bordered surfaces.

Let S^* be a bordered surface, $p \in S^*$, and h a homeomorphism of an open neighborhood of p onto an open subset of $\bar{H}$. Let us assume that $h(p)$ is an interior point of $\bar{H}$. If k is another homeomorphism of a neighborhood of p onto an open set in $\bar{H}$, we can choose a disc D with center at $h(p)$ and with closure in $\bar{H}$, such that $k \circ h^{-1}$ is homeomorphic in D. Now $k \circ h^{-1}$ is an injection of D into the plane and it is continuous in the topology of the plane. Since D is an open neighborhood of $h(p)$ in the topology of the plane, it follows from the invariance of open sets under continuous injections that $k(p)$ must be an interior point of $\bar{H}$. Hence, if the image of p under one local parameter of S^* is an interior point of $\bar{H}$, then it is so under all local parameters of S^* defined at p.

It follows that we can write $S^* = S \cup B$, where S is the subset of S^* whose points map into the interior of $\bar{H}$ under any local parameter, and B the set whose points map on the boundary of H. Clearly, S is open in S^*, and so B is closed. Each neighborhood of every point of S^* contains points of S. Consequently, S^* is the closure of S. It is easy to show that S is connected, whence it follows that S is a surface. The set B is called the *border* of S^*. Each point $p \in B$ has a neighborhood U in S^* which is homeomorphic to the half-disc $\{z = x + iy \,|\, |z| < 1, y \geq 0\}$ under a parameter which maps p to the origin and $B \cap U$ onto the line segment $(-1, 1)$. We conclude that B is a real one-dimensional manifold.

A *bordered Riemann surface* is a bordered surface $S^* = S \cup B$ with a conformal structure on S determined by an atlas on S^*. It follows that S is a Riemann surface.

1.4. Differentials on Riemann Surfaces

Let S be a Riemann surface whose conformal structure is determined by the local parameters z_i with domains U_i, $i = 1, 2, \ldots$.

Let f be a complex-valued holomorphic function on S. Suppose that z_i and z_j have overlapping domains, and write $f_i = f \circ z_i^{-1}$, $f_j = f \circ z_j^{-1}$. It is customary to regard z_i and z_j as complex variables. If we do so and differentiate $f_i(z_i) = f_j(z_j)$, we arrive at the invariance

$$f_i' \, dz_i = f_j' \, dz_j. \tag{1.1}$$

It follows that while an invariant derivative cannot be defined for f, we can speak of the invariant differential df, defined locally by (1.1).

Let us now generalize the notion of a differential. A collection φ of complex-valued functions φ_i defined on U_i, $i = 1, 2, \ldots$, is said to be an (m, n)-differential on S if

$$\varphi_i \left(\frac{dz_i}{dz_j}\right)^m \left(\overline{\frac{dz_i}{dz_j}}\right)^n = \varphi_j \tag{1.2}$$

in $U_i \cap U_j$. The function element φ_i is a representation of φ in terms of the local coordinate z_i. The differential φ is said to be holomorphic if all functions φ_i are holomorphic. Meromorphic differentials are defined similarly.

Two (m, n)-differentials φ and ψ on S are equal if their local representations in the same local coordinate always agree. In case φ and ψ are meromorphic, we conclude that $\varphi = \psi$ if their representations agree for some local coordinate.

From the definition it is clear that the set of (m, n)-differentials on S and the subset of all holomorphic (m, n)-differentials form linear spaces over the complex numbers. We can also form the product of an (m, n)-differential and a (p, q)-differential in an obvious manner and so obtain an $(m + p, n + q)$-differential.

Particularly important in the following is the case $m = 2$, $n = 0$. Then φ is called a *quadratic differential*. Such differentials will be studied in sections 5, 6, and 7 of this chapter, as a preparation for the applications in Chapter V. A holomorphic or meromorphic $(1, 0)$-differential is called an Abelian differential. Its square is a quadratic differential. Deeper connections between holomorphic Abelian and quadratic differentials will be investigated in 5.5.

Quasiconformal mappings between Riemann surfaces will be introduced in V.1. It turns out that in this setting the complex dilatation generalizes to a $(-1, 1)$-differential. For a $(-1, 1)$-differential φ it follows from the invariance (1.2) that we can speak of the function $|\varphi|$ on S. If φ is Lebesgue measurable in all local coordinates and $\|\varphi\|_\infty < 1$, then the $(-1, 1)$-differential φ is called a *Beltrami differential*.

In applications $(1, 1)$-differentials are often important, because they can be integrated with respect to the two-dimensional Lebesgue measure. This follows from the fact that if $z \to w$ is a change of local coordinates, then

$$dw \, d\bar{w} = |w'(z)|^2 \, dz \, d\bar{z},$$

where $|w'|^2$ is the Jacobian of the mapping $z \to w$. If φ is a quadratic differential, then $|\varphi_i \, dz_i^2| = |\varphi_i| \, dz_i \, d\bar{z}_i$ for all local representations φ_i. Thus the absolute value of a quadratic differential is a $(1, 1)$-differential. Another important observation is that the product of a quadratic differential and a $(-1, 1)$-differential is a $(1, 1)$-differential.

1.5. Isothermal Coordinates

The natural question of how to make a concrete surface in $\mathbb{R}^3$ into a Riemann surface leads us to quasiconformal mappings. Let S be an orientable C^1-surface in $\mathbb{R}^3$, and $f = (f_1, f_2, f_3)$ the inverse of a local parameter of S. The metric on S is defined locally by the line element ds, where

$$ds^2 = \sum_{i=1}^{3} \left(\frac{\partial f_i}{\partial x} dx + \frac{\partial f_i}{\partial y} dy \right)^2 = E \, dx^2 + 2F \, dx \, dy + G \, dy^2. \tag{1.3}$$

Here

$$E = \sum_{i=1}^{3} \left(\frac{\partial f_i}{\partial x} \right)^2, \qquad F = \sum_{i=1}^{3} \frac{\partial f_i}{\partial x} \frac{\partial f_i}{\partial y}, \qquad G = \sum_{i=1}^{3} \left(\frac{\partial f_i}{\partial y} \right)^2$$

are the classical Gaussian quantities. The expression (1.3) is invariant, i.e., independent of the choice of the local parameter.

Using the complex notation $dz = dx + i \, dy$, $d\bar{z} = dx - i \, dy$, we obtain from (1.3)

$$ds = \lambda |dz + \mu \, d\bar{z}|, \tag{1.4}$$

with

$$\lambda^2 = \frac{1}{4}(E + G + 2\sqrt{EG - F^2}), \qquad \mu = \frac{E - G + 2iF}{E + G + 2\sqrt{EG - F^2}}.$$

Note that

$$|\mu|^2 = \frac{E + G - 2\sqrt{EG - F^2}}{E + G + 2\sqrt{EG - F^2}} < 1. \tag{1.5}$$

It is a classical result (and not difficult to prove by direct calculation) that f is conformal in the sense that it preserves angles if and only if $E = G$, $F = 0$. This condition is equivalent to μ being identically zero. In this case

$$ds = \lambda |dz|.$$

Local coordinates z of S with this property are called *isothermal*.

Let us consider another local parameter of S which defines the local coordinates w. If the coordinates z and w are both isothermal $(ds = \lambda_1 |dw|)$ and if the induced mapping $z \rightarrow w$ is defined in some non-empty open set of the plane, then

$$\lambda_1 |dw| = \lambda |dz|.$$

This shows that $z \rightarrow w$ is conformal or indirectly conformal. Since S is orientable, we may assume that $z \rightarrow w$ is conformal. We conclude that *isothermal coordinates define a natural conformal structure for an orientable C^1-surface, which thus becomes a Riemann surface.*

1.6. Riemann Surfaces and Quasiconformal Mappings

We are thus led to the new problem of finding isothermal coordinates for a surface. Without bothering about minimal conditions, we show how a solution is obtained with the aid of the existence theorem for Beltrami equations.

Theorem 1.1. *Every orientable C^2-surface in $\mathbb{R}^3$ can be made into a Riemann surface.*

PROOF. Let S be an orientable C^2-surface. Consider an arbitrary local parameter of S inducing local coordinates z in a domain A of the complex plane. The theorem follows if we can transform the z-coordinates diffeomorphically so that the new coordinates are isothermal.

Expressed in terms of z, the line element of S is of the form (1.4). Here μ is continuously differentiable and by (1.5), we have $\sup |\mu(z)| < 1$ in every relatively compact subdomain of A. Let $z \rightarrow w$ be a quasiconformal mapping of such a subdomain with complex dilatation μ. By the Existence theorem I.4.4 such a mapping w exists, and by the remark in I.4.5, w is continuously differentiable and $\partial w(z) \neq 0$ everywhere. Comparison of

$$|dw| = |\partial w\, dz + \bar{\partial} w\, d\bar{z}| = |\partial w| |dz + \mu\, d\bar{z}|$$

with (1.4) shows that

$$ds = \frac{\lambda}{|\partial w|} |dw|.$$

We see that the w-coordinates are isothermal, and the theorem is proved.

$\square$

In essence, Theorem 1.1 is due to Gauss [1]. Since every sense-preserving diffeomorphism is locally quasiconformal, there is no undisputed criterion to determine the first appearance of quasiconformal mappings in analysis. But in developing the theory of surfaces, Gauss realized the importance of finding locally injective solutions for a Beltrami equation (i.e., quasiconformal mappings with a given complex dilatation) and he actually constructed such solutions. Therefore, it is not without justification to say that quasiconformal mappings entered analysis around 1825, in connection with the problem of how to map a plane domain conformally onto a portion of a surface imbedded in euclidean three-space.

For Gauss conformal mappings were just a tool in differential geometry. It was Riemann who recognized the fundamental connection between conformal mappings and complex analysis. Later, the concrete method of Theorem 1.1 to generate Riemann surfaces was used by Klein [1], whose work paved the way for Weyl's monograph [1] cited in 1.2.

Theorem 1.1 and its proof mark the first indication of the intimate relationship between Riemann surfaces and quasiconformal mappings. Later we shall uncover plenty of additional evidence of the depth of this connection, and Theorem 1.1 will be generalized in various ways.

2. Topology of Covering Surfaces

2.1. Lifting of Paths

The unifying link between the theory of abstract Riemann surfaces and complex analysis in the plane is provided by covering surfaces. In this section, we shall discuss topological properties of covering surfaces.

A *smooth covering surface* of a surface S is a pair (W, f), where W is a surface and $f: W \to S$ is a local homeomorphism. The mapping f is called a projection, and the inverse images of a point $p \in S$ are said to lie over p. Being locally a homeomorphism, f is both continuous and open.

Let γ be a path on S, more precisely a continuous map of the closed unit interval $I = \{t | 0 \le t \le 1\}$ into S. A path γ' on W with initial point $a = \gamma'(0)$ and with the property $f \circ \gamma' = \gamma$ is called a *lift* of γ from a. It is easy to prove that *on a smooth covering surface the lift from a fixed initial point is unique* (Ahlfors–Sario [1], p. 28).

From the local injectiveness of f it is clear that a part of γ can always be lifted from an arbitrary point a lying over its initial point. If the whole of γ cannot be lifted, there is a t_0, $0 < t_0 \le 1$, such that γ restricted to any closed subinterval of $[0, t_0)$ can be lifted from a but $\gamma | [0, t_0]$ cannot be so lifted.

A smooth covering surface (W, f) of S is said to be *unlimited* (regular, in the terminology of Ahlfors–Sario [1]) if every path on S has a lift to W from each point lying over its initial point. In this case $f: W \to S$ is surjective, and the cardinality of the set $f^{-1}\{p\}$ is the same for all points $p \in S$.

Unlimited covering surfaces have an important topological property (Ahlfors–Sario [1], p. 30).

Theorem 2.1 (Monodromy Theorem). *Let (W, f) be an unlimited covering surface of a surface S, and γ_0 and γ_1 homotopic paths on S. Then the lifts of γ_0 and γ_1 on W from the same initial point have the same terminal point and they are homotopic.*

Suppose, in particular, that the surface S is simply connected, i.e., that the fundamental group of S is trivial. In this case the monodromy theorem yields an interesting corollary:

If (W, f) is an unlimited covering surface of a simply connected surface S, then the mapping $f: W \to S$ is a homeomorphism.

For since the projection f is continuous, open and surjective, it is enough to show that f is injective. Assume that there are two points a and b of W such that $f(a) = f(b)$. A path γ from a to b then has a projection on S which is a closed curve. This is homotopic to zero, since S is simply connected. By the monodromy theorem, γ terminates at the same point as the constant path $t \to a$. Hence $a = b$.

2.2. Covering Surfaces and the Fundamental Group

Let S be a surface, (W, f) an unlimited covering surface of S, and a a point of S. We denote by F_a the fundamental group of S whose elements are homotopy classes $[\gamma]$ of closed paths γ on S from a. Fix a point $a' \in W$ over a, and consider a homotopy class $[\gamma]$ containing a path whose lift from a' is closed. It follows from the Monodromy theorem that the lifts of all paths of $[\gamma]$ from a' are then closed, and such elements $[\gamma]$ form a subgroup of F_a. We denote this subgroup, determined by the triple (W, f, a'), by $\Gamma_{a'}$.

The choice of $a \in S$ is immaterial. For given another point $b \in S$, consider a path σ on S from a to b; let b' be the terminal point of the lift of σ from a'. If $[\gamma] \in F_a$, then $[\gamma] \to [\sigma^{-1}\gamma\sigma]$ is a group isomorphism of F_a onto F_b which carries $\Gamma_{a'}$ onto the subgroup $\Gamma_{b'}$ determined by (W, f, b'). Also, if $a' \in W$ is replaced by another point a'' of W over a, then the groups determined by (W, f, a') and (W, f, a'') are conjugate subgroups of F_a. This can be easily verified.

An unlimited covering surface (W, f) of S is said to be *normal* if the triple (W, f, a') determines a normal subgroup of F_a. This notion is well defined, because the property of a covering surface being normal does not depend on the choice of the points $a \in S$ and $a' \in W$. If (W, f) is normal, then all triples (W, f, a') with $a' \in f^{-1}\{a\}$ determine the same subgroup of F_a.

There is an important connection between the topologies of W and S:

Let (W, f) be an unlimited covering surface of S. Then the fundamental group of W is isomorphic to the subgroup of F_a determined by a triple (W, f, a').

The proof is easy. We consider closed paths γ of W from a' and check that $[\gamma] \to [f \circ \gamma]$ is a required isomorphism (Ahlfors–Sario [1], p. 36).

The connection between the triples (W, f, a') and the subgroups of F_a makes it possible to partially order the unlimited covering surfaces of S according to strength. The strongest covering surfaces are those which determine the trivial subgroup of F_a. They are called *universal covering surfaces* of S. By what we just proved, *a universal covering surface is simply connected.* It is of course a normal covering surface.

Every surface possesses universal covering surfaces. This can be shown by direct construction. Given a surface S and a point $a \in S$, consider all paths γ of S from a to a point p. We define the set

$$W = \{p' = (p, [\gamma]) | p \in S\}$$

and the mapping $f: W \to S$ by the requirement $f(p') = p$. Then a topology can be introduced on W so that (W, f) is a universal covering surface of S. For the details of the proof we refer to Ahlfors–Sario [1], p. 35. There the more general result is proved that, given any subgroup of F_a, there exists an unlimited covering surface of S which determines this subgroup.

The notion of a universal covering surface is due to H. A. Schwarz, who noticed its importance in the theory of Riemann surfaces (see Theorem 3.4 in the next section).

2.3. Branched Covering Surfaces

In elementary function theory, Riemann surfaces are first encountered in connection with the mapping $z \to z^n$. This defines the plane as a covering of itself, but in such a way that the projection mapping has branch points at 0 and ∞. More generally, if $f: W \to S$ is a non-constant analytic mapping between the Riemann surfaces W and S, then (W, f) need not be a smooth covering surface of S. This state of affairs leads us to generalize the notion of smooth covering surface.

A *covering surface* of a surface S is a pair (W, f), where W is a surface, $f: W \to S$ is a continuous mapping, and every point $p \in W$ has a neighborhood U such that $(U \setminus \{p\}, f | U \setminus \{p\})$ is a smooth covering surface of $S \setminus \{f(p)\}$.

A smooth covering surface is trivially a covering surface. A covering surface which is not smooth is called *branched*.

We can deduce from the definition that the projection mapping $f: W \to S$ behaves locally like the mapping $z \to z^n$ for suitable n. More precisely, the following result is true:

Lemma 2.1. *Let (W, f) be a covering surface of S. For every $p \in W$, there are parameter discs $U \ni p$ and $f(U)$, with local parameters k and h normalized by $k(p) = h(f(p)) = 0$, such that in U,*

$$h \circ f = k^n, \tag{2.1}$$

where n is a natural number.

The proof is given in Ahlfors–Sario [1], p. 40. Conversely, if $f: W \to S$ is a continuous mapping and the above condition holds, we conclude immediately that (W, f) is a covering surface of S. Thus this condition characterizes covering surfaces.

We mention here that there are other non-trivially equivalent characterizations of covering surfaces, even though we shall not be using them in what follows. Let S and W be surfaces and $f: W \to S$ a continuous mapping. Then (W, f) is a covering surface of S if and only if f is locally homeomorphic with the possible exception of a discrete set, or if and only if f is light (the preimage of a point is totally disconnected) and open. (A function which is continuous, light and open is called an interior mapping. It is a famous theorem of Stoïlov that an interior mapping f of a plane domain is of the form $f = \varphi \circ h$, where h is homeomorphic and φ analytic.)

2.4. Covering Groups

Let S be a surface and (W, f) its smooth covering surface. A *cover transformation* g of W over S is a homeomorphism $g: W \to W$ such that $f \circ g = f$. All such mappings g form a group G which is called the *covering group* of W over S.

Two points of W equivalent under G have the same projection on S. If conversely, any two points lying over the same point of S are equivalent under G, then G is said to be *transitive*.

Let us consider the quotient space W/G and furnish it with the quotient topology. Under certain conditions, W/G is a surface which is homeomorphic to the surface S.

Theorem 2.2. *If the projection mapping $f: W \to S$ is surjective and the covering group G of W over S is transitive, then W/G and S are homeomorphic.*

PROOF. We write $[p] \in W/G$ for the equivalence class containing the point $p \in W$ and prove that

$$[p] \to f(p) \qquad\qquad (2.2)$$

is a homeomorphism of W/G onto S. First, it follows from $f = f \circ g$, $g \in G$, that (2.2) is well defined in W/G. It is surjective, because $f: W \to S$ is onto, and injective, because G is transitive. Its continuity follows from the continuity of $f: W \to S$, and the continuity of its inverse from the fact that $f: W \to S$ is locally homeomorphic.   □

The points of W are isolated with respect to G in the following sense:

Each point of the surface W has a neighborhood in which no two points are equivalent under the action of the covering group.

In fact, it follows from the definition of a cover transformation that an open set in which the projection mapping $f: W \to S$ is injective cannot contain points equivalent modulo G. From this observation we can draw another conclusion:

Except for the identity mapping, a cover transformation has no fixed points.

For assume that q is a fixed point for a transformation $g \in G$. Since g is continuous, it maps a point p near q onto a point $g(p)$ near $g(q) = q$. Because p and $g(p)$ are equivalent under G, we conclude that for all p in a sufficiently small neighborhood of q, we have $g(p) = p$. It follows that the set in which $g(p) = p$ is open. It is also closed and nonvoid. Since a surface is connected, we see that g is the identity mapping.

For the most part, we shall be dealing with the covering groups corresponding to universal covering surfaces.

Theorem 2.3. *The covering group of a universal covering surface W over a surface S is transitive.*

PROOF. Suppose that a and a' are points of W which lie over the same point of S. Choose a point $p \in W$, join a to p by a path on W, project this path onto S, and lift the projection back, but from the point a'. Let p' be the terminal point of this lift. We define g by the condition $g(p) = p'$, and check that g is well defined and a cover transformation of W over S. Hence a and a' are equivalent under the covering group.   □

Combined with Theorem 2.3, Theorem 2.2 says that *for a universal covering surface W of S, the space W/G is always homeomorphic to S.* The following result sheds additional light on this connection.

Theorem 2.4. *The covering group of a universal covering surface of S is isomorphic to the fundamental group of S.*

PROOF. Given a point $a \in W$, let γ be a closed path on S from $f(a)$, and $b \in W$ the terminal point of the lift of γ from a. Then a and b both lie over $f(a)$. By Theorem 2.3, there is a unique cover transformation g_γ with the property $g_\gamma(a) = b$. It is easy to verify that $[\gamma] \to g_\gamma$ is the desired group isomorphism (cf. Ahlfors–Sario [1], p. 38). $\qquad\qquad\square$

2.5. Properly Discontinuous Groups

Starting with a given surface S, we arrived via a covering surface (W, f) of S at the covering group G of W over S. Theorem 2.2 tells that, under very general conditions, the circle from S to W to G closes, in the sense that the quotient W/G is homeomorphic to S.

We shall now take a different starting point and prescribe directly a surface W together with a group G of homeomorphic self-mappings of W. Again, we form the quotient space W/G, furnish it with the quotient topology, and want to impose a condition on G which makes W/G a surface.

For a covering group, every point has a neighborhood in which no two points are equivalent. However, this property does not characterize covering groups. In fact, examples can be given of groups G which possess this property but for which W/G is not even a Hausdorff space. We need a stronger condition on G.

A group G acting on W is said to be *properly discontinuous* if for any two compact sets $A, B \subset W$, the intersection $g(A) \cap B$ is void, except for finitely many $g \in G$. Unlike a covering group, a properly discontinuous group need not be fixed point free.

A point $p \in W$ is a *limit point* of a group G acting on W if there are distinct mappings $g_n \in G$, $n = 1, 2, \ldots$, such that $p = \lim g_n(q)$ for some point $q \in W$. A *properly discontinuous group has no limit points*. This follows immediately from the definition of proper discontinuity.

A fixed point free properly discontinuous group G shares the property of covering groups that every point $p \in W$ has a neighborhood in which no two points are equivalent modulo G. For assume that there are two sequences of different points a_n, b_n, $n = 1, 2, \ldots$, in a compact neighborhood A of p such that $\lim a_n = \lim b_n = p$ and $b_n = g_n(a_n)$ for mappings $g_n \in G$. Then $g_n(A) \cap A \neq \varnothing$. If there are infinitely many different mappings g_n, then G is not properly discontinuous. If there are only finitely many different transformations g_n, then at least one of them appears infinitely many times in the sequence. For such a mapping p is a fixed point.

Theorem 2.5. *A transitive covering group is properly discontinuous.*

PROOF. Let (W, f) be a covering surface of S, and suppose that the covering group G of W over S is transitive. Given two compact sets A and B in W, we consider the subset $C = \{(p,q) | f(p) = f(q)\}$ of $A \times B$. The complement of C is open, because f is continuous and two different points $f(p)$ and $f(q)$ have disjoint open neighborhoods in S. It follows that C is closed and hence compact.

For $g \in G$ fixed, we write $U_g = \{(p,q) \in C | q = g(p)\}$. Each U_g is open in C, because f is a local homeomorphism. By the transitivity of G, the union of the disjoint sets U_g, where g runs through all elements of G, agrees with C. Since C is compact, we conclude that only finitely many of the sets U_g are non-empty. Consequently, G is properly discontinuous. □

The following result is a converse to Theorem 2.2.

Theorem 2.6. *Let W be a surface, G a properly discontinuous fixed point free group of homeomorphisms of W onto itself, and $f: W \to W/G$ the canonical projection. Then*

1. *W/G is a surface,*
2. *(W, f) is an unlimited covering surface of W/G,*
3. *G is the (transitive) covering group of W over W/G.*

PROOF. By definition, f is continuous. If $A \subset W$, then $f^{-1}(f(A)) = \cup g(A)$, $g \in G$, from which we conclude that f is open.

In order to prove that W/G is a Hausdorff space we consider two different points $f(a)$ and $f(b)$ of W/G. Since G is properly discontinuous, there exists a compact neighborhood B of b which does not contain any point $g(a)$, $g \in G$. After this we conclude the existence of a compact neighborhood A of a such that $A \cap g(B)$ is empty for every $g \in G$. Then $g_1(A) \cap g_2(B) = \varnothing$ for all $g_1, g_2 \in G$, and it follows that $f(A)$ and $f(B)$ are disjoint neighborhoods of $f(a)$ and $f(b)$.

Clearly W/G is connected and has a countable base for topology. In order to find local parameters we fix a point $p \in W$. Since G is properly discontinuous and fixed point free, there exists an open neighborhood U of p such that $g(U) \cap U = \varnothing$ for all mappings $g \in G$ different from the identity. Then $f|U$ is injective, and if U is so small that it lies in the domain of a local parameter h of W, then $h \circ (f|U)^{-1}$ maps the open set $f(U)$ in W/G homeomorphically onto an open set in the plane. Since $f: W \to W/G$ is surjective, it follows that W/G is a surface. Also, (W, f) is a smooth covering surface of W/G.

From the definition it is clear that every $g \in G$ is a cover transformation. Conversely, let w be a cover transformation and $p \in W$. Then there is a $g \in G$ such that $g(p) = w(p)$, for otherwise we would have $f(w(p)) \neq f(p)$. Hence $w = g$.

Since G is a transitive covering group, it is not difficult to show that (W, f) is an unlimited covering surface of W/G (cf. Ahlfors–Sario [1], p. 29). □

3. Uniformization of Riemann Surfaces

3.1. Lifted and Projected Conformal Structures

Let us now apply the results of section 2 to Riemann surfaces.

Theorem 3.1. *Let S be a Riemann surface and (W, f) a smooth covering surface of S. Then W carries a unique conformal structure which makes the projection mapping f analytic.*

PROOF. Let H be the conformal structure of S. For every point $p \in W$ we choose a neighborhood U of p such that $f|U$ is injective and $f(U)$ is contained in the domain of some $h \in H$. Then the atlas $\{h \circ (f|U)|p \in W\}$ defines a conformal structure for W, and f is analytic with respect to this structure.

We say that this conformal structure of W is obtained by lifting the conformal structure of S. If the projection $f: W \to S$ is analytic with respect to a conformal structure of W, then the condition which expresses this fact shows directly that this structure is the same as the lifted structure. Thus the uniqueness assertion in the theorem follows.   □

Using the characterization (2.1) of a covering surface, we could show without difficulty that Theorem 3.1 holds also in the case where (W, f) is a branched covering surface of S (cf. Ahlfors–Sario [1], p. 119).

In the sequel, a covering surface of a Riemann surface is always regarded as the Riemann surface with the lifted conformal structure.

The following observation is immediate: Let S be a Riemann surface and (W, f) a smooth covering surface of S. *Then the cover transformations of W over S are conformal.* For locally a cover transformation g is of the form $(f|g(U))^{-1} \circ f|U$, and hence conformal.

Theorem 2.6 can be refined in the setting of Riemann surfaces.

Theorem 3.2. *Let W be a Riemann surface, G a properly discontinuous fixed point free group of conformal self-mappings of W, and $f: W \to W/G$ the canonical projection. Then the surface W/G carries a unique conformal structure which lifts to the original conformal structure of W.*

This follows immediately from the way the local parameters of W/G were defined in the proof of Theorem 2.6. In the situation of Theorem 3.2, the conformal structure of W is said to have been projected to W/G. If W is a given Riemann surface, we always regard the quotient W/G as the Riemann surface with the projected structure.

Suppose that G is a properly discontinuous group of conformal self-mappings of a Riemann surface W, but not fixed point free. We can still conclude that W/G is a surface and that W/G carries a conformal structure which makes the projection mapping $f: W \to W/G$ analytic. However, W is now a covering surface of W/G which is branched at the fixed points of G (Ahlfors–Sario [1], p. 121).

3.2. Riemann Mapping Theorem

Our results in 2.4, 2.5 and 3.1 lead to a fundamental representation of Riemann surfaces if they are combined with the following generalization of the Riemann mapping theorem for plane domains.

Theorem 3.3 (Riemann Mapping Theorem). *Every simply connected Riemann surface is conformally equivalent to one and only one of the following plane domains: the unit disc, the complex plane, or the extended plane.*

This is a deep result, and a complete proof requires lengthy preparations. We content ourselves, therefore, with sketching the main lines of a proof based on the use of subharmonic functions. (Subharmonic functions are defined on Riemann surfaces with the aid of local parameters. This is possible because subharmonicity is a local and conformally invariant property.)

First of all, a classification of Riemann surfaces into compact, parabolic, and hyperbolic surfaces is needed. A non-compact Riemann surface S is parabolic if every negative subharmonic function on S is constant; otherwise S is hyperbolic.

Using subharmonic functions and Perron families, we can define Green's functions for Riemann surfaces just as it is done for the case of plane domains. The Green's function g_p of a Riemann surface S with singularity at the point $p \in S$ is a function which is positive and harmonic on $S - \{p\}$. To describe its singularity, we consider a local parameter z mapping a neighborhood of p onto the unit disc such that $z(p) = 0$. Then it is required that $g_p + \log|z|$ be harmonic at p; this is an invariant definition not depending on the choice of the local parameter. The Green's function is characterized by the property that among all functions positive and harmonic on $S - \{p\}$ and possessing the same singularity at p as g_p, the function g_p is the smallest. If a Green's function exists for some $p \in S$, then it exists for every $p \in S$. By a theorem of Ohtsuka, the Green's function exists if and only if S is hyperbolic.

If S is parabolic or compact, Green's functions do not exist but it is possible to prove the existence of a function $u_{p,q}$ with the following properties: $u_{p,q}$ is harmonic in $S - \{p\} - \{q\}$; if $z(p) = 0$, then $u_{p,q} - \log|z|$ is harmonic at p, and if $z(q) = 0$, then $u_{p,q} + \log|z|$ is harmonic at q; outside parameter discs (preimages of discs under z) containing p and q, the function $u_{p,q}$ is bounded (Nevanlinna [1], p. 212).

Suppose now that S is simply connected. If S is hyperbolic, we take a Green's function g_p, form its conjugate g_p^* in a parametric disc, and extend

$\exp(-(g_p + ig_p^*))$ by analytic continuation to S. Using the Monodromy theorem, we conclude that the extended function is single-valued on S. Finally, application of the maximum principle shows that it maps S conformally onto the unit disc (Nevanlinna [1], p. 204).

If S is parabolic or compact, a conformal mapping of S into the extended plane can be constructed with the aid of the function $u_{p,q}$ (Nevanlinna [1], p. 213). In case S is parabolic the boundary of the image consists of one point, whereas the boundary is empty if S is compact.

Theorem 3.3 can also be proved by a method which is based on the use of quasiconformal mappings (Bers [5]). The idea is to first construct a topological and locally quasiconformal mapping of the given Riemann surface S into the plane, then apply the existence theorem of Beltrami equations to obtain a conformal mapping of S (as Gauss did; cf. 1.6), and complete the proof with the aid of the Riemann mapping theorem for plane domains.

Theorem 3.3 occupies a central position in the theory of Riemann surfaces. It is often called the general uniformization theorem. The first proofs are attributed to Koebe (in 1907) and Poincaré.

3.3. Representation of Riemann Surfaces

Let S be an arbitrary Riemann surface and (W, f) its universal covering surface. Since W is simply connected, we conclude from Riemann's mapping theorem the existence of a conformal mapping $w: W \to D$, where D is the unit disc, the finite plane or the extended plane. But then $(D, f \circ w^{-1})$ is also a universal covering surface of the Riemann surface S, and we have proved the following important result:

> *Every Riemann surface admits as its universal covering surface the unit disc, the finite plane, or the extended plane.*

This makes possible a far-reaching normalization of universal covering surfaces. A consequence of basic importance is the fact that the elements of the covering group of D over S, being conformal self-mappings of D, are Möbius transformations.

Summarizing the topological results in 2.4–5 and the analytical results in 3.1–2, we obtain the basic representation theorem for Riemann surfaces.

Theorem 3.4. *Given an arbitrary Riemann surface S, let D be its universal covering surface, and G the covering group of D over S. Then S is conformally equivalent to the Riemann surface D/G.*

PROOF. It follows from Theorems 2.3, 2.5, and 3.2 that the quotient D/G is a Riemann surface with the projected conformal structure. By Theorem 2.2, the mapping (2.2) is a homeomorphism of D/G onto S. It is conformal, because the conformal structure of S is also obtained by projection from D.   □

Theorem 3.4 is the analytic counterpart of the topological result expressed in 2.4 that every surface S is topologically equivalent to the quotient W/G, where W is a universal covering surface of S.

Theorem 3.4 can be supplemented as follows: *Let G be an arbitrary properly discontinuous group of conformal mappings acting on D. Then D/G is a Riemann surface.* This follows from Theorem 3.2 and the remark made after it.

3.4. Lifting of Continuous Mappings

We have seen above that, under certain circumstances, a mapping between universal covering surfaces projects to a mapping between the underlying surfaces. Here we take the opposite view and show how to lift mappings between Riemann surfaces to mappings between their universal covering surfaces.

Let φ be a continuous mapping of a Riemann surface S_1 into another Riemann surface S_2, and (D_i, π_i) a universal covering surface of S_i, $i = 1, 2$. Here we consider only the case $D_1 = D_2$. By 3.3, we can choose this common surface, which we denote by D, so that D is the unit disc, the complex plane or the extended plane.

In the extended plane every Möbius transformation has a fixed point. Thus, if the extended plane is the universal covering surface of a Riemann surface, the covering group over this surface is trivial. By Theorem 3.4, the surface itself is then the extended plane up to conformal equivalence. Therefore, we can exclude this trivial case in what follows.

The continuous mapping $\varphi\colon S_1 \to S_2$ always induces a continuous mapping f of D into itself. More precisely, there is a continuous $f\colon D \to D$ such that

$$\varphi \circ \pi_1 = \pi_2 \circ f. \tag{3.1}$$

The construction of f, which is called a *lift* of φ, is as follows: Fix first $z_0 \in D$ and $w_0 \in \pi_2^{-1}\{\varphi(\pi_1(z_0))\}$. If γ is a path in D from z_0 to z, we define $f(z)$ as the endpoint of the path which we obtain by lifting $\varphi \circ \pi_1 \circ \gamma$ from w_0. If γ' is another path in D from z_0 to z, then $\varphi \circ \pi_1 \circ \gamma$ and $\varphi \circ \pi_1 \circ \gamma'$ are homotopic, and it follows from the Monodromy theorem that $f(z)$ is well defined. From the definition it is clear that f is continuous.

The mapping φ induces a mapping of the covering group G_1 of D over S_1 into the covering group G_2 of D over S_2: We shall show that the relation

$$\theta(g) \circ f = f \circ g \tag{3.2}$$

defines a mapping θ which is a homomorphism of G_1 into G_2.

In order to prove that (3.2) defines a homomorphism $\theta\colon G_1 \to G_2$, we choose an element $g \in G_1$. From (3.1) it follows that

$$\pi_2 \circ f \circ g = \varphi \circ \pi_1 \circ g = \varphi \circ \pi_1 = \pi_2 \circ f.$$

For every $z \in D$ we thus have an element $h \in G_2$ such that $f(g(z)) = h(f(z))$.

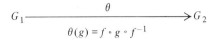

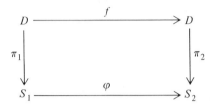

Figure 8. Mappings induced by a homeomorphic φ.

The mapping h depends only on g, not on the point z. For if γ is a path in D from z to z', then $f \circ g \circ \gamma$ and $h \circ f \circ \gamma$ start at the same point and have the same projection $\varphi \circ \pi_1 \circ \gamma$. It follows that they have the same terminal point, i.e., the element of G_2 corresponding to z' is also h. Hence (3.2) holds for $\theta(g) = h$. From

$$\theta(g_2) \circ \theta(g_1) \circ f = \theta(g_2) \circ f \circ g_1 = f \circ g_2 \circ g_1 = \theta(g_2 \circ g_1) \circ f,$$

we see that the mapping θ is a homomorphism.

The mappings f and θ induced by φ are not unique. If f is a lift of φ, then $f \circ g$ is also a lift of φ for every $g \in G_1$. From $f \circ g = \theta(g) \circ f$ it follows that such lifts are of the form $h \circ f$, where $h = \theta(g)$ is an element of G_2. The mappings

$$h \circ f, \qquad h \in G_2,$$

represent all possible lifts of φ. We see this by repeating the reasoning which showed that (3.2) defines an element $\theta(g)$ of G_2.

If f induces the group homomorphism θ, then $h \circ f$ induces the homomorphism $g \to h \circ \theta(g) \circ h^{-1}$, which differs from θ by an inner automorphism of G_2. We call two such homomorphisms equivalent and conclude that all homomorphisms induced by $\varphi: S_1 \to S_2$ are equivalent.

Suppose that $\varphi_0: S_1 \to S_2$ and $\varphi_1: S_1 \to S_2$ determine equivalent homomorphisms. If φ_0 induces θ, then φ_1 can be so lifted that it also induces θ. In fact, a lift f_1 of φ_1 induces a homomorphism $g \to h \circ \theta(g) \circ h^{-1}$, $h \in G_2$, and so $h^{-1} \circ f_1$, which is a lift of φ_1, induces θ.

If $\varphi: S_1 \to S_2$ is a homeomorphism, then so is every lift $f: D \to D$. In this case $g \to \theta(g) = f \circ g \circ f^{-1}$ is an isomorphism of G_1 onto G_2. (Fig. 8.)

3.5. Homotopic Mappings

Lifting of mappings is closely related to the topological notion of homotopy. Let $\varphi_0: S_1 \to S_2$ be a homeomorphism and $I = \{t \mid 0 \le t \le 1\}$ the unit interval. A homeomorphism $\varphi_1: S_1 \to S_2$ is said to be homotopic to φ_0 if there is a continuous mapping $h: S_1 \times I \to S_2$, such that $h(.,0) = \varphi_0$, $h(.,1) = \varphi_1$. The mapping h is called a homotopy from φ_0 to φ_1.

Suppose that h is a homotopy from φ_0 to φ_1. Given a lift f_0 of φ_0, the mappings $h(.,t): S_1 \to S_2$, $0 \le t \le 1$, can then be so lifted that we obtain a homotopy from f_0 to a lift f_1 of φ_1. This homotopy lifting property follows easily from the definitions.

Theorem 3.5. *Two homeomorphisms* $\varphi_i: S_1 \to S_2$, $i = 0$, 1, *induce the same group isomorphisms if and only if they are homotopic.*

PROOF. Assume first that φ_0 is homotopic to φ_1. Let h be a homotopy from φ_0 to φ_1 and f_t a lift of $h(.,t)$, $0 \le t \le 1$, such that f_t is a homotopy between f_0 and f_1.

Choose $g \in G_1$ and $z \in D$, and consider the two paths $t \to f_t(g(z))$ and $t \to (f_0 \circ g \circ f_0^{-1})(f_t(z))$. Both have the same initial point $f_0(g(z))$ and the same projection $t \to \pi_2(f_t(z))$ on S_2. Hence they agree, and for $t = 1$ we obtain the desired result

$$f_0 \circ g \circ f_0^{-1} = f_1 \circ g \circ f_1^{-1}. \tag{3.3}$$

Assume, conversely, that φ_0 and φ_1 have lifts f_0 and f_1 such that (3.3) holds for every $g \in G_1$. If D is the unit disc, we define $f_t(z)$, $0 < t < 1$, as follows: $f_t(z)$ is the point of the hyperbolic geodesic arc joining $f_0(z)$ and $f_1(z)$ which divides the hyperbolic length of this arc in the ratio $t: (1 - t)$. Then f_t is a homotopy between f_0 and f_1.

Under the mapping $\theta(g) = f_0 \circ g \circ f_0^{-1} (= f_1 \circ g \circ f_1^{-1})$ the endpoints of the arc map to $f_0(g(z))$ and $f_1(g(z))$. But since $\theta(g)$ leaves hyperbolic distances invariant, $\theta(g)$ maps the point $f_t(z)$ to $f_t(g(z))$. Hence, $\theta(g) \circ f_t = f_t \circ g$. In other words, all mappings f_t, $0 \le t \le 1$, determine the same group homomorphism. It follows that $\pi_2 \circ f_t \circ \pi_1^{-1}$ is a well defined mapping, and it is a homotopy between φ_0 and φ_1.

If D is the finite plane, all cover transformations are translations $z \to z + b$. Therefore, the above reasoning remains valid if the hyperbolic metric is replaced by the euclidean. $\square$

3.6. Lifting of Differentials

Let S and W be Riemann surfaces and $f: W \to S$ a non-constant analytic mapping. It follows from the definition of a covering surface in 2.3 that (W, f) is a covering surface of S (cf. Ahlfors–Sario [1], p. 119).

Let the conformal structures of W and S be determined by the local parameters h_i and k_j, respectively. Given an arbitrary point of W, we consider an open neighborhood V of this point which is contained in the domain of a local parameter h_i and is so small that $f(V)$ lies in the domain of a local parameter k_j. For $p \in V$, we write $z_i = h_i(p)$, $w_j = k_j(f(p))$.

Let φ be an (m, n)-differential on S (cf. 1.4), and let φ_j denote its representation in the local coordinate w_j. Set

$$\psi_i = \varphi_j \left(\frac{dw_j}{dz_i}\right)^m \left(\overline{\frac{dw_j}{dz_i}}\right)^n. \tag{3.4}$$

By formula (1.2), the function element ψ_i does not depend on the particular choice of the parameter made in $f(V)$. We now require that (3.4) remains valid when we change the parameter of W in V. Then (3.4) defines an (m, n)-differential on V, and since we started with an arbitrary point of W, on the whole surface W. It is called the *lift* of φ to W.

We can say a little more: Formula (3.4) shows that $f: W \to S$ induces a linear mapping of the space of holomorphic (m, n)-differentials of S into the space of holomorphic (m, n)-differentials of W. If f is a conformal mapping of W onto S, the induced mapping is bijective.

Suppose, in particular, that $W = D$ is the universal covering surface of S, where D is the unit disc or the complex plane. In both cases the conformal structure of S is determined by the local inverses of the projection mapping of D onto S. The conformal structure of D is of course given by the identity mapping $z \to z$.

What we gain from the use of the universal covering surface D is that we now possess a global coordinate $z \in D$ for the representation of φ. In other words, in (3.4) we can put $z_i = w_j = z$. Thus $\psi_i = \varphi_j$, and formula (1.2) shows that every φ_j is the restriction of a function globally defined in D. We denote this function by φ, i.e., we identify it with the collection of its restrictions. It follows that *the lift of the differential φ of S to the universal covering surface D is a global representation of φ in terms of the coordinate $z \in D$.*

Let g be an arbitrary element of the covering group G of D over S. Suppose $p \to z$ is a local parameter in an open subset U of S defined by the inverse of a suitable restriction of the projection mapping. Then $p \to g(z)$ is a local parameter in U such that z and $g(z)$ have the same preimage in U. Hence, it follows from the invariance (1.2) that

$$\varphi(g(z))g'(z)^m \overline{g'(z)^n} = \varphi(z) \tag{3.5}$$

for every $g \in G$.

Conversely, if φ is a complex-valued function in D with the property (3.5), then φ defines an (m, n)-differential of S. A function φ satisfying (3.5) is said to be an (m, n)-*differential for the group* G. Thus there is no difference whether we interpret φ to be a differential on the Riemann surface S or for the covering group G of D over S.

As one application, we introduce the hyperbolic metric, which we have so far considered in plane domains conformally equivalent to a disc, to Riemann surfaces. Let S be a Riemann surface which has the unit disc D as its universal covering surface. The Poincaré density $z \to \eta(z) = 1/(1 - |z|^2)$ of D satisfies the condition

$$(\eta \circ g)|g'| = \eta \tag{3.6}$$

for every cover transformation g. It follows that η is a $(1/2, 1/2)$-differential on S, and $\eta(z) = 1/(1 - |z|^2)$ is its global representation in the coordinate z of D.

Now let γ be a rectifiable arc on S. We define its hyperbolic length to be equal to the hyperbolic length of its lift to D. Because of the invariance (3.6) it does not matter how we choose the point of D lying over the initial point of γ.

In order to study the geodesics on S, we take two points p and q of S and join them by an arc γ. Let z_0 be a point of D over p, and let z_1 denote the terminal point of the lift of γ from z_0. Then the projection of the hyperbolic geodesic from z_0 to z_1 has the shortest hyperbolic length in the homotopy class of γ. The infimum of these shortest lengths over all homotopy classes is the hyperbolic distance between p and q. The infimum is attained. In fact, the hyperbolic geodesic in D joining z_0 to a "closest" point of the preimage of q projects to a geodesic between p and q.

4. Groups of Möbius Transformations

4.1. Covering Groups Acting on the Plane

Let S be an arbitrary Riemann surface. We again normalize its universal covering surface D so that D is the unit disc, the complex plane or the extended plane, and denote by G the covering group of D over S. In view of the representation $S = D/G$ modulo conformal equivalence, the theory of Riemann surfaces can be regarded as essentially equivalent with the theory of discontinuous groups of Möbius transformations acting on D. Lehner ([1], Chapter I) gives an interesting survey of the historical development of the theory of Möbius groups.

The points of D/G are called orbits of G. A subdomain of D is said to be a *fundamental domain* of G if it contains at most one point of every orbit of G and its closure in D meets every orbit of G.

In the cases where the universal covering surface D is the extended plane or the complex plane, all possible covering groups of D over S can be readily listed. We know already that if D is the extended plane, the covering group G is trivial and S is conformally equivalent to D.

Suppose next that D is the complex plane. Since the elements of G have their fixed point at ∞, they are translations $z \to z + a$. Here three possible types of discontinuous groups G arise. First, G may be trivial, in which case S is conformally equivalent to the finite plane. Second, G can be infinite, generated by a transformation $z \to z + \omega$, $\omega \neq 0$. A fundamental domain of such a group is the interior of the parallel strip bounded by straight lines through 0 and through ω and perpendicular to the vector from 0 to ω. Topologically, D/G is an infinite cylinder. The function $z \to \exp(2\pi i z/\omega)$, which is invariant under G, shows that D/G is conformally equivalent to the finite plane punctured at 0.

A third possibility is that G has two generators $z \to z + \omega_1$, $z \to z + \omega_2$, $\text{Im}(\omega_1/\omega_2) \neq 0$. A fundamental domain is now the interior of the parallelogram P with vertices at 0, ω_1, $\omega_1 + \omega_2$, and ω_2. In this case the Riemann surface $S = D/G$ is compact. For the closure $\bar{P}$ is compact and S is the image of $\bar{P}$ under the continuous projection mapping $D \to D/G$.

Since the opposite sides of P are equivalent under G, it follows that topologically S is a torus. Two different tori obtained in this fashion are generally not conformally equivalent. The conformal structures on a torus will be studied in V.6.

A simple geometric argument shows that there are no other ways to form groups of translations which are properly discontinuous in the finite plane.

4.2. Fuchsian Groups

Let us now consider the case in which the Riemann surface S admits a disc D as its universal covering surface. It follows from the results of section 3 that the covering group G of D over S is a properly discontinuous fixed point free group of Möbius transformations which keeps the disc D invariant. We call such groups *Fuchsian groups*. (In the literature, fixed points are usually allowed.) Conversely, every Fuchsian group G acting on D is the covering group of D over the Riemann surface D/G.

An arbitrary Möbius transformation $z \to w$ with two finite fixed points z_1 and z_2 has the representation

$$\frac{w - z_1}{w - z_2} = \rho e^{i\theta} \frac{z - z_1}{z - z_2}.$$

If $z_2 = \infty$, we have $w - z_1 = \rho e^{i\theta}(z - z_1)$. The geometric action of a Möbius transformation is best seen from this representation, which also gives rise to the division of Möbius transformations (different from the identity) into four classes. If $\rho = 1$, $\theta \neq 0$, the transformation is elliptic, if $\rho \neq 1$, $\theta = 0$, it is hyperbolic, and if $\rho \neq 1$, $\theta \neq 0$, it is loxodromic ($0 \le \theta < 2\pi$). A Möbius transformation with only one fixed point is parabolic. The class of a Möbius transformation g remains the same when g is conjugated by an arbitrary Möbius transformation h, i.e., when g is replaced by $h \circ g \circ h^{-1}$.

A loxodromic transformation does not keep any disc invariant. If an elliptic transformation g maps a disc D onto itself, then one of the fixed points of g lies inside D while the other is its mirror image with respect to ∂D.

Let a Fuchsian group act on a disc D. Since it has no fixed points in D, it follows from the reflection principle that the elements of the group do not have fixed points in the complement of the closure of D either (with the identity mapping excluded of course). We conclude that the elements of a Fuchsian group are hyperbolic or parabolic, with fixed points on ∂D.

In studying a Fuchsian group G, we choose the invariant disc on which G acts to be the upper half-plane. The elements of G are then of the form

$$z \to g(z) = \frac{az + b}{cz + d},$$

where the coefficients a, b, c, d are real and $ad - bc > 0$. Groups of Möbius transformations with real coefficients are said to be real.

The conformally invariant hyperbolic metric $ds = |dz|/2 \operatorname{Im} z$ of H is a natural tool for the study of the geometric properties of G. As before, we use the notation $h(z_1, z_2)$ for the distance between the points z_1 and z_2 of H in this metric. We fix a point $z_0 \in H$ and denote by $G(z_0)$ the orbit of z_0. For every $z_j \in G(z_0), j = 0, 1, \ldots,$ we write

$$N_j = \{z \in H \,|\, h(z, z_j) < h(z, \zeta) \quad \text{for all} \quad \zeta \in G(z_0), \zeta \neq z_j\}.$$

The sets N_j are non-empty, open and mutually disjoint, and the union of the closures in H of all N_j is H. If $z_k = g(z_j)$ for $g \in G$, then $N_k = g(N_j)$. It follows that all sets N_j are congruent in the non-euclidean geometry of H. In studying the properties of the sets N_j we may therefore restrict ourselves to one of them, say to N_0, for which we also use the shorter notation N.

A point $z \in H$ lies on the boundary of N if and only if $h(z, z_0) \leq h(z, z_k)$ for all $z_k \in G(z_0)$ and equality holds for at least one z_k, $k \neq 0$. The set $\{z \in H \,|\, h(z, z_0) = h(z, z_k)\}$ is the non-euclidean line which is the perpendicular bisector of the non-euclidean segment joining z_0 and z_k. It follows that N is a convex polygon; in particular, N is connected. N is called the *Dirichlet region* of G with center at z_0.

From the definition of N we conclude that N is a fundamental domain of G. Its boundary arcs lie either on the real axis, in which case they are said to be free sides, or they are situated in H, with the possible exception of endpoints on $\mathbb{R}$, and are called inner sides of N.

The inner sides of N are pairwise equivalent under G, whereas inner points of a free side have no equivalent points in the closure of N. These properties of N can be deduced without difficulty from the definition. More careful analysis is required to prove the following fundamental result:

Theorem 4.1. *The elements of a Fuchsian group which map the inner sides of a Dirichlet region onto each other generate the whole group.*

Dirichlet regions are studied in detail in Springer [1]; for the proof of Theorem 4.1, see p. 237.

4.3. Elementary Groups

A group of Möbius transformations can of course be regarded as acting on the extended plane. We shall now adopt this point of view.

The *limit set* L of a group G of Möbius transformations is the set of the limit points of G. From the definition of a limit point, which was given in 2.5, it follows that $g(L) = L$ for every $g \in G$. Also, L is closed (cf. Lehner [1], p. 88). If L contains at most two points, the group G is said to be *elementary*.

All elementary groups can be listed. (A comprehensive treatment of elementary groups is given in Ford [1], Chapter VI.) First of all, a finite group is necessarily elementary, its limit set being empty. If the Riemann sphere is rotated so that a regular solid remains invariant, then stereographic projection leads to Möbius transformations which form a finite group. All non-cyclic finite groups of Möbius transformations are obtained from such groups by conjugation.

The elements of a finite group are elliptic transformations. They are of finite period, i.e., there is a natural number n such that the nth iterate of the transformation is the identity mapping. A cyclic group generated by an elliptic transformation which is not of finite period is very different: Every point of the plane is a limit point of such a group (Lehner [1], p. 87).

There is a second type of elementary groups G all of whose elements are elliptic or parabolic transformations sharing a common fixed point. Then G has this common fixed point as its sole limit point (Lehner [1], p. 93). A simple example is the group generated by the elliptic transformation $z \to -z$ and the parabolic transformation $z \to z + 1$. In this case $L = \{\infty\}$.

Any other infinite group is elementary if and only if it is cyclic and the generator is not elliptic. The limit set L then agrees with the set of the fixed points of the generator (Lehner [1], p. 87). Thus L consists of a single point if the generator is parabolic and of two points if the generator is hyperbolic or loxodromic.

An example of an elementary Fuchsian group acting on the upper half-plane H is the real cyclic group

$$G = \{z \to a^n z \,|\, n = 0, \pm 1, \pm 2, \ldots\}, \qquad a > 1.$$

In this case $L = \{0, \infty\}$. The function

$$z \to e^{-2\pi i \log z / \log a} \tag{4.1}$$

is invariant under G. By studying the image of the fundamental domain $\{z \in H \,|\, 1 < |z| < a\}$ under (4.1) we deduce that the annulus

$$A = \{w \,|\, 1 < |w| < e^{2\pi^2/\log a}\}$$

is a model of H/G. In other words, G is the covering group of the upper half-plane H over the annulus A.

4.4. Kleinian Groups

Returning to an arbitrary group G of Möbius transformations, we denote by Ω the complement of the limit set L with respect to the plane. A point of Ω is called an ordinary point of G, and Ω is said to be the *set of discontinuity* of G.

The set Ω can be empty. A trivial example is the group of all Möbius transformations. But Ω can be empty even for a cyclic group: We pointed out in 4.3 that this is always the case if the group is generated by an elliptic transformation which is not of finite period.

Since L is closed, Ω is open. The set Ω need not be connected. From the invariance of L under G it follows that $g(\Omega) = \Omega$ for every $g \in G$.

If Ω is not empty, G is called a *Kleinian group*. A Kleinian group is countable (Lehner [1], p. 90). Fuchsian groups and elementary groups are of course special cases of Kleinian groups.

Let g be a Möbius transformation and $g(z) = (az + b)/(cz + d)$ its unimodular representation, i.e., $ad - bc = 1$. All 2×2-matrices with determinant 1 form a group $SL(2)$ under matrix multiplication. The mapping

$$g \to \begin{pmatrix} a & b \\ c & d \end{pmatrix} \tag{4.2}$$

is an isomorphism of the group M of all Möbius transformations onto the quotient group $SL(2)/\pm I$, where I is the identity matrix.

If the distance of the matrices (a_{ij}) and (b_{ij}) is defined by

$$\max\{|a_{ij} - b_{ij}| \, | \, i, j = 1, 2\},$$

then $SL(2)$ becomes a topological group. Via the mapping (4.2), the topological structure is transferred to M. A subgroup G of M is called *discrete* if its elements are isolated in the topology of M. It is not difficult to prove that G is discrete if and only if it does not contain infinitesimal transformations, i.e., if and only if there is no sequence of distinct elements $g_n \in G$, $n = 1, 2, \ldots$, such that $\lim g_n(z) = z$ for every z (Lehner [1], p. 96). From this characterization of discreteness we conclude that if G is not discrete, then every point of the plane is a limit point of G. It follows that *a Kleinian group is discrete*. The converse is not true.

The set of discontinuity Ω can be characterized by means of normal families. Let A be a domain of the plane and G a Kleinian group. *The family $\{g|A|g \in G\}$ is normal if and only if A is a subdomain of Ω* (Lehner [1], p. 98). In particular, if Ω is connected, then Ω is the largest domain in which the mappings $g \in G$ constitute a normal family.

4.5. Structure of the Limit Set

The normal family criterion for sets of discontinuity can be used to proving the following result, which reveals several properties of the limit set (Lehner [1], p. 103).

Lemma 4.1. *For a Kleinian group G, every point $\zeta \in L$ is the cluster point of each orbit $G(z)$, with the possible exception of $z = \zeta$ and one other point $z \in L$.*

We first deduce from this lemma that if G is not elementary, every point of L is the cluster point of other limit points. Hence, L is then always a perfect set. It follows that for Möbius groups there is a striking dichotomy: *Either the limit set contains at most two points or else it contains uncountably many points.*

A second conclusion from Lemma 4.1 is that *the limit set of a Kleinian group agrees with the boundary of the set of discontinuity.* For we have trivially $\partial\Omega = \bar{\Omega} \cap L$, so that $\partial\Omega \subset L$. On the other hand, we infer from Lemma 4.1 that $L \subset \bar{\Omega}$. Hence, $L \subset \bar{\Omega} \cap L = \partial\Omega$, and we obtain the desired result

$$L = \partial\Omega. \tag{4.3}$$

Third, Lemma 4.1 (or (4.3)) shows that *the limit set of a Kleinian group is nowhere dense in the plane.* For to every $\zeta \in L$ there is a point $z \in \Omega$ and mappings $g_n \in G$, such that $g_n(z) \to \zeta$. For Fuchsian groups the same reasoning gives the following result:

The limit set of a Fuchsian group acting on a disc D is either the whole boundary ∂D or a nowhere dense subset of ∂D.

If the limit set of a Fuchsian group G agrees with the boundary of the invariant disc, G is said to be *of the first kind* (or horocyclic). Otherwise, G is *of the second kind.* It follows that for groups of the first kind, Ω has two components, whereas Ω is connected if the group is of the second kind.

For a Kleinian group G, let F denote the set of the fixed points of its elements (other than the identity). If $g \in G$ and z is a fixed point of $g_0 \in G$, then $g(z)$ is a fixed point of $g \circ g_0 \circ g^{-1}$. Hence $g(F) = F$ for every $g \in G$.

For the group generated by $z \to z + 2$ and $z \to -1/z$, the set Ω is the union of the upper and lower half-planes while the point i belongs to F. This shows that F need not be a subset of L even though the group is infinite. But *if a Kleinian group does not contain elliptic transformations of finite period, then the closure of the set of its fixed points coincides with its limit set.*

In particular, if G is a Fuchsian group of the first kind acting on the upper half-plane, then the fixed points of G are everywhere dense on the real axis.

There are even sharper relations between F and L. Let F_h, F_l, and F_p denote the subsets of F consisting of the fixed points of the hyperbolic, loxodromic and parabolic elements of the Kleinian group G. Then

$$L = \bar{F}_h, \qquad L = \bar{F}_l, \qquad L = \bar{F}_p, \tag{4.4}$$

whenever G contains an element from the class in question. The relations (4.4) can be proved with the aid of Lemma 4.1 (cf. Lehner [1], p. 104).

Suppose that G is a Fuchsian group of the second kind acting on H. Then

$$S^* = (H \cup (\mathbb{R} \setminus L))/G$$

is a bordered Riemann surface with $(\mathbb{R} \setminus L)/G$ as its border (cf. the definition given in 1.3).

From this representation of a bordered Riemann surface S^* we see that S^* can be imbedded in a larger Riemann surface. In fact, interpreting G as acting on the plane, we form the quotient Ω/G. By Theorem 3.2, it is a Riemann surface, and clearly it contains S^*. It is called the *double* of $S = H/G$. The Riemann surface $(\Omega \setminus (H \cup \mathbb{R}))/G$ is called the *mirror image* of $S = H/G$, no matter whether G be of the first or of the second kind.

4.6. Invariant Domains

The components of the set of discontinuity Ω of a Kleinian group G are disjoint domains. A component which is mapped onto itself by every element of G is called an *invariant domain* of G.

A Kleinian group G with no fixed points in Ω is Fuchsian if it has a disc in Ω as an invariant domain, and it is said to be *quasi-Fuchsian* if it has a Jordan domain in Ω as an invariant component.

The following result makes it possible to analyze invariant domains.

Lemma 4.2. *Let G be a Kleinian group such that Ω has an invariant component A which is a Jordan domain different from a disc. Then ∂A does not have a tangent at a fixed point of a loxodromic element of G.*

PROOF. Assume that the tangent exists at a fixed point of a loxodromic element $g \in G$. We may suppose without loss of generality that the fixed point of g lies at $z = 0$, that the tangent at $z = 0$ is the real axis and that ∞ is the repulsive fixed point of g. Then $g(z) = re^{i\theta}z$, where $0 < r < 1$ and $0 < \theta < 2\pi$.

Suppose first that $\theta \neq \pi$, and set

$$a = \min(\theta/2, |\pi - \theta|/2, (2\pi - \theta)/2);$$

then $0 < a \leq \pi/4$. Consider the two angles $V_a = \{\rho e^{i\varphi} | \varphi \in (-a, a) \text{ or } \varphi \in (\pi - a, \pi + a), \rho \geq 0\}$. Since the real axis is a tangent, we have for every $a > 0$ a disc D_a centered at the origin, such that

$$\partial A \cap D_a \subset V_a \cap D_a. \tag{4.5}$$

Now choose $z \in \partial A \cap D_a \cap V_a$, $z \neq 0$. Then $g(z) \in \partial A \cap D_a$. On the other hand, it follows from the definition of a that $g(z) \notin V_a$. This contradicts (4.5).

If $z \to g(z) = -rz$ belongs to G, then $g \circ g$ is a hyperbolic transformation with the same fixed points as g. A modification of the above proof shows that ∂A does not have a tangent at a fixed point of a hyperbolic element of G. This proves the lemma. $\qquad \square$

Combined with our previous results on Kleinian groups, Lemma 4.2 yields the following result.

Theorem 4.2. *The boundary of an invariant component of a quasi-Fuchsian group is either a circle or a Jordan curve which fails to have a tangent on an everywhere dense set.*

PROOF. First, if A denotes an invariant component, we clearly have $\partial A \subset \partial \Omega$. From (4.3) we then conclude that $\partial A \subset L$. If the group is not Fuchsian, it always contains loxodromic elements (Lehner [1], p. 107). By (4.4), we have in this case $\partial A \subset \bar{F}_l$. Hence, the theorem follows from Lemma 4.2.   □

It was Klein who first noticed that such weird invariant Jordan domains exist. He obtained such domains by direct construction. For details, interesting pictures and almost philosophical comments on this unexpected phenomenon we refer to Fricke–Klein [1]. Here we shall only briefly explain Klein's method. In V.3.4 we shall arrive at invariant quasidiscs in a completely different manner.

Let $\{D_i | i = 1, 2, \ldots\}$ be a closed chain of discs, i.e., the discs D_i are disjoint but for every i, the closure of the union $\bigcup D_j, j \neq i$, is connected. We form the group G whose elements are compositions of an even number of reflections in the circles ∂D_i. The elements of G are Möbius transformations. We say that G is generated by the chain $\{D_i\}$.

If the number of the discs in the chain is one, G is trivial, if it is two, G is cyclic, and if it is three, G is Fuchsian. In case the complement of the closure of the union $\bigcup D_i$ of all discs is not empty, it is easy to see that this complement is contained in the set of discontinuity Ω of G. In other words, G is then always a Kleinian group.

The importance of Klein's method derives from the fact that under very general conditions, Ω has two invariant components which are complementary Jordan domains. This is the case, for instance, if the chain has only a finite number (> 2) of discs. The boundary of the invariant domain passes through the points at which the closed discs touch each other and through their reflected images (Fig. 9).

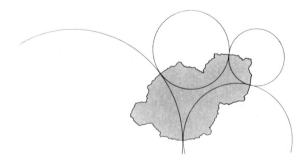

Figure 9. Klein's method of generating invariant domains.

5. Compact Riemann Surfaces

5.1. Covering Groups over Compact Surfaces

Let G be a Fuchsian group acting on the upper half-plane H. We recall that all Dirichlet regions of G are congruent in the non-euclidean geometry induced by the hyperbolic metric on H. The Dirichlet regions clearly have a bounded hyperbolic diameter if and only if their closures lie in H. This property characterizes compact Riemann surfaces.

Theorem 5.1. *Let S be a Riemann surface and G the covering group of the upper half-plane H over S. Then S is compact if and only if the Dirichlet regions of G are bounded in the hyperbolic metric of H.*

PROOF. Suppose first that S is compact. Let N be a Dirichlet region with center a. We consider the hyperbolic discs $D_n = \{z | h(z, a) < n\}$, $n = 1, 2, \ldots$. Their projections on $S = H/G$ form an open covering of S. Since S is compact, there is an n such that the projection of D_n alone covers S. In other words, for every $z \in H$ there exists a mapping $g \in G$ for which $h(g(z), a) < n$. Now if $z \in N$, then $h(z, a) \leq h(g(z), a)$ for every $g \in G$. It follows that $N \subset D_n$.

Assume, conversely, that the closure of a Dirichlet region of G lies in H. Then S is the image of a compact set under a continuous mapping and hence compact. □

Theorem 5.1 admits interesting conclusions.

Theorem 5.2. *The covering group of the upper half-plane over a compact Riemann surface is finitely generated and of the first kind.*

PROOF. Let S be a compact Riemann surface and G the covering group of H over S. The vertices of a Dirichlet region of G cannot have a limit point in H. Hence, by Theorem 5.1, a Dirichlet region for G has a finite number of sides. We conclude using Theorem 4.1 that G is finitely generated.

In order to determine the limit set L of G, we consider an arbitrary point x of the real axis and set $U = \{\zeta \in H | |\zeta - x| < r\}$. The hyperbolic distance from the point $x + iy \in U$ to the semicircle $|\zeta - x| = r$ tends to ∞ as $y \to 0$. On the other hand, by Theorem 5.1 the Dirichlet region containing $x + iy$ has a uniformly bounded hyperbolic diameter for every $y > 0$. It follows that U contains a Dirichlet region for every $r > 0$. Consequently, $x \in L$, and so L is the whole real axis. □

5.2. Genus of a Compact Surface

Let S be a compact Riemann surface and G the covering group of H over S. A Dirichlet region N for G is then a non-euclidean polygon with finitely many sides, which are pairwise equivalent under G. Now it is possible to transform N to another non-euclidean polygon which is also a fundamental domain of G and whose sides follow each other according to the pattern

$$a_1 b_1 a_1' b_1' \ldots a_p b_p a_p' b_p', \tag{5.1}$$

the sides a_i and b_i being equivalent to a_i' and b_i', respectively. The number p is at least 2. Such a polygon is called a *normal polygon* for G. The transformation of a Dirichlet region to a normal polygon is described in Nevanlinna [1], pp. 229–230.

The pattern (5.1), which generalizes the pattern $a_1 b_1 a_1' b_1'$ of a torus (see 4.1), can be regarded as a representation of the compact Riemann surface S. But it is in fact of a more general character. Let S be an arbitrary compact orientable surface. A fixed (necessarily finite) triangulation of S leads in a natural manner to a polygon representing S, in which identified sides either have the simple pattern $a_1 a_1'$ or else the pattern (5.1), with $p = 1, 2, \ldots$ (Springer [1], p. 117). Here p does not depend on the triangulation by way of which we arrived at it. The pattern $a_1 a_1'$ occurs if and only if S is a topological sphere.

The number p in (5.1) is called the *genus* of the compact surface S. A sphere is said to have the genus $p = 0$. It follows that a torus has the genus $p = 1$ and every compact Riemann surface which has the upper half-plane as a universal covering surface has a genus $p > 1$.

The genus characterizes the topology: *Two orientable compact surfaces are homeomorphic if and only if they have the same genus.* In analogy with the case of a torus, the topological type of a surface can be read from the pattern (5.1): *A compact orientable surface of genus p is a topological sphere with p handles.*

These two results are proved in Springer [1], which contains a detailed discussion of the relations of genus to such topological invariants as the fundamental group, homology groups, and the Euler characteristic. We shall study here certain analytic implications of genus.

5.3. Function Theory on Compact Riemann Surfaces

Let f be a non-constant meromorphic function on a compact Riemann surface S of genus p. If $p = 0$, the surface S can be identified with the extended plane, and f is a rational function. In particular, f takes on every complex value and ∞, each of them the same number of times, provided of course that multiple values are counted according to their multiplicities.

If $p = 1$, the study of f amounts to function theory on a torus. (Its elements are described in Springer [1], p. 34.) The lift of f to the universal covering surface $\mathbb{C}$ is double-periodic, i.e., an elliptic function.

Regardless of p, the property of rational functions remains true as indicated above (Springer [1], p. 176):

Theorem 5.3. *On a compact Riemann surface, a non-constant meromorphic function assumes every value the same finite number of times.*

For topological reasons, injective meromorphic functions can exist only if $p = 0$. The theory of meromorphic functions on compact Riemann surfaces is classical analysis, which is intimately connected with the theory of algebraic functions (see, e.g., Springer [1], p. 286). In a way, the theory was born before the notion of a Riemann surface existed in any form.

A striking example of the interaction between topology and analysis on compact Riemann surfaces is provided by the complex vector space consisting of holomorphic Abelian differentials. The study of the periods of regular harmonic differentials on a homology basis yields the following result (Springer [1], p. 252):

On a compact Riemann surface of genus p, the dimension of the space of holomorphic Abelian differentials is equal to p.

We shall use this result in determining the dimension of the complex vector space formed by holomorphic quadratic differentials. This linear space will play an important role in the theory of Teichmüller spaces.

5.4. Divisors on Compact Surfaces

Let S be a compact Riemann surface. A *divisor* D on S is a mapping of S whose values are integers and which is non-zero only at finitely many points of S. Addition of two divisors D_1 and D_2 is defined by $(D_1 + D_2)(p) = D_1(p) + D_2(p)$. The *degree* of D, $\deg D$, is the sum of its values. We write $D_1 \geq D_2$ if $D_1(p) \geq D_2(p)$ for every point $p \in S$.

Let φ be a holomorphic differential of an arbitrary type on S. Fix a point $p \in S$ and consider two local parameters z_1 and z_2 in a neighborhood of p, both mapping p to zero. The mapping $z_1 \to z_2$ is conformal at the origin and has, therefore, a non-zero derivative at 0. We conclude from formula (1.2) in 1.4 that if the representation of φ in z_1 has a zero of order n (≥ 0) at the origin, then the representation of φ in z_2 also has a zero of order n at the origin. Thus the zeros of φ and their orders are well defined, being independent of local parameters. Similarly, we infer that the poles of a meromorphic differential and their orders can be defined in an invariant manner.

Let φ be a meromorphic differential on S with the zeros of order m_i at the points p_i and the poles of order n_j at the points q_j. The divisor D_φ of the differential φ is the function which takes the value m_i at p_i, the value $-n_j$ at q_j, and vanishes elsewhere. Hence, $\deg D_\varphi = \sum m_i - \sum n_j$.

Suppose, in particular, that φ is a function on S. By Theorem 5.3, we then have $\deg D_\varphi = 0$. If φ_1 and φ_2 are differentials of the same type, the quotient φ_1/φ_2 is a function. From the definition of the divisor of a differential we see that $\deg D_{\varphi_1/\varphi_2} = \deg D_{\varphi_1} - \deg D_{\varphi_2}$. It follows that *the divisors of differentials of the same type all have the same degree.*

Given a divisor D, consider the family which consists of all meromorphic functions f with $D_f \geq D$, together with the function which is identically zero. This family is a complex linear space. Its dimension is called the *dimension of the divisor D* and denoted by $\dim D$.

If $D = 0$, i.e., $D(p) = 0$ for every $p \in S$, the space consists of the constants, and so

$$\dim 0 = 1. \tag{5.2}$$

We also conclude that

$$\dim D = 0 \quad \text{if} \quad \deg D > 0, \tag{5.3}$$

because the space then contains only the zero function.

5.5. Riemann–Roch Theorem

We denote here by Q the complex vector space of all holomorphic quadratic differentials on S. We fix a non-zero $\psi \in Q$ and write $D_\psi = D_2$. If φ is an arbitrary meromorphic quadratic differential, $f = \varphi/\psi$ is a meromorphic function. From $D_\varphi = D_f + D_2$ we see that φ is holomorphic if and only if $D_f \geq -D_2$. It follows that

$$\dim Q = \dim(-D_2). \tag{5.4}$$

Exactly the same reasoning can be applied to $(1, 0)$-differentials. If D_1 is the divisor of a holomorphic Abelian differential, we conclude that the space of these differentials has the dimension $\dim(-D_1)$. On the other hand, by what was said in 5.3, this dimension is equal to the genus p of S. Hence

$$\dim(-D_1) = p. \tag{5.5}$$

The results (5.4) and (5.5) can be put together with the help of a classical result on compact surfaces (Springer [1], p. 264).

Theorem 5.4 (Riemann–Roch Theorem). *On a compact Riemann surface of genus p, every divisor D satisfies the equation*

$$\dim D = \dim(-D - D_1) - \deg D - p + 1. \tag{5.6}$$

Let us first apply (5.6) for $D = -D_1$. Then, by (5.5) and (5.2), $p = 1 + \deg D_1 - p + 1$, so that $\deg D_1 = 2p - 2$. By our previous remark, we have

$$\deg D_{\varphi_1} = 2p - 2$$

for every meromorphic $(1, 0)$-differential φ_1.

Now let φ_2 be a meromorphic quadratic differential. Then φ_2/φ_1 is a $(1, 0)$-differential. From $\deg D_{\varphi_1} + \deg D_{\varphi_2/\varphi_1} = \deg D_{\varphi_2}$ it thus follows that

$$\deg D_{\varphi_2} = 4p - 4. \tag{5.7}$$

In particular, *every non-zero holomorphic quadratic differential on a Riemann surface of genus p has $4p - 4$ zeros.*

The dimension of Q can now be readily determined.

Theorem 5.5. *On a compact Riemann surface of genus p, the space of holomorphic quadratic differentials has dimension 1 if $p = 1$ and $3p - 3$ if $p > 1$.*

PROOF. In the case $p = 1$, the Riemann–Roch theorem is not needed to determine the dimension of Q. We saw in 4.1 that cover transformations are translations $z \to z + m\omega_1 + n\omega_2$, m, $n \in \mathbb{Z}$. Formula (3.5) shows, therefore, that φ is a holomorphic quadratic differential for the covering group G if and only if $\varphi(z + m\omega_1 + n\omega_2) = \varphi(z)$ for all m and n. It follows that $\varphi \in Q$ is a bounded holomorphic function in the complex plane and hence a constant. Conversely, every constant is a quadratic differential for G. We see that $\dim Q = 1$.

Next suppose that $p > 1$. We fix a holomorphic quadratic differential and denote its divisor by D_2. After this, we choose $D = -D_2$ in (5.6). Then, by (5.4) and (5.7),

$$\dim Q = \dim(D_2 - D_1) + 3p - 3. \tag{5.8}$$

Now $\deg(D_2 - D_1) = \deg D_2 - \deg D_1 = 2p - 2 > 0$. Hence the desired result $\dim Q = 3p - 3$ follows from (5.8) and (5.3). $\qquad\square$

If $p = 0$, the space Q reduces to zero.

6. Trajectories of Quadratic Differentials

6.1. Natural Parameters

Let φ be a holomorphic quadratic differential on a Riemann surface S. We assume that φ is not identically zero, and regard φ as fixed throughout this section. A point $p \in S$ is said to be *regular* if $\varphi(p) \neq 0$, and *critical* if $\varphi(p) = 0$. We showed in 5.4 that these are invariant definitions, independent of the representation of φ. Critical points form a discrete set, and on a compact Riemann surface there are only finitely many of them.

Let p be a regular point and $q \to h(q) = z$ a local parameter in a neighborhood of p mapping p to the origin. Since $\varphi(0) \neq 0$, there is a simply connected domain containing the origin in which the two branches of $z \to \sqrt{\varphi(z)}$ are single-valued. For a fixed branch of $\sqrt{\varphi}$, every integral function

$$z \to \Phi(z) = \int \sqrt{\varphi(z)}\, dz$$

is then also single-valued in this neighborhood of the origin and uniquely determined up to an additive constant. From the invariance of $\sqrt{\varphi(z)}\, dz$ under changes of parameter it follows that every Φ is a function on S near p.

From $\Phi'(0) = \sqrt{\varphi(0)} \neq 0$ we conclude that there is a disc around the origin which $z \to \Phi(z)$ maps injectively into the complex plane. It follows that $q \to w = \Phi(z)$ is a local parameter near p. From

$$dw^2 = \varphi(z)\, dz^2$$

we see that with respect to w, the function representing the quadratic differential φ is the function which is identically equal to 1.

We call $w = \Phi(z)$ a *natural parameter* at p. An arbitrary natural parameter at p is of the form $\pm w + \text{constant}$. We see that in each case, *near a regular point the local representation of φ in terms of a natural parameter is the constant function* 1.

There are natural parameters at a critical point also. Suppose that $p \in S$ is a zero of order n of φ. Again, let $q \to h(q) = z$ be a local parameter near p which maps p to the origin. Then there is a disc $D(0, r)$ around the origin in which $\varphi(z) = z^n \psi(z)$ with $\psi(z) \neq 0$. We fix a single-valued branch of $\sqrt{\psi}$ in $D(0, r)$. If n is odd, we cut $D(0, r)$ along its positive radius $I = \{x \mid 0 \leq x < r\}$, and fix a branch of $z \to z^{n/2}$ in $D(0, r) \setminus I$; if n is even, no such cut is needed. In either case

$$z \to \Phi(z) = \int_0^z \sqrt{\varphi(z)}\, dz = z^{(n+2)/2}(c_0 + c_1 z + \cdots), \qquad c_0 \neq 0,$$

is single-valued in $D(0, r) \setminus I$. Moreover, $z \to \omega(z) = \Phi(z) z^{-(n+2)/2}$ is single-valued and $\neq 0$ in a disc $D(0, r_1) \subset D(0, r)$. (Note that the cut I is no longer needed in the definition of ω.)

In $D(0, r_1)$, the function

$$z \to \Phi(z)^{2/(n+2)} = z \omega(z)^{2/(n+2)}$$

is single-valued. Since it has the non-zero derivative $\omega(0)^{2/(n+2)}$ at the origin, it is injective in a disc $D(0, r_2) \subset D(0, r_1)$. It follows that

$$q \to \zeta = \Phi(z)^{2/(n+2)}$$

is a local parameter near p. We now call ζ a natural parameter at p.

From $\varphi = (\Phi')^2$ we obtain

$$\varphi(z)\,dz^2 = \left(\frac{n+2}{2}\right)^2 \zeta^n\,d\zeta^2. \tag{6.1}$$

In other words, *near a critical point of order n, a holomorphic quadratic differential has the representation $\zeta \to (1 + n/2)^2\zeta^n$ in terms of the natural parameter ζ.*

The idea of associating natural parameters with a quadratic differential is due to Teichmüller [1].

6.2. Straight Lines and Trajectories

A continuously differentiable mapping γ of an open interval I into the Riemann surface S with a non-zero derivative on I is called a regular path on S. Near a point $p \in \gamma(t)$ we introduce a local parameter $q \to h(q) = z$ and write, with a slight abuse of notation, $z(t) = (h \circ \gamma)(t)$. We assume that γ does not pass through any critical point of φ. The function $t \to \arg(\varphi(z(t))z'(t)^2)$ is then well defined on I, of course modulo 2π.

If

$$\arg(\varphi(z(t))z'(t)^2) = \theta = \text{constant} \tag{6.2}$$

at every point $t \in I$, we say that γ is a *straight line* (in the geometry induced by the quadratic differential φ). The condition for γ to be a straight line is often expressed in the form

$$\arg(\varphi(z)\,dz^2) = \text{constant}$$

along γ. It follows from the definition that a straight line does not pass through a zero of φ.

We say that a straight line is horizontal if $\theta = 0$, and vertical if $\theta = \pi$. The straight line (6.2) is a horizontal line for the quadratic differential $e^{-i\theta}\varphi$. A straight line (6.2) is called maximal if it is not properly contained in any regular curve on which (6.2) is true. A *horizontal trajectory* is a maximal horizontal straight line. Similarly, a vertical trajectory is a maximal vertical straight line.

In natural parameters, the trajectories have simple representations. Let w be a natural parameter near a point of γ. From $dw^2 = \varphi(z)\,dz^2$ and from (6.2) it follows that locally, a horizontal straight line is a euclidean horizontal line segment in the w-plane. Similarly, a vertical straight line is locally a euclidean vertical segment.

Near a critical point $p \in S$ the behavior of trajectories is more complicated. Let $\zeta = w^{2/(n+2)}$, $w = \Phi(z)$, be a natural parameter in a neighborhood of p. Horizontal lines near p are horizontal line segments in the w-plane, whereas in the ζ-plane, they are located in $n + 2$ different sectoral domains. Consider, in particular, a horizontal line segment in the w-plane which contains the

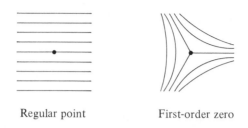

Regular point   First-order zero

Figure 10. Horizontal trajectories under natural parameters.

origin. In view of the relation $\zeta = w^{2/(n+2)}$, we conclude that its preimage on S consists of $n + 2$ rays emanating from the critical point p such that the angle between two adjacent rays is equal to $2\pi/(n + 2)$ (Fig. 10).

A trajectory with an endpoint which is a zero of φ is called critical. The number of critical horizontal trajectories is countable, and on a compact Riemann surface there are only finitely many such trajectories.

From the definition it is clear that given a regular point of S, there exists exactly one horizontal trajectory passing through that point.

6.3. Orientation of Trajectories

A sufficiently small subarc of a horizontal trajectory is mapped by a natural parameter onto a segment of the real axis. The orientation of the real axis can thus be transferred locally to horizontal trajectories.

The global situation is more complicated. Let $S_0 = S \setminus \{\text{zeros of } \varphi\}$. For any two natural parameters w_1 and w_2 of S_0 for which $w_2 \circ w_1^{-1}$ is defined, we have $w_2 \circ w_1^{-1}(z) = \pm z + \text{constant}$. The trajectory structure of φ is said to be *orientable* if S_0 has an atlas of natural parameters w_i, such that every change of variables is of the form

$$w_j \circ w_i^{-1}(z) = z + \text{constant}.$$

More briefly, we then say that φ is orientable.

Let us assume that φ is not orientable. We shall show that S then has a two-sheeted covering surface $\tilde{S}$, branched over the zeros of φ of odd order, such that the lift of φ to $\tilde{S}$ is orientable.

In order to prove this, we consider for a moment φ on the punctured surface S_0. Suppose that φ is a collection of holomorphic functions φ_i, $i = 1$, $2, \ldots$, defined in simply connected domains $U_i \subset S_0$ with local parameters z_i. Then $\varphi_i \, dz_i^2 = \varphi_j \, dz_j^2$ on $U_i \cap U_j$.

Consider all triples (p, z_i, α_i), where $p \in U_i$ and α_i is holomorphic and satisfies $\alpha_i^2 = \varphi_i$ on U_i. We identify (p, z_i, α_i) with (q, z_j, α_j) if $p = q$ and $\alpha_i \, dz_i = \alpha_j \, dz_j$ at $p = q$. Let $\tilde{S}_0$ denote the set of identified triples and $\pi: \tilde{S}_0 \to S_0$ the projection which maps (p, z_i, α_i) to p.

We introduce the standard topology on $\tilde{S}_0$. It follows that $(\tilde{S}_0, \pi)$ is an unlimited covering surface of S_0. It becomes a Riemann surface with the conformal structure lifted from S_0.

Let $\tilde{S}$ be the natural extension of $\tilde{S}_0$ to the holes over the points of $S \setminus S_0$. More precisely, if the lift of a closed path on S_0 around a zero p of φ terminates at the initial point, we have two points of $\tilde{S}$ over p (as we have over all points of S_0). Otherwise $\tilde{S}$ has a branch point of order 2 over p. The latter alternative occurs if and only if p is a zero of odd order. If $\pi: \tilde{S} \to S$ is the natural extension of $\pi: \tilde{S}_0 \to S_0$, then $(\tilde{S}, \pi)$ is a two-sheeted branched covering surface of S.

Finally, let $\tilde{\varphi}$ and $\tilde{\alpha}_i$ denote the lifts of φ and α_i to $\tilde{S}$. From the construction it is clear that the functions $\tilde{\alpha}_i$ form a holomorphic Abelian differential $\tilde{\alpha}$ and that $\tilde{\varphi} = \tilde{\alpha}^2$. As the square of an Abelian differential $\tilde{\varphi}$ is orientable, as we wished to show.

6.4. Trajectories in the Large

In order to study the global structure of trajectories, we choose a regular point $p_0 \in S$, fix a branch of Φ_0 in a neighborhood of p_0, and normalize it so that $\Phi_0(p_0) = 0$. Let $r \leq \infty$ be the largest number, such that the analytic continuation f_0 of the local inverse of Φ_0 maps the disc $D_0 = D(0, r)$ injectively into S. The image $V_0 = f_0(D_0)$ is called a maximal disc around p_0, and $r = r(p_0)$ is said to be its radius. The maximal disc V_0 is uniquely determined by φ, i.e., V_0 does not depend on the choice of the integral function of $\sqrt{\varphi}$. The function $p \to r(p)$ is continuous in p.

Next we choose a point $u_1 \in D_0$ which lies on the real axis $\mathbb{R}$; then $f_0(u_1) = p_1$ is a regular point. Hence, there is a conformal mapping f_1 of $D_1 = D(u_1, r(p_1))$ onto the maximal disc V_1 around p_1, such that $f_1 = f_0$ in $D_0 \cap D_1$ and that f_1 is the inverse of Φ_1. By continuing this procedure, we obtain connected chains of discs $D_0, D_1, \ldots, D_n$ with centers on $\mathbb{R}$, such that $f_i = (\Phi_i | V_i)^{-1}$ is a conformal mapping of D_i into S and that f_{i+1} is a direct analytic continuation of f_i.

Let G be the union of all the discs of such chains which we obtain by starting from D_0. Since the intersection of two chains is connected and contains D_0, the analytic continuation f of f_0 to G is single-valued. We write $I = G \cap \mathbb{R}$ and deduce that $f(I)$ is the horizontal trajectory which passes through the point p_0.

If $[a, b]$ is a closed subinterval of I on which f is injective, there is a rectangle $\{u + iv \mid a \leq u \leq b, -\delta \leq v \leq \delta\}$ which f maps injectively into S. The image of every horizontal line segment in this rectangle is a subarc of some horizontal trajectory.

We shall now show that the character of the horizontal trajectory $f(I)$ varies, depending on whether f is injective or not on the whole interval I.

6.5. Periodic Trajectories

Suppose first that there are points a, $b \in I$ such that $f(a) = f(b)$. Among the pairs of points of $[a, b]$ at which f takes the same value, there is a pair u_0, u_1 with a minimal distance from each other. It is not difficult to show that f is then periodic in G, with the primitive period $\omega = u_1 - u_0$ (cf. Strebel [6], p. 39). In this case f can be continued analytically to the whole real axis $\mathbb{R}$ by periodicity. The horizontal trajectory $\alpha = f(\mathbb{R}) = f([u_0, u_1]) = f([0, \omega])$ is a closed curve. There is a maximal rectangle $R_0 = \{u + iv | 0 \le u \le \omega, v_1 \le v \le v_2\}$ in whose interior f is analytic.

If f is injective in R_0, then

$$\zeta \to f\left(\frac{\omega}{2\pi i} \log \zeta\right)$$

maps the annulus $A = \{\zeta | e^{-2\pi v_2/\omega} < |\zeta| < e^{-2\pi v_1/\omega}\}$ conformally onto a ring domain on S. The image is called the maximal annulus around the horizontal trajectory α. Every circle $|\zeta| = $ constant in A maps onto a closed horizontal trajectory of S. It follows that the maximal annulus around α is swept out by closed horizontal trajectories, freely homotopic to α and all of the same length ω in the w-plane. From the maximality it is clear that if α_1 and α_2 are closed horizontal trajectories of S, their maximal annuli are either disjoint or identical.

If f is not injective in the rectangle R_0, simple reasoning shows that f has another primitive period $\omega' \notin \mathbb{R}$ and that f has an analytic extension throughout the complex plane $\mathbb{C}$ (cf. Strebel [6], p. 41). The parallelogram with vertices at the points 0, ω, $\omega + \omega'$ and ω' and with the opposite sides identified is mapped by f bijectively onto S. We conclude that S is a torus. The pair $(\mathbb{C}, f)$ is a universal covering surface of S. From the global representation $dw^2 = \varphi(z) dz^2$ it follows that the straight lines on S are images under f of euclidean straight lines in the plane. All horizontal trajectories are closed curves on S.

6.6. Non-Periodic Trajectories

Let us now assume that the function f, obtained by analytic continuation of a germ of Φ^{-1}, is injective on the interval $I = G \cap \mathbb{R} = (u_{-\infty}, u_\infty)$. Then $f: I \to S$ is a parametric representation of the horizontal trajectory α passing through the point $p_0 \in S$ with which we started. The trajectory is now an open arc, and we define its length to be the same as the length of I. (The metric induced by φ will be studied in the next section 7.) The two parts into which p_0 divides α are called trajectory rays from p_0. We denote $\alpha^+ = f([0, u_\infty))$, $\alpha^- = f((u_{-\infty}, 0])$.

Let L be the limit set of the ray α^+, i.e., L is the set of points $p \in S$ for which there exists a sequence of points $u_n \in I$ tending to u_∞ such that $f(u_n) \to p$. The limit set L is contained in the closure of α, and it does not depend on the choice of the initial point p_0 of α.

Regarding the set L, there are three essentially different possibilities. First, L may be empty. We then say that α^+ tends to the boundary of S, and call α^+ a boundary ray. If both α^+ and α^- are boundary rays, the trajectory α is said to be a *cross-cut* of S. On a compact surface, there are no boundary rays.

Second, L might consist of a single point p. The point p cannot be regular, because f could then be continued on $\mathbb{R}$ past u_∞. Hence, $f(u)$ tends towards the critical point p along α^+ as $u \to u_\infty$ on I. In this case, α^+ is called a critical ray.

The remaining case is that L contains more than one point. Suppose $p \in L$ is a regular point, and consider the horizontal trajectory α_p through p. This cannot be closed, because it could then be covered by an open annulus which contains only closed trajectories. Choose another point $q \in \alpha_p$, and construct an open rectangle R_0 such that $f_p | R_0$ is injective and that the image of the middle line of R_0 contains the trajectory arc from p to q. Since $p \in L$, we have points $p_n \in \alpha$, such that $p_n \to p$. If $f_p^{-1}(p_n) \in R_0$, the horizontal line segment in R_0 through $f_p^{-1}(p_n)$ maps on a subarc of α. There are infinitely many such subarcs, and we conclude that α has infinite length.

The same reasoning shows that if the subarc of α_p from p to q has length a and if $p_n = f(u_n) \to p$, then $q_n = f(u_n \pm a) \in \alpha$ with properly chosen signs converge to q. Here $q \in \alpha_p$ was arbitrarily chosen. We conclude that if $p \in L$, then the whole trajectory through p belongs to L.

The trajectory α has infinite length also in the case when $p \in L$ is a critical point. The ray α^+ cannot end at p, because L would then reduce to the single point p. Therefore, at least one of the finitely many sectors into which the horizontal trajectory rays emanating from p divide a neighborhood of p contains infinitely many points $p_n \in \alpha$. After this, the reasoning used in the case where p was a regular point can be modified so as to yield the desired result (Strebel [6], p. 44).

Consequently, if the limit set L of the ray α^+ contains more than one point, then α^+ is always of infinite length. The ray α^+ is then said to be divergent.

Suppose that the initial point p_0 of α^+ belongs to the set L. From what we just proved it follows that the whole trajectory α is then contained in L. Since L lies in the closure $\bar{\alpha}$ of α, we conclude that in this case

$$L = \bar{\alpha}.$$

A trajectory ray α^+ with $p_0 \in L$ is called recurrent. A trajectory both rays of which are recurrent is said to be a *spiral*, and its limit set L is called a spiral set (Fig. 11).

On a compact Riemann surface, every divergent ray is recurrent and all non-periodic trajectories are spirals, save for finitely many exceptions. (For two simple examples, see Figs. 11 and 12. For a detailed account of trajectories on compact surfaces, we refer to Strebel [6], § 11.)

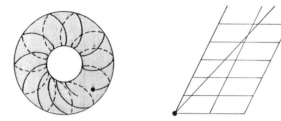

Figure 11. A torus and a part of its universal covering surface. A straight line projects on a spiral.

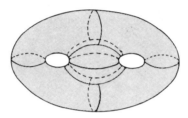

Figure 12. Horizontal trajectories on a surface of genus 2. $4p - 4 = 4$ critical points.

7. Geodesics of Quadratic Differentials

7.1. Definition of the Induced Metric

As in the previous section, φ is a holomorphic quadratic differential on a Riemann surface S, and not identically zero. The invariant differential

$$|\varphi(z)|^{1/2}|dz| \tag{7.1}$$

is called the line element of the metric induced by φ.

Let γ be a curve on S locally rectifiable with respect to the euclidean metric in any parametric plane. The length of γ in the metric induced by φ can be obtained with the aid of the following geometric reasoning. First, if γ lies in a maximal disc around a point p, the length of γ is equal to the euclidean length of the image of γ under a natural parameter defined in a neighborhood of p. An arbitrary γ not passing through any critical points can be subdivided into parts each one lying in a maximal disc. The length $l(\gamma)$ of γ is the sum of the lengths of these parts; it is independent of the subdivision of γ.

Since the differential (7.1) is an invariant on S, we can also define directly

$$l(\gamma) = \int_\gamma |\varphi(z)|^{1/2}|dz|,$$

in terms of arbitrary local parameters. The length $l(\gamma)$ is then well defined also in the case where γ passes through critical points.

The invariant $|\varphi(z)| \, dx \, dy$ is called the area element of the φ-metric. Hence, the total area of S is the L^1-norm of φ.

In the following, φ is fixed, and notions like distance, length, and area, refer to the φ-metric, unless otherwise stated.

The area of a compact surface S is always finite. If the genus of S is 1, i.e., if S is a torus, its universal covering surface is the complex plane. The lifted φ is then bounded in the plane and hence constant. It follows that the φ-metric is euclidean, a result at which we arrived in a different manner in 6.5.

7.2. Locally Shortest Curves

The existence of a unique shortest curve joining two given points of S can be established without difficulty if the points are close to each other. The curve itself can be described geometrically by use of natural parameters.

Theorem 7.1. *Every point of a Riemann surface has a neighborhood in which any two points can be joined by a unique shortest curve.*

PROOF. Let a point $p \in S$ be given and suppose first that p is regular. Let V be the maximal disc around p and $\{w \,|\, |w| < r\}$ its image under a natural parameter $w = \Phi(z)$. Let $V_0 \subset V$ be the preimage of $|w| < r/2$, and p_1, p_2 arbitrary points of V_0. Then the preimage γ_0 of the line segment connecting $\Phi(p_1)$ and $\Phi(p_2)$ is the unique shortest curve which joins p_1 and p_2 on S. For let $\gamma \,(\neq \gamma_0)$ be an arbitrary curve on S which joins p_1 and p_2. If γ stays in V, then clearly $l(\gamma_0) < l(\gamma)$. If γ leaves V, then $l(\gamma) \geq r > l(\gamma_0)$.

Suppose next that p is a zero of φ of order n. We proved in 6.1 that if ζ is a natural parameter near p, then

$$\varphi(\zeta) = \left(\frac{n+2}{2}\right)^2 \zeta^n, \qquad w = \Phi(\zeta) = \zeta^{(n+2)/2}, \tag{7.2}$$

in a disc $|\zeta| < r$. Let V_0 now be the preimage of the disc $|\zeta| < 2^{-2/(n+2)}r$ on S. Then any two points p_1 and p_2 in V_0 can be connected by a unique shortest curve. This is either a straight line segment in the w-plane, or it is composed of two radii in the ζ-plane which emanate from the origin. The former case occurs if and only if $|\arg \zeta_1 - \arg \zeta_2| < 2\pi/(n+2)$, where ζ_1 and ζ_2 are the ζ-images of p_1 and p_2. These conclusions can be drawn from (7.2); for the details we refer to Strebel [6], p. 35.   □

It follows from the above that if the shortest curve is the union of two radii, both angles θ between these rays satisfy the inequality

$$\theta \geq \frac{2\pi}{n+2}. \tag{7.3}$$

This "angle condition" will be utilized in the study of globally shortest curves.

7.3. Geodesic Polygons

A curve γ on S is called a *geodesic* if it is locally shortest, i.e., every point $p \in \gamma$ has a neighborhood V on S such that for any two points $p_1, p_2 \in \gamma \cap V$, no arc joining p_1 and p_2 in V is shorter than the subarc of γ from p_1 to p_2. From the results of 7.2, we know the local structure of a geodesic. It will turn out that a geodesic is also globally the unique shortest curve in its homotopy class.

A geodesic polygon is a curve which consists of open intervals of straight lines (in the geometry of φ) and of their endpoints. The endpoints can be zeros of φ. In case the polygon is a Jordan curve, it is called simple and closed.

Assuming the existence of a geodesic, we shall first prove that it is uniquely extremal in its homotopy class. The proof uses the argument principle in its generalized form, in which the holomorphic function considered in a subdomain of the complex plane is allowed to have zeros on the boundary of that domain.

Argument Principle. *Let f be holomorphic in the closure of a plane domain A bounded by finitely many piecewise regular curves. Let γ_j denote the arcs into which the zeros $z_j \in \partial A$ of f divide the boundary, and θ_j the interior angle at z_j between the arcs γ_{j-1} and γ_j. Then*

$$\int_{\partial A} d \arg f(z) = \sum \int_{\gamma_j} d \arg f(z) = 2\pi \sum m_i + \sum \theta_j n_j,$$

where m_i are the orders of the zeros of f in A and n_j the orders of those on ∂A.

If $f(z) \neq 0$ on ∂A, this is the standard principle of argument. The refinement says that the zeros z_j on the boundary have the weights $\theta_j/2\pi$.

Teichmüller ([1], p. 162) drew the following conclusion from the Argument principle.

Lemma 7.1 (Teichmüller's Lemma). *Let φ be holomorphic in the closure of a domain A in the complex plane which is bounded by a simple closed polygon in the φ-metric, whose sides γ_j form the angles θ_j at the vertices. If m_i and n_j denote the orders of the zeros of φ in A and on ∂A, respectively, then*

$$\sum \left(1 - (n_j + 2)\frac{\theta_j}{2\pi} \right) = 2 + \sum m_i. \tag{7.4}$$

PROOF. On γ_j we have $\arg(\varphi(z)\,dz^2) = $ constant, and so

$$d\arg\varphi(z) + 2d(\arg dz) = 0. \tag{7.5}$$

The argument of the tangent vector dz increases by $2\pi - \sum(\pi - \theta_j)$ after a full turn along ∂A. This observation, coupled with (7.5) and the Argument principle, yields

$$2\pi\sum m_i + \sum\theta_j n_j = -4\pi + 2\sum(\pi - \theta_j),$$

which is (7.4).   □

It follows from (7.4) that

$$\sum\left(1 - (n_j + 2)\frac{\theta_j}{2\pi}\right) \geq 2. \tag{7.6}$$

We conclude that there are at least three angles θ_j so small that

$$\theta_j < \frac{2\pi}{n_j + 2}. \tag{7.7}$$

Hence, these angles do not satisfy the angle condition (7.3).

7.4. Minimum Property of Geodesics

In order to prove that a geodesic is globally the unique shortest curve in its homotopy class, we need two auxiliary results. As before, we assume that we have a fixed φ-metric.

Lemma 7.2. *Let $S = G$ be a simply connected domain in the complex plane and z_1 and z_2 points of G. Then there exists at most one geodesic from z_1 to z_2.*

PROOF. Let us assume that there are two geodesics joining z_1 and z_2 in G. If they do not coincide we can find two subarcs, both from a point a to a point b, which form a simple closed polygon. The angle condition (7.3) is satisfied at the vertices, except possibly at the two points a and b. This is in contradiction with the fact that (7.7) holds for at least three angles.   □

A geodesic is called maximal if it is not a proper subset of any other geodesic.

Lemma 7.3. *In a simply connected subdomain of the complex plane every maximal geodesic is a cross-cut.*

PROOF. Let γ be a maximal geodesic in a simply connected plane domain $S = G$. Fix a point $z_0 \in \gamma$ and represent a ray of γ with the initial point z_0 by using its arclength u as parameter, $0 \leq u < u_\infty$. Assume that $\gamma(u)$ does not tend to ∂G as $u \to u_\infty$. Then there is a sequence of points $u_n \to u_\infty$ such that $z_n = \gamma(u_n) \to z \in G$. By Theorem 7.1, there is a disc U around z in which any two points can be joined by a unique shortest curve. Consider the maximal subarc γ_k of γ which contains the point z_k and lies in U. There is a point $z_n \in U$, $n > k$, which is not on γ_k. Otherwise γ would terminate at z, which contradicts the fact that every geodesic arc in U can be continued to ∂U. Therefore, the part of γ from z_k to z_n is a geodesic which leaves U. On the other hand, there is a shortest curve and hence a geodesic from z_k to z_n inside U. This is in contradiction with Lemma 7.2. $\qquad\square$

With the aid of the above two lemmas, the unique extremality of geodesics can now be established.

Theorem 7.2. *Let S be a Riemann surface and p and q points of S. Then a geodesic arc from p to q is strictly shorter than any other curve in its homotopy class.*

PROOF. Let γ_1 and γ_2 be homotopic curves on S joining p and q. Let z_0 be a point of the universal covering surface D of S over the point p. By Theorem 2.1 (Monodromy theorem), the lifts of γ_1 and γ_2 from z_0 terminate at the same point z over q. Since lifting does not change lengths, we may assume that S is a simply connected domain D in the complex plane.

Let γ be a geodesic from z_0 to z; by Lemma 7.2, γ is unique. Consider an arbitrary curve γ' in D which connects z_0 to z. Replacing subarcs of γ' by locally shortest arcs with the same endpoints does not make γ' longer. We may thus assume that γ' is a geodesic polygon. Also, it is not difficult to construct a Jordan domain G, $\bar{G} \subset D$, which is bounded by a geodesic polygon and which contains γ and γ'.

Suppose first that γ is a straight arc; we may assume that it is horizontal. Let $z_1, z_2, \ldots, z_{n-1}$ denote the points of γ which lie on a critical vertical arc with respect to G. We pick an arbitrary point z' of an open interval (z_{i-1}, z_i) of γ, where i can take any value $1, 2, \ldots, n$; $z_n = z$. By Lemma 7.3, the maximal vertical arc β through z' is a cross-cut of G. It follows that γ' intersects β.

Now let z' run through all points of (z_{i-1}, z_i). The vertical arcs β then sweep out a simply connected domain A_i, which is mapped by Φ onto a vertical parallel strip. If a_i denotes the width of the strip, then the length of $\gamma' \cap A_i$ is at least a_i, with equality if and only if $\gamma' \cap A_i$ is a single arc parallel to γ. The domains A_i are disjoint so that $l(\gamma') \geq \sum a_i = l(\gamma)$. Equality can hold only if $l(\gamma' \cap A_i) = a_i$ for every i. Starting with $i = 1$, we first deduce that then $\gamma' \cap A_1 = \gamma \cap A_1$, and continuing the reasoning we conclude that $\gamma' = \gamma$.

After this, let γ be an arbitrary geodesic. Then γ is the union of straight arcs γ_j. For every γ_j, we construct the orthogonal strips A_{ij} as before. An arc β_j orthogonal to γ_j does not meet γ again; this follows from Lemma 7.2. Neither can a β_j intersect a β_k orthogonal to γ_k, $k \neq j$. For if it would, we would get a simple closed geodesic polygon with two interior angles equal to $\pi/2$, one positive angle at the intersection of β_j and β_k, and the other angles satisfying the condition (7.3). This would contradict the inequality (7.6). Therefore, the strips A_{ij} are disjoint, and since γ' goes through every A_{ij}, we conclude as before that $l(\gamma') \geq l(\gamma)$ and that equality can hold only if $\gamma' = \gamma$.   □

It follows from Theorem 7.2 that every horizontal arc is uniquely length minimizing in its homotopy class.

7.5. Existence of Geodesics

If the Riemann surface S and the φ-metric are arbitrary, it is not always possible to connect two given points p and q with a geodesic. A simple example is the case in which S is a non-convex plane domain and the metric is euclidean. The existence of geodesics can be shown if the distance between the lifts of the points p and q to the universal covering surface is smaller than their distances to the boundary.

In view of our applications, we shall restrict ourselves to the case in which S is a compact surface. Then every point of the universal covering surface D has an infinite distance to the boundary of D (Ahlfors [1]).

This is trivially true if D is the complex plane. For then $\partial D = \{\infty\}$, and we know that the φ-metric is euclidean (cf. 6.5 and 7.1).

If D is the unit disc, the Dirichlet regions of the covering group of D over S are relatively compact (Theorem 5.1). Hence, given a point $\zeta \in D$, there is an $r_0 < 1$ such that the disc $|z| < r_0$ covers the Dirichlet region with center at ζ. Pick an r_1 such that $r_0 < r_1 < 1$, and let d denote the φ-distance between the circles $|z| = r_0$ and $|z| = r_1$. The circle $|z| = r_1$ can be covered by the images of $|z| < r_0$ under finitely many cover transformations. The images of the disc $|z| < r_1$ under these finitely many transformations are contained in a disc $|z| < r_2$ with $r_1 < r_2 < 1$. From the invariance of φ-distances under the covering group we conclude that the distance from ζ to the circle $|z| = r_2$ is $\geq 2d$. A repetition of the argument shows that the distance from ζ to ∂D is infinite.

We can now prove the existence of geodesics on compact surfaces.

Theorem 7.3. *Let S be a compact Riemann surface and p and q points of S. Then each homotopy class of curves joining p and q on S contains a unique shortest (hence geodesic) arc.*

PROOF. As in the proof of Theorem 7.2, we may replace S by its universal covering surface D. Let two points z_1 and z_2 of D be given. Since the distance from z_1 and z_2 to ∂D is infinite, we can find a Jordan domain G, $\bar{G} \subset D$, such that $z_1, z_2 \in G$ and that any arc connecting z_1 and z_2 in D and leaving G cannot be length minimizing. If a denotes the infimum of the lengths of the curves in D which join z_1 and z_2, we then obtain the same infimum a if we restrict attention only to curves which lie in G.

Let (γ_i) be a minimal sequence of curves in G from z_1 to z_2, i.e., $l(\gamma_i) \to a$. Subdivide the parameter interval $[0, l(\gamma_i)]$ into n equal parts and take n so large that the endpoints of the resulting subarcs of γ_i can be joined by a unique shortest arc in D. That this is possible follows from Theorem 7.1, combined with a standard compactness argument. For a subsequence (γ_{i_k}), these $n + 1$ endpoints converge. By joining the limit points with shortest arcs in D we obtain a shortest arc γ from z_1 to z_2. Being globally shortest γ is also locally shortest, i.e., a geodesic.

The uniqueness follows from Theorem 7.2. $\square$

7.6. Deformation of Horizontal Arcs

A horizontal arc α of S is shortest in its homotopy class. We shall prove that this is asymptotically true if the competing curves are images of α under deformations of S (Teichmüller [1], p. 159).

Lemma 7.4. *Let S be a compact Riemann surface, $f: S \to S$ a homeomorphism homotopic to the identity, and α a horizontal arc. Then there is a constant M, which does not depend on α, such that*

$$l(f(\alpha)) \geq l(\alpha) - 2M.$$

PROOF. Let $h: S \times [0,1] \to S$ be a homotopy from the identity mapping to f. Fix a point $p \in S$ and denote by $\tilde{\gamma}_p$ the path $t \to h(p, t)$. Let γ_p be the (unique) geodesic in the homotopy class of $\tilde{\gamma}_p$. If p' is close to p, the difference $|l(\gamma_p) - l(\gamma_{p'})|$ is majorized by the sum of the distances between p and p' and $f(p)$ and $f(p')$. Hence, the function $p \to l(\gamma_p)$ is continuous. Since S is compact, it follows that

$$M = \max_{p \in S} l(\gamma_p) < \infty.$$

Now let p be the initial point and q the terminal point of the horizontal arc α. If γ_q^{-1} denotes the path $t \to \gamma_q(1 - t)$, then $\gamma_p f(\alpha) \gamma_q^{-1}$ is homotopic to α. By Theorem 7.2, the geodesic α is shortest in its homotopy class. Therefore,

$$l(\alpha) \leq l(f(\alpha)) + 2M,$$

as we wished to show. $\square$

On a spiral trajectory we can take α as long as we please. If α is a part of a closed trajectory, we may allow α to cover itself. Thus we can always let $l(\alpha) \to \infty$, and have then $\liminf l(f(\alpha))/l(\alpha) \geq 1$.

CHAPTER V

Teichmüller Spaces

Introduction to Chapter V

In this chapter we introduce the theory of Teichmüller spaces of Riemann surfaces by utilizing the results in all four preceding chapters.

In section 1 we define the notion of a quasiconformal mapping between Riemann surfaces and prove that the existence and uniqueness theorems for Beltrami equations generalize from the plane to Riemann surfaces. The complex dilatation turns out to be a $(-1, 1)$-differential on a surface. The uniqueness theorem shows that every such Beltrami differential determines a conformal structure for the surface.

The Teichmüller space of a Riemann surface is defined in section 2 as a set of equivalence classes of quasiconformal mappings. A metric is introduced in this space in the same manner as in the universal Teichmüller space. The use of the complex dilatation leads to a characterization of the Teichmüller space in terms of different conformal structures.

In sections 3, 4 and 5 we consider Riemann surfaces which have a half-plane as a universal covering surface. In section 3, the quasiconformal mappings used in the definition of a Teichmüller space are lifted to mappings of the half-plane onto itself. This gives a clear picture of an arbitrary Teichmüller space as a subset of the universal space, and makes it possible to generalize many previous results.

In III.4 we mapped the universal Teichmüller space homeomorphically onto an open set in the space of Schwarzian derivatives. In section 4 of this chapter, we consider the restriction of this mapping to the Teichmüller space of a Riemann surface. The image is then contained in the subspace consisting of the Schwarzians which are holomorphic quadratic differentials for the covering group of the half-plane over the "mirror image" of the given surface.

The unifying link with the earlier results concerning the universal space is provided by a theorem which says that the image of an arbitrary Teichmüller space is the intersection of the image of the universal Teichmüller space with the space of the quadratic differentials for the covering group. It follows that the image is an open set in the space of quadratic differentials. The distance from a point of this image to its boundary can be estimated. Using Schwarzian derivatives, we also obtain simple estimates relating the metric of a Teichmüller space and the metric it inherits from the universal space, showing that these two metrics are topologically equivalent.

The results of section 4 are used in section 5, where a complex analytic structure is introduced into Teichmüller spaces. Quasiconformally equivalent Riemann surfaces turn out to have isometrically and biholomorphically isomorphic Teichmüller spaces.

While sections 2–5 deal with the general theory of Teichmüller spaces, the remaining sections 6–9 are primarily concerned with Teichmüller spaces of compact surfaces. Section 6 is devoted to the study of the Teichmüller space of a torus, which is shown to be isomorphic to the upper half-plane furnished with the hyperbolic metric.

In section 7 we consider extremal quasiconformal mappings of Riemann surfaces, which determine the distance in the Teichmüller space, i.e., which have the smallest maximal dilatation in their homotopy class. A necessary condition for the extremal complex dilatation is derived in the general case. If the surface is compact, we can conclude that the extremal is always a Teichmüller mapping, i.e., its complex dilatation is of the form $k\bar{\varphi}/|\varphi|$, where $0 \leq k < 1$ and φ is a holomorphic quadratic differential of the surface.

In section 8 we prove Teichmüller's famous theorem that on compact Riemann surfaces of genus > 1, every Teichmüller mapping is a unique extremal in its homotopy class. In section 9 we show how this result leads to a mapping of the Teichmüller space of a compact surface onto the open unit ball in the space of holomorphic quadratic differentials. If the surface is of genus $p \, (> 1)$, the Teichmüller space is proved to be homeomorphic to the euclidean space $\mathbb{R}^{6p-6}$. Finally, the connection between the Teichmüller metric and the complex analytic structure is discussed briefly, and some remarks are made on the Teichmüller spaces of Riemann surfaces of finite type.

1. Quasiconformal Mappings of Riemann Surfaces

1.1. Complex Dilatation on Riemann Surfaces

A homeomorphism f between two Riemann surfaces S_1 and S_2 is called K-quasiconformal if for any local parameters h_i of an atlas on S_i, $i = 1, 2$, the mapping $h_2 \circ f \circ h_1^{-1}$ is K-quasiconformal in the set where it is defined. The mapping f is quasiconformal if it is K-quasiconformal for some finite $K \geq 1$.

Suppose that the local parameters h_1, k_1 of S_1 have overlapping domains U_1, V_1, and that $f(U_1 \cap V_1)$ lies in the domains of the local parameters h_2, k_2 of S_2. Using the notation $g = h_1 \circ k_1^{-1}$, $h = k_2 \circ h_2^{-1}$, we then have in $k_1(U_1 \cap V_1)$,

$$k_2 \circ f \circ k_1^{-1} = h \circ (h_2 \circ f \circ h_1^{-1}) \circ g.$$

The mappings h and g are conformal and, therefore, do not change the maximal dilatation. It follows that K-quasiconformal mappings between Riemann surfaces are well defined.

Let (D, π_1) be a universal covering surface of S_1, where D is the unit disc or the complex plane. (Here and in what follows the trivial case in which D is the extended plane is excluded.) For the quasiconformal mapping $f: S_1 \to S_2$, consider all mappings $w = h_2 \circ f \circ h_1^{-1}$, where we choose h_1^{-1} to be a suitable restriction of π_1. Then the complex dilatations of the mappings w define a function μ on D. Set $k_1 = g^{-1} \circ h_1$, where g is an arbitrary cover transformation of D over S_1. Then $h_2 \circ f \circ k_1^{-1} = w \circ g$. From this and formula (4.4) in I.4.2 it follows that μ satisfies the condition

$$\mu = (\mu \circ g)\frac{\overline{g'}}{g'} \tag{1.1}$$

for every cover transformation g. Consequently, a quasiconformal mapping f of a Riemann surface S_1 determines a Beltrami differential for the covering group G or, what is the same, a Beltrami differential on the surface S_1 (cf. IV.1.4 and IV.3.6). This differential is called the complex dilatation of f.

We can also arrive at the complex dilatation of a quasiconformal mapping of a Riemann surface in a slightly different manner; namely, by lifting the given quasiconformal mapping to a mapping between the universal covering surfaces. Let (D, π_i) be a universal covering surface of S_i, $i = 1, 2$, and G_i the covering group of D over S_i. Consider a lift $w: D \to D$ of the given quasiconformal mapping $f: S_1 \to S_2$. Since the projections π_1 and π_2 are analytic local homeomorphisms, w is quasiconformal. Let μ be the complex dilatation of w. Because $w \circ g \circ w^{-1}$ is conformal for every $g \in G_1$ (cf. IV.3.4), the mappings w and $w \circ g$ have the same complex dilatation. Hence, we again obtain (1.1). Clearly this μ is the complex dilatation of f.

We assumed that S_1 and S_2 admit the same universal covering surface D. But if S_1 and S_2 are quasiconformally equivalent, the same is true of their universal covering surfaces. Therefore, it is not possible that the universal covering surface of one of the surfaces is a disc and of the other the complex plane.

The existence theorem for Beltrami equations (Theorem I.4.4) can be easily generalized to Riemann surfaces.

Theorem 1.1. *Let μ be a Beltrami differential on a Riemann surface S. Then there is a quasiconformal mapping of S onto another Riemann surface with complex dilatation μ. The mapping is uniquely determined up to a conformal mapping.*

PROOF. We consider μ as a Beltrami differential for the covering group G of D over S. By Theorem I.4.4, there is a quasiconformal mapping $f: D \to D$ with complex dilatation μ. Since (1.1) holds, f and $f \circ g$ have the same complex dilatation for every $g \in G$. Then $f \circ g \circ f^{-1}$ is conformal, and we conclude that f induces an isomorphism of G onto the Fuchsian group $G' = \{f \circ g \circ f^{-1} | g \in G\}$. If π and π' denote the canonical projections of D onto S and $S' = D/G'$, then $\varphi \circ \pi = \pi' \circ f$ defines a quasiconformal mapping φ of S onto S'. This mapping has the complex dilatation μ.

Let ψ be another quasiconformal mapping of S with complex dilatation μ and $w: D \to D$ its lift. Then $w \circ f^{-1}: D \to D$ is conformal, and so its projection $\psi \circ \varphi^{-1}$ is also conformal. $\qquad\square$

1.2. Conformal Structures

Let μ be a Beltrami differential on a Riemann surface S, and h an arbitrary local parameter on S with domain V. From the existence theorem for Beltrami equations it follows that there is a complex-valued quasiconformal mapping w of $h(V)$ with complex dilatation $\mu \circ h^{-1}$. (For this conclusion we can use the plane version, Theorem I.4.4.) Then $f = w \circ h$ is a quasiconformal mapping of V into the plane with complex dilatation μ. If f_1 and f_2 are two such mappings with intersecting domains V_1 and V_2, then by the Uniqueness theorem (Theorem I.4.2), $f_2 \circ f_1^{-1}$ is conformal in $f_1(V_1 \cap V_2)$. This allows an important conclusion:

A Beltrami differential of S defines a conformal structure on S.

If H is the original conformal structure and H_μ the structure induced by μ, then H_μ is determined by all quasiconformal mappings of open subsets of (S, H) into the plane whose complex dilatations are restrictions of μ. These mappings are conformal with respect to the structure H_μ.

We can relax the conditions on μ slightly and still obtain conformal structures. In fact, the above reasoning works if μ is a $(-1, 1)$-differential of S and $\|\mu\|_\infty < 1$ in every compact subset of S.

1.3. Group Isomorphisms Induced by Quasiconformal Mappings

Let us now assume that the universal covering surface D of S is the unit disc. The lifts of homeomorphisms of Riemann surfaces need not possess limits at the boundary of D. However, if the homeomorphism is quasiconformal, then a lift always admits a homeomorphic extension to the boundary. This makes it possible to rephrase Theorem IV.3.5 in terms of the boundary behavior of lifted mappings.

Before formulating the theorem, we make a remark on the transformation of the limit sets of covering groups. Let $f: D \to D$ be a lift of a quasiconformal mapping of a Riemann surface with the covering group G onto a Riemann surface with the covering group G'. If z is a fixed point of $g \in G$ and $f(z) = \zeta$, then $(f \circ g \circ f^{-1})(\zeta) = f(g(z)) = \zeta$. We conclude that f maps the fixed points of G onto the fixed points of G'. Since the limit set is the closure of the set of the fixed points (IV.4.5), it follows that f maps the limit set L of G onto the limit set L' of G'.

Theorem 1.2. *Let S and S' be Riemann surfaces with non-elementary covering groups G and G', $\varphi_i: S \to S'$, $i = 0, 1$, two quasiconformal mappings, and f_0 a lift of φ_0. Then φ_0 and φ_1 induce the same group isomorphism between G and G' if and only if there is a lift f_1 of φ_1 which agrees with f_0 on the limit set of G.*

PROOF. Suppose first that there is a lift f_1 of φ_1 such that $f_1 = f_0$ on the limit set L of G. Because f_0 and f_1 map L onto the limit set L' of G' and because L is invariant under G, we then have

$$f_0 \circ g \circ f_0^{-1} = f_1 \circ g \circ f_1^{-1}, \qquad g \in G, \tag{1.2}$$

at every point of L'. Both sides are Möbius transformations. Since they are equal on a set with at least three points, they agree everywhere.

In order to prove the necessity of the condition, we now assume that (1.2) is true in D. Setting $h = f_0^{-1} \circ f_1$, we rewrite (1.2) in the form

$$g \circ h = h \circ g.$$

If z is a fixed point of some g, then $g(h(z)) = h(z)$, i.e., $h(z)$ is also a fixed point of g. If z is an attractive fixed point and $\zeta \in D$, then for the nth iterate g_n of g, $g_n(h(\zeta)) \to z$ as $n \to \infty$. On the other hand, $g_n(h(\zeta)) = h(g_n(\zeta)) \to h(z)$. Hence $h(z) = z$ for all fixed points of G. Since these fixed points comprise a dense subset of L (see IV.4.5), it follows that $f_0(z) = f_1(z)$ for all z in L. $\qquad \square$

Theorem 1.2, combined with Theorem IV.3.5, plays an important role in the theory of Teichmüller spaces. The following special case deserves particular attention.

Theorem 1.3. *Let S be a Riemann surface with a non-elementary covering group. If $f: S \to S$ is a conformal mapping homotopic to the identity, then f is the identity mapping.*

PROOF. By Theorem IV.3.5, f and the identity mapping of S induce the same group isomorphism of the covering group of D over S. By Theorem 1.2, f has a lift which is the identity mapping of D. Hence, the projection f itself is the identity mapping. $\qquad \square$

1.4. Homotopy Modulo the Boundary

For covering groups of the first kind, the existence of homotopy between two mappings φ_0 and φ_1 is equivalent to the existence of lifts which agree on the whole boundary of D. In the theory of Teichmüller spaces this is a very satisfactory state of affairs. For analogous behavior to occur when covering groups are of the second kind, which would make it possible to develop a unified theory, we need a stronger form of homotopy.

Let us consider a Riemann surface $S = D/G$, where D is the unit disc and G is of the second kind. We denote by B the non-void complement of the limit set L of G with respect to the unit circle. Then $S^* = D/G \cup B/G$ is a bordered Riemann surface (cf. IV.1.3 and IV.4.5).

If we use the quotient representation D/G for Riemann surfaces under consideration, an amazingly strong result can be easily proved: *A quasiconformal mapping φ of $S = D/G$ onto $S' = D/G'$ can always be extended to a homeomorphism of S^* onto $(S')^*$.*

In order to prove this, we consider a lift $f: D \to D$ of φ. We continue f by reflection to a quasiconformal mapping of the plane. The extended f then induces the isomorphism $g \to f \circ g \circ f^{-1}$ between G and G' in the whole plane. We proved in 1.3 that f maps the set of discontinuity Ω of G onto the set of discontinuity Ω' of G'.

We assumed that G is of the second kind, in which case G' also is of the second kind. Extend the canonical projections $\pi: D \to D/G$ and $\pi': D \to D/G'$ to the domains Ω and Ω'. Then

$$\varphi^* \circ \pi = \pi' \circ f$$

defines a quasiconformal mapping φ^* of the double Ω/G of S onto the double Ω'/G' of S'. Its restriction to $S^* = (D \cup B)/G$ is the desired extension of φ.

Let $\varphi_i: S \to S'$, $i = 0, 1$, be two quasiconformal mappings between the Riemann surfaces $S = D/G$ and $S' = D/G'$. We just proved that φ_0 and φ_1 can be extended to mappings of S^* onto $(S')^*$. We say that φ_0 is *homotopic to φ_1 modulo the boundary* if $\varphi_0 = \varphi_1$ on the border and there is a homotopy from φ_0 to φ_1 which is constant on the border.

Theorem 1.4. *Two quasiconformal mappings $\varphi_i: S \to S'$, $i = 0, 1$, are homotopic modulo the boundary if and only if they can be lifted to mappings of D which agree on the boundary.*

PROOF. Assume first that φ_0 and φ_1 are homotopic modulo the boundary. If f_0 is a lift of φ_0, then the lift f_1 of φ_1 homotopic to f_0 through the lifted homotopy agrees with f_0 on the set B. The mappings f_0 and f_1 determine the same group isomorphism (Theorem IV.3.5). From the proof of Theorem 1.2 it follows that $f_0 = f_1$ on L.

Conversely, if $f_0 = f_1$ on the boundary of D, we construct a homotopy f_t from f_0 to f_1 as in the proof of Theorem IV.3.5, and conclude again that it can be projected to produce a homotopy from φ_0 to φ_1. Since f_t keeps every point of B fixed, the projected homotopy is constant on the border of S. □

1.5. Quasiconformal Mappings in Homotopy Classes

Not every sense-preserving homeomorphism between two given Riemann surfaces is homotopic to a quasiconformal mapping. A trivial counterexample is the case where one of the surfaces is a disc and the other the complex plane. These are homeomorphic Riemann surfaces but not quasiconformally equivalent.

In the case of compact surfaces, the situation is different (Teichmüller [2]).

Theorem 1.5. *Let S and S' be compact, topologically equivalent Riemann surfaces. Then every homotopy class of sense-preserving homeomorphisms of S onto S' contains a quasiconformal mapping.*

PROOF. Let $f: S \to S'$ be a sense-preserving homeomorphism. Since S is compact, it has a finite covering by domains $U_1, U_2, \ldots, U_n$, such that U_k is conformally equivalent to the unit disc and ∂U_k is an analytic curve. Set $f_0 = f$, and define inductively a sequence of mappings f_k, $k = 1, 2, \ldots, n$, as follows: $f_k = f_{k-1}$ in $S \backslash U_k$, while in U_k, the mapping f_k is the Beurling–Ahlfors extension of the boundary values $f_{k-1} | \partial U_k$. More precisely, we map U_k and $f_{k-1}(U_k)$ conformally onto the upper half-plane H. Since U_k and $f_{k-1}(U_k)$ are Jordan domains, these conformal transformations of U_k and $f_{k-1}(U_k)$ onto H have homeomorphic extensions to the boundary (see I.1.2). We normalize the mappings so that the induced self-mapping w of H keeps ∞ fixed. After that, we form the Beurling–Ahlfors extension of $w | \mathbb{R}$ as in I.5.3. By transferring this extension to S we obtain $f_k | U_k$. The mapping $f_k | U_k$ is a diffeomorphism and hence locally quasiconformal. Moreover, if f_{k-1} is quasiconformal at a point $z \in \partial U_k$, then $f_k | U_k \cap V$ is quasiconformal for some neighborhood V of z (cf. [LV], pp. 84–85). Hence, f_k is quasiconformal at z, because ∂U_k is a removable singularity (cf. Lemma I.6.1). It follows that f_n is a quasiconformal mapping of S, since S is compact.

The mapping $(p, t) \to t f_k(p) + (1 - t) f_{k-1}(p)$ is a homotopy between f_{k-1} and f_k. It follows that f_n is homotopic to f. □

Theorem 1.5 is not true for arbitrary Riemann surfaces S and S', not even in cases in which S and S' each admits a disc as its universal covering surface.

We shall prove later (Theorems 4.5 and 6.3) that if S and S' are arbitrary Riemann surfaces which are quasiconformally equivalent, then every homotopy class of quasiconformal mappings of S onto S' contains a real analytic quasiconformal mapping. This result can be regarded as a generalization of Theorem III.1.1.

2. Definitions of Teichmüller Space

2.1. Riemann Space and Teichmüller Space

We shall now generalize the notion of the universal Teichmüller space introduced in III.1 and define the Teichmüller space for an arbitrary Riemann surface.

Let us consider all quasiconformal mappings f of a Riemann surface S onto other Riemann surfaces. If two such mappings f_1 and f_2 are declared to be equivalent whenever the Riemann surfaces $f_1(S)$ and $f_2(S)$ are conformally equivalent, the collection of equivalence classes forms the *Riemann space* R_S of S.

In the classical case where S is a compact Riemann surface we could equally well start with homeomorphic mappings of S: by Theorem 1.5, every homotopy class of homeomorphisms contains quasiconformal mappings. The study of R_S is called Riemann's problem of moduli. In the case where S is the upper half-plane, the equivalence relation is so weak that all mappings f are equivalent, and so R_S reduces to a single point.

Teichmüller [1] observed that even in the case of a compact surface, a space simpler than R_S is obtained if we use a stronger equivalence relation. Let f_1 and f_2 be quasiconformal mappings of a Riemann surface S. Suppose that the universal covering surface of S is the extended plane or the complex plane or a disc with a covering group of the first kind. Then f_1 and f_2 are said to be equivalent if $f_2 \circ f_1^{-1}$ is homotopic to a conformal mapping of $f_1(S)$ onto $f_2(S)$. If the universal covering surface of S is a disc and the covering group is of the second kind, i.e., if S is bordered, "homotopic" in this definition of equivalence is to be replaced by "homotopic modulo the boundary".

Definition. The Teichmüller space T_S of the Riemann surface S is the set of the equivalence classes of quasiconformal mappings of S.

Teichmüller restricted his interest to compact Riemann surfaces and, a little more generally, to certain cases in which T_S is finite-dimensional (cf. 9.7). Ahlfors [1] seems to have been the first to use the name "Teichmüller space", this in 1953. The above definition applying to all Riemann surfaces is due to Bers ([7], [8]).

It is not difficult to see that if $S = H$, then T_S agrees with the universal Teichmüller space T_H. In applying the above definition of T_S to $S = H$, we first note that all quasiconformal images of H are conformally equivalent. It follows that we may consider without loss of generality only the normalized quasiconformal self-mappings of H which we denoted in III.1 by f^μ. By Theorem 1.4, the condition that $f^{\mu_2} \circ (f^{\mu_1})^{-1}$ be homotopic modulo the boundary to a conformal mapping is fulfilled if and only if $f^{\mu_2} \circ (f^{\mu_1})^{-1}$ agrees with the identity mapping on the real axis $\mathbb{R}$. Consequently, f^{μ_1} is equivalent to

f^{μ_2} by the above definition if and only if $f^{\mu_1}|\mathbb{R} = f^{\mu_2}|\mathbb{R}$. By the definition in III.1.1, this is the condition for f^{μ_1} and f^{μ_2} to determine the same point in the universal Teichmüller space T_H.

A connection is obtained between T_S and the universal Teichmüller space if we lift the mappings between S and other Riemann surfaces to mappings between the universal covering surfaces. In this way we are able to transfer many results associated with the universal Teichmüller space to the general case. This will be done in sections 3–5.

By slightly changing the definition of T_S we arrive at the *reduced Teichmüller space* of S. Its points are also equivalence classes of quasiconformal mappings of S, but now two such mappings f_1 and f_2 are declared equivalent if $f_2 \circ f_1^{-1}$ is just homotopic (not necessarily homotopic modulo the boundary) to a conformal mapping.

The reduced Teichmüller space differs from the Teichmüller space T_S only if S is bordered. In what follows, we shall not deal with reduced Teichmüller spaces. For this reason we henceforth use the term "homotopy" to mean "homotopy modulo the boundary" in the case of bordered surfaces. This simplifies the language and, if this convention is kept in mind, should not cause confusion.

2.2. Teichmüller Metric

Exactly as in the case of the universal Teichmüller space, we define the distance

$$\tau(p, q) = \tfrac{1}{2}\inf\{\log K_{g \circ f^{-1}} | f \in p, g \in q\} \tag{2.1}$$

between the points p and q of the Teichmüller space T_S (cf. III.2.1). In the proof that τ defines a metric in T_S, the only non-trivial step is again to show that $\tau(p, q) = 0$ implies $p = q$. This can be deduced from the following result.

Theorem 2.1. *Let $f_0 \colon S \to S'$ be a quasiconformal mapping and F the class of all quasiconformal mappings of S onto S' homotopic to f_0. Then F contains an extremal mapping, i.e., one with smallest maximal dilatation.*

PROOF. Let D be a universal covering surface of S. The theorem is trivial if D is the extended plane or if D is the complex plane and S is non-compact. In the case where D is the complex plane and S is compact, the theorem will be proved in 6.4. Hence, we may assume that $D = H$ is the upper half-plane (cf. IV.4.1).

By Theorem 1.4, we can lift each $f \in F$ to a self-mapping w_f of H such that all mappings w_f agree on the real axis. The class $W = \{w_f | f \in F\}$ contains its quasiconformal limits. Hence, there exists a mapping $w \in W$ with smallest maximal dilatation (cf. I.5.7). The projection of w is the extremal sought in F. $\square$

Given the points $p, q \in T_S$, we fix the mappings $f_0 \in p$, $g_0 \in q$, and let F now be the class of all quasiconformal mappings of $f_0(S)$ onto $g_0(S)$ homotopic to $g_0 \circ f_0^{-1}$. Again mimicking what was done in the case of the universal Teichmüller space (cf. III.2.1) we conclude that

$$\tau(p, q) = \tfrac{1}{2} \inf\{\log K_f | f \in F\}. \tag{2.2}$$

In other words, in determining the distance between two points of T_S we can always take the infimum of maximal dilatations in a homotopy class of quasiconformal mappings between two fixed Riemann surfaces.

Theorem 2.1 says that the inf in (2.2) can be replaced by min. Consequently, if $\tau(p, q) = 0$, the class F contains a conformal mapping, and so $p = q$. After this, it is clear that (T_S, τ) is a metric space.

The point which is defined by the identity mapping of S is called the origin of T_S. The origin contains all conformal mappings of S.

2.3. Teichmüller Space and Beltrami Differentials

The definition of the Teichmüller space T_S can also be formulated in terms of the Beltrami differentials on S. Every quasiconformal mapping of S determines a Beltrami differential on S, namely, its complex dilatation. Conversely, if μ is a Beltrami differential of S, then by Theorem 1.1 there is a quasiconformal mapping of S whose complex dilatation is μ, and by the uniqueness part of Theorem 1.1, all such mappings determine the same point of T_S. Two Beltrami differentials are said to be equivalent if the corresponding quasiconformal mappings are equivalent. Hence, a point of T_S can be thought of as a set of equivalent Beltrami differentials. The Teichmüller distance (2.1) can be expressed in terms of Beltrami differentials:

$$\tau(p, q) = \frac{1}{2} \inf\left\{ \log \left| \frac{1 + \|(\mu - v)/(1 - \bar{\mu}v)\|_\infty}{1 - \|(\mu - v)/(1 - \bar{\mu}v)\|_\infty} \right| \mu \in p, v \in q \right\}. \tag{2.3}$$

Let S admit the half-plane as its universal covering surface. Then geodesics in T_S allow the same description as in the universal Teichmüller space (see Theorem III.2.2): *If μ is an extremal complex dilatation for the point $p \in T_S$, then*

$$\mu_t = \frac{(1 + |\mu|)^t - (1 - |\mu|)^t}{(1 + |\mu|)^t + (1 - |\mu|)^t} \cdot \frac{\mu}{|\mu|}, \qquad 0 \le t \le 1, \tag{2.4}$$

is extremal for the point $p_t = [\mu_t]$. The arc $t \to p_t$ is a geodesic from 0 to p, and $\tau(p_t, 0) = t\tau(p, 0)$.

By using Theorem 3.1, to be established in subsection 3.1, we can merely repeat the proof of Theorem III.2.2. We only have to make the additional verification that μ_t represents a point of T_S. Since μ represents a point of T_S, it is a Beltrami differential for the covering group G of the universal covering surface over S, i.e., $(\mu \circ g)\overline{g'}/g' = \mu$ for every $g \in G$. From (2.4) we see immediately that μ_t also satisfies this condition. Hence $[\mu_t]$ is a point of T_S.

The following generalization of Theorem III.2.1 is immediate.

Theorem 2.2. *The Teichmüller space T_S is pathwise connected.*

PROOF. The geodesic $t \to [\mu_t]$ is a path joining the origin to the point p in T_S; the path $t \to [t\mu]$ of T_S also has this property. □

2.4. Teichmüller Space and Conformal Structures

Let S be a Riemann surface with the conformal structure H and $\{h\}$ an atlas of local parameters belonging to H. If f is a homeomorphism of S onto itself, then $\{h \circ f^{-1}\}$ is an atlas which determines another conformal structure of S. We denote this structure by $f_*(H)$ and note that $f_*(H)$ does not depend on the particular choice of the atlas on H.

It follows from the definition that $f: (S, H) \to (S, f_*(H))$ is a conformal mapping. Conversely, if H and H' are conformal structures of S and $f: (S, H) \to (S, H')$ is conformal, then $H' = f_*(H)$.

We say that *two conformal structures H and H' of S are deformation equivalent if H can be deformed conformally to H'*, i.e., if there is a conformal mapping of (S, H) onto (S, H') which is homotopic to the identity.

In 1.2 we showed that every Beltrami differential μ on the Riemann surface (S, H) defines a new conformal structure H_μ. Given two conformal structures H and H' of S, suppose that there exists a quasiconformal mapping $f: (S, H) \to (S, H')$. Let μ denote the complex dilatation of f. Then

$$H' = f_*(H_\mu). \tag{2.5}$$

This follows directly from the definitions, because now $f: (S, H_\mu) \to (S, H')$ is conformal.

There is a simple connection between different structures H_μ and points of the Teichmüller space of S.

Theorem 2.3. *The conformal structures H_1 and H_2 induced by the Beltrami differentials μ_1 and μ_2 on the Riemann surface S are deformation equivalent if and only if μ_1 and μ_2 determine the same point in the Teichmüller space T_S.*

PROOF. Let f_i, $i = 1, 2$, be quasiconformal mappings of S with complex dilatations μ_i. If $\varphi: (S, H_1) \to (S, H_2)$ is a conformal mapping homotopic to the identity, we first conclude that the mapping

$$h = f_2 \circ \varphi \circ f_1^{-1}: f_1(S) \to f_2(S)$$

is conformal. Also, we see that $f_2 \circ f_1^{-1}$ is homotopic to h. It follows that μ_1 and μ_2 are equivalent.

Conversely, if μ_1 and μ_2 are equivalent, there is a conformal map $h: f_1(S) \to f_2(S)$ such that $\varphi = f_2^{-1} \circ h \circ f_1: S \to S$ is homotopic to the identity. In addition, $\varphi: (S, H_1) \to (S, H_2)$ is conformal, and so H_1 is equivalent to H_2.   □

We conclude that *the Teichmüller space T_S can be characterized as the set of equivalence classes of conformal structures H_μ on S modulo deformation*. Note that a conformal structure H' on S is of the form H_μ if and only if id: $(S, H) \to (S, H')$ is quasiconformal.

2.5. Conformal Structures on a Compact Surface

In considering different conformal structures on a surface S, we assume here that for any two structures H and H', the identity mapping of (S, H) onto (S, H') is sense-preserving. The above results can then be supplemented if S is a compact surface.

Theorem 2.4. *On a compact Riemann surface S, every conformal structure is deformation equivalent to a structure induced by a Beltrami differential of S.*

PROOF. Let H be the given and H' an arbitrary conformal structure on S. By Theorem 1.5, there is a quasiconformal mapping $f: (S, H) \to (S, H')$ which is homotopic to the identity. Let f have the complex dilatation μ. Then $H' = f_*(H_\mu)$ (formula (2.5)). But $f: (S, H_\mu) \to (S, f_*(H_\mu))$ is a conformal mapping homotopic to the identity. Consequently, $H' = f_*(H_\mu)$ is deformation equivalent to H_μ.   □

Theorems 2.3 and 2.4 yield an important characterization of T_S.

Theorem 2.5. *The Teichmüller space of a compact Riemann surface is isomorphic to the set of equivalence classes of conformal structures modulo deformation.*

This result can also be expressed in somewhat different terms. Let $\mathscr{H}(S)$ denote the set of all conformal structures of S. The group Homeo$^+(S)$ consisting of all sense-preserving homeomorphic self-mappings of S acts on $\mathscr{H}(S)$: If $H \in \mathscr{H}(S)$ and $f \in \text{Homeo}^+(S)$, then $f_*(H) \in \mathscr{H}(S)$.

Let Homeo$_0(S)$ be the subgroup of Homeo$^+(S)$ whose mappings are homotopic to the identity. Then $H, H' \in \mathscr{H}(S)$ are deformation equivalent if and only if there is an $f \in \text{Homeo}_0(S)$ such that $f_*(H) = H'$. It follows, therefore, that for a compact surface S we have the isomorphism

$$T_S \simeq \mathscr{H}(S)/\text{Homeo}_0(S).$$

In section 8 we shall prove that every class of equivalent complex dilatations contains a unique dilatation of the form $k\bar{\varphi}/|\varphi|$, where $0 \leq k < 1$ and φ is a holomorphic quadratic differential of S. It thus follows from Theorems 2.4 and 2.5 that every conformal structure of a compact Riemann surface is deformation equivalent to a structure induced by a Beltrami differential $k\bar{\varphi}/|\varphi|$. Since φ is uniquely determined up to a multiplicative positive constant (except of course for the case $k = 0$), we see that there is a simple relationship between the normalized quadratic differentials and the equivalence classes of conformal structures on S.

The role of holomorphic quadratic differentials in the Teichmüller theory of compact Riemann surfaces will be studied in more detail in sections 7–9.

2.6. Isomorphisms of Teichmüller Spaces

In III.5.2 we proved that the universal Teichmüller spaces associated with different quasidiscs are all isomorphic. Again, it is a trivial consequence of the definition that we have a counterpart of this result in the general case.

Theorem 2.6. *The Teichmüller spaces of two quasiconformally equivalent Riemann surfaces are isometrically bijective.*

PROOF. Let S and S' be Riemann surfaces and h a quasiconformal mapping of S onto S'. The mapping $f \to f \circ h^{-1}$ is a bijection of the family of all quasiconformal mappings f of S onto the family of all quasiconformal mappings of S'. If $w_i = f_i \circ h^{-1}$, we have $w_2 \circ w_1^{-1} = f_2 \circ f_1^{-1}$. We first conclude that f_1 and f_2 determine the same point of T_S if and only if w_1 and w_2 determine the same point in $T_{S'}$, i.e.,

$$[f] \to [f \circ h^{-1}] \tag{2.6}$$

is a bijective mapping of T_S onto $T_{S'}$. It also follows that (2.6) is an isometry, i.e., it leaves all Teichmüller distances invariant.   □

Under (2.6) the point $[h]$ of T_S is mapped to the origin of $T_{S'}$. We shall later utilize this simple method of moving an arbitrary point of one Teichmüller space to the origin of another isometric Teichmüller space.

If S and S' are compact Riemann surfaces of the same genus, they are homeomorphic (IV.5.2). By Theorem 1.5, they are also quasiconformally equivalent. We conclude from Theorem 2.6 that *all Teichmüller spaces of compact surfaces of the same genus are isomorphic*.

In sections 5 and 6 we shall introduce complex analytic structure in Teichmüller spaces. We can then enhance Theorem 2.6 and prove that the Teichmüller spaces of quasiconformally equivalent Riemann surfaces are even biholomorphically isomorphic.

2.7. Modular Group

Let h be a quasiconformal self-mapping of S. Then (2.6) defines a bijective isometry of T_S onto itself. The group Mod(S) of all such isomorphisms $[f] \rightarrow [f \circ h^{-1}]$ of T_S is called the *modular group* of T_S.

If $S = H$, in which case T_S is the universal Teichmüller space, Mod(S) is the universal modular group introduced in III.1.2.

The modular group Mod(S) is also a generalization of the classical modular group Γ of Möbius transformations acting on the upper half-plane H, in the following sense: If S is a torus, the Teichmüller space T_S can be identified with H and the group Mod(S) with Γ. This will be explained in 6.7.

In section 5 we shall prove that even in the general case, the elements of the modular group are biholomorphic self-mappings of T_S.

Let $Qc(S)$ be the group of all quasiconformal self-mappings of S and $Qc_0(S)$ the normal subgroup of $Qc(S)$ whose mappings are homotopic to the identity. We associate with every $h \in Qc(S)$ the element $[f] \rightarrow [f \circ h^{-1}]$ of Mod(S). This rule defines a mapping of the quotient group $Qc(S)/Qc_0(S)$ into Mod(S). In fact, if $h_2^{-1} \circ h_1 \in Qc_0(S)$, then $[f \circ h_1^{-1}] = [f \circ h_2^{-1}]$. Clearly, this mapping of $Qc(S)/Qc_0(S)$ into Mod(S) is surjective and a group homomorphism. We remark that the mapping is injective if S admits no conformal self-mappings other than the identity transformation. It follows that in this case the modular group Mod(S) is isomorphic with the quotient group $Qc(S)/Qc_0(S)$. This is also true of all Riemann surfaces quasiconformally equivalent to such an S.

The following result illustrates the homotopy condition which makes R_S a quotient space of T_S.

Theorem 2.7. *The Riemann space is the quotient of the Teichmüller space by the modular group.*

PROOF. Assume first that the points $[f]$ and $[g]$ of T_S are equivalent under Mod(S). We then have a quasiconformal mapping $h: S \rightarrow S$ such that $f \circ h^{-1}$ is equivalent to g. But this means that there is a conformal mapping of $f(S)$ onto $g(S)$, i.e., f and g determine the same point of R_S.

Conversely, let f and g represent the same point of R_S. Then a conformal mapping $\varphi: f(S) \rightarrow g(S)$ exists, and $h = g^{-1} \circ \varphi \circ f$ is a quasiconformal self-mapping of S. From $g = \varphi \circ (f \circ h^{-1})$ we see that g and $f \circ h^{-1}$ determine the same point of T_S.   □

Theorem 2.7 says that two points $[f_1]$ and $[f_2]$ of the Teichmüller space T_S are equivalent under Mod(S) if and only if the Riemann surfaces $f_1(S)$ and $f_2(S)$ are conformally equivalent. In other words, the modular group is transitive if and only if the Riemann space R_S reduces to a singleton. This occurs only in the exceptional cases where quasiconformal equivalence of Riemann surfaces implies their conformal equivalence. The universal Teichmüller space $(S = H)$ is such an exception.

Not only the Riemann space but even the Teichmüller space may reduce to a single point. This occurs if S is a sphere (a compact surface of genus zero) or a sphere from which 1, 2 or 3 points are removed. All quasiconformal images of S are then conformally equivalent; this can be seen if Theorem I.4.4 is combined with the fact that three points of the extended plane can be moved to arbitrary positions by a Möbius transformation. After this we conclude that every point of T_S contains a conformal mapping, i.e., that T_S is a singleton. If S is the sphere minus three points, S has the disc as a universal covering surface (cf. IV.4.1), and the covering group is of the first kind.

3. Teichmüller Space and Lifted Mappings

3.1. Equivalent Beltrami Differentials

For a Riemann surface S, we defined the Teichmüller space T_S by means of quasiconformal mappings of S onto Riemann surfaces. Lifting these mappings to mappings between the universal covering surfaces leads to new characterizations of T_S and makes it possible to see better the connection between the general space T_S and the universal Teichmüller space.

We impose on the Riemann surface S the sole restriction that it has a half-plane as its universal covering surface. Since we try to follow as closely as possible the reasoning applied in III.1 and III.2 in the case of the universal Teichmüller space, we take here the lower half-plane H' as the universal covering surface of S. The cases in which the universal covering surface of S is the complex plane will be discussed in section 6.

Given a Riemann surface S, we consider a Beltrami differential μ on S or, what is the same, a function μ defined in H' which is a Beltrami differential for the covering group of H' over S. As before, we denote by f^μ the uniquely determined quasiconformal self-mapping of H' which has the complex dilatation μ and which keeps fixed the points 0, 1 and ∞ on the real axis $\mathbb{R}$, and by f_μ the quasiconformal mapping of the plane which has the complex dilatation μ in H', is conformal in the upper half-plane H and fixes the points 0, 1 and ∞. Theorem III.1.2 has an exact counterpart:

Theorem 3.1. *The Beltrami differentials μ and v of S are equivalent if and only if $f^\mu|\mathbb{R} = f^v|\mathbb{R}$ or if and only if $f_\mu|H = f_v|H$.*

PROOF. Let us first assume that μ and v are equivalent. Let φ and ψ be quasiconformal mappings of S which lift to f^μ and f^v, respectively. Then there is a conformal map $\eta: \varphi(S) \to \psi(S)$ such that $\eta \circ \varphi$ is homotopic to ψ. By Theorem 1.4, we have $f^v = h \circ f^\mu$ on the real axis $\mathbb{R}$, where h, as a lift of η, is a Möbius transformation. Since f^μ and f^v both fix 0, 1, ∞, it follows that h is the identity.

Suppose, conversely, that $f^\mu = f^\nu$ on the boundary $\mathbb{R}$. Then f^μ and f^ν induce the same isomorphism of the covering group of H' over S onto a Fuchsian group G'. The projections of f^μ and f^ν map S onto the same Riemann surface H'/G', and by Theorem 1.4, these projections are homotopic. It follows that μ and ν are equivalent.

After this we can show that $f^\mu|\mathbb{R} = f^\nu|\mathbb{R}$ if and only if $f_\mu|H = f_\nu|H$ by repeating the proof of Theorem III.1.2 verbatim. $\square$

3.2. Teichmüller Space as a Subset of the Universal Space

Theorem 3.1 says that

$$[\mu] \to f^\mu|\mathbb{R} \quad \text{and} \quad [\mu] \to f_\mu|H$$

are well defined injective mappings of the Teichmüller space. In particular, T_S can be characterized as the set of equivalence classes $[f^\mu]$, two mappings being equivalent if they agree on $\mathbb{R}$. We have thus arrived at the situation which was our starting point in III.1 when we defined the universal Teichmüller space. In the general case the complex dilatations of the mappings f^μ are Beltrami differentials for the covering group G. If G is trivial, then T_S is the universal Teichmüller space T (cf. also the remarks in 2.1).

This characterization of T_S shows that the family of Teichmüller spaces admits a partial ordering. Let S_1 and S_2 be Riemann surfaces and G_1 and G_2 the covering groups of H' over S_1 and S_2. If G_1 is a subgroup of G_2, then $T_{S_2} \subset T_{S_1}$. In particular, *every Teichmüller space T_S can be regarded as a subset of the universal Teichmüller space T.*

Let τ and τ_S denote the Teichmüller metrics in the spaces T and T_S. Then the restriction $\tau|T_S$ is also a metric in T_S. From the definitions of τ and τ_S it follows immediately that

$$\tau|T_S \leq \tau_S. \tag{3.1}$$

It was for many years an open question whether the metrics τ_S and $\tau|T_S$ actually agree. We now know that T_S *does not inherit its metric from the universal Teichmüller space*: The metrics τ_S and $\tau|T_S$ need not be the same.

This was proved by Strebel [4] who gave two examples of surfaces S for which $\tau|T_S$ is strictly less than τ_S. In one case S is a punctured torus, in the other a compact surface of genus 2; cf. also 3.7 and 7.6.

Even though (3.1) does not always hold as an equality, the metrics $\tau|T_S$ and τ_S are topologically equivalent. In other words, the inclusion $(T_S, \tau_S) \to (T, \tau)$ is a homeomorphism onto its image. This will be proved in 4.6.

3.3. Completeness of Teichmüller Spaces

Lemma III.2.2 is true in every Teichmüller space T_S:

A Cauchy sequence in (T_S, τ_S) contains always a subsequence whose points have representatives μ_n such that $\lim \mu_n(z) = \mu(z)$ exists almost everywhere,

$[f_{\mu_n}] \to [f_\mu]$ in the τ_S-metric, $f_{\mu_n}(z) \to f_\mu(z)$ uniformly in the spherical metric, and $f^{\mu_n}(z) \to f^\mu(z)$ locally uniformly in H' in the euclidean metric.

The proof is word for word the same as in Lemma III.2.2. In this case every μ_n is a Beltrami differential for G, i.e., $(\mu_n \circ g)\overline{g'}/g' = \mu_n$. From $\mu_n(z) \to \mu(z)$ almost everywhere it follows that the limit μ also is a Beltrami differential for G, i.e., $[\mu]$ is a point of T_S.

From this observation we obtain a generalization of Theorem III.2.3.

Theorem 3.2. *The Teichmüller space* (T_S, τ_S) *is complete.*

For an application in section 9 we need the following result.

Lemma 3.1. *Let* $[\mu_n] \to [\mu]$ *in* T_S, $\|\mu_n\|_\infty \le k < 1$, *and* $\mu_n \to \nu$ *a.e. Then* $[\mu] = [\nu]$ *in* T_S.

PROOF. Let $\lambda_n \in [\mu_n]$ be an extremal complex dilatation for which $\tau_S([\mu_n], [\mu]) = \operatorname{artanh}\|(\lambda_n - \mu)/(1 - \bar\mu\lambda_n)\|_\infty$ (formula (2.3)). The hypothesis $[\mu_n] \to [\mu]$ then implies that $\lambda_n \to \mu$ in L^∞. By Theorem I.4.6, $f_{\mu_n} \to f_\nu$ and $f_{\lambda_n} \to f_\mu$. Since $f_{\mu_n}|H = f_{\lambda_n}|H$ it follows that $f_\mu|H = f_\nu|H$, and so $[\mu] = [\nu]$. □

3.4. Quasi-Fuchsian Groups

The mappings f_μ lead to discontinuous groups of Möbius transformations with an invariant domain different from a disc.

Theorem 3.3. *The mapping* $g \to f_\mu \circ g \circ f_\mu^{-1}$ *defines an isomorphism of the covering group* G *onto a group* G_μ *of Möbius transformations acting on the quasidisc* $f_\mu(H')$.

PROOF. Consider the quasiconformal mapping $f_\mu \circ g \circ f_\mu^{-1}$, $g \in G$, of the plane. It is conformal in $f_\mu(H)$, because $f_\mu|H$ is conformal. Since μ is a Beltrami differential for G, the mappings f_μ and $f_\mu \circ g$ have the same complex dilatation. It follows that $f_\mu \circ g \circ f_\mu^{-1}$ is conformal in $f_\mu(H')$ also. The common boundary of $f_\mu(H)$ and $f_\mu(H')$, being the image of the real axis under f_μ, is a quasicircle. We conclude, therefore, from Lemma I.6.1 that $f_\mu \circ g \circ f_\mu^{-1}$ is a Möbius transformation. □

By the terminology we adopted in IV.4.6, the group

$$G_\mu = \{f_\mu \circ g \circ f_\mu^{-1} | g \in G\}$$

is quasi-Fuchsian. A quasi-Fuchsian group of this special type is called a quasiconformal deformation of the Fuchsian group G. Such groups were discovered by Bers [4].

The invariant domain $f_\mu(H')$ is a half-plane if and only if μ is a *trivial* complex dilatation, i.e., μ is equivalent to the complex dilatation which is

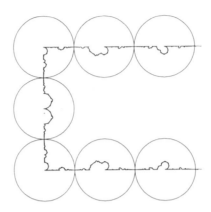

Figure 13. Invariant Jordan domain which is not a quasidisc.

identically zero. If G is of the first kind, the limit set of G_μ is the whole
boundary of $f_\mu(H')$. By Theorem IV.4.2, only two strikingly different cases are
then possible: Either the boundary is a straight line or else it is a quasicircle
which fails to have a tangent at each point of a set dense in the curve.

Not all simply connected invariant domains of properly discontinuous
groups of Möbius transformations are necessarily quasidiscs. A simple example is obtained if we consider a countable set of circles, all of diameter 1,
of which one has the center at 0 and the others at the points $\pm i + n, n = 0, 1,$
$2, \dots$. The method of Klein, which we described at the end of IV.4.6, applied
to this family of circles yields an invariant Jordan domain whose boundary
has a cusp at the origin. Thus the boundary curve violates the condition
(6.11) of Theorem I.6.7 at the origin ($z_2 = 0$, z_1 and $z_3 \to 0$) and cannot be a
quasicircle (Fig. 13).

There can even be invariant domains whose boundary has positive area
(Abikoff [1]). In 4.3 we shall exhibit a general (albeit implicit) method for
producing invariant domains which are not quasidiscs.

A point $[\mu] \in T_S$ uniquely determines the domains $A_\mu = f_\mu(H)$ and $A'_\mu =
f_\mu(H')$. Like the universal Teichmüller space, T_S can be regarded as a collec-
tion of the quasidiscs A_μ (cf. III.1.5). In III.4 we defined a distance between
two such domains by using Schwarzian derivatives. The relation of this
distance to the Teichmüller distance will be studied in section 4.

3.5. Quasiconformal Reflections Compatible with a Group

The point $[\mu] = p \in T_S$ determines uniquely the quasicircle $f_\mu(\mathbb{R})$. We show
here that $f_\mu(\mathbb{R})$ admits always quasiconformal reflections which are compatible with the covering group G.

In order to make this statement more precise, we choose a $\mu \in p$ and
consider the mapping

$$\lambda = f_\mu \circ j \circ f_\mu^{-1}; \tag{3.2}$$

as before j denotes the reflection $z \to \bar{z}$. This mapping is a quasiconformal reflection in $f_\mu(\mathbb{R})$.

By Theorem 3.3, the mapping f_μ induces an isomorphism of the group G acting on H' onto the quasi-Fuchsian group $G_\mu = \{f_\mu \circ g \circ f_\mu^{-1} | g \in G\}$ acting on A'_μ. But since the elements of G and G_μ are restrictions of Möbius transformations, we may consider g and $g_\mu = f_\mu \circ g \circ f_\mu^{-1}$ in H and A_μ also. If $z = f_\mu(\zeta)$, then

$$\lambda(g_\mu(z)) = f_\mu(\bar{g}(\zeta)) = f_\mu(g(\bar{\zeta})) = g_\mu(\lambda(z)).$$

We conclude that the sense-reversing quasiconformal mappings λ and $\lambda \circ g_\mu$ have the same complex dilatation. Therefore, if $\kappa_\lambda = \partial\lambda/\bar{\partial}\lambda$ denotes this complex dilatation, then

$$(\kappa_\lambda \circ g_\mu)(g'_\mu/\overline{g'_\mu}) = \kappa_\lambda \tag{3.3}$$

for every $g_\mu \in G_\mu$. In other words, *the complex conjugate of the complex dilatation of λ is a Beltrami differential for the quasi-Fuchsian group G_μ*.

Conversely, we prove that if λ is a quasiconformal reflection in $f_\mu(\mathbb{R})$ whose complex dilatation κ_λ satisfies the relation (3.3), then λ is of the form (3.2) (cf. Lemma I.6.2). Set $f = f_\mu$ in the closure of H, and $f = \lambda \circ f_\mu \circ j$ in H'. Then f is a quasiconformal mapping in the plane which is conformal in H. For $g \in G$, we have in $f_\mu(H')$,

$$f \circ g \circ f^{-1} = \lambda \circ f_\mu \circ g \circ f_\mu^{-1} \circ \lambda^{-1} = \lambda \circ g_\mu \circ \lambda^{-1}.$$

Because of (3.3), $\lambda \circ g_\mu \circ \lambda^{-1}$ is conformal. Hence $f \circ g \circ f^{-1}$ is a conformal self-mapping of $f_\mu(H')$. Since $f \circ g \circ f^{-1} = g_\mu$ in $f_\mu(H)$, it follows that $f \circ g \circ f^{-1} = g_\mu$ everywhere. We see that f is a mapping f_ν equivalent to f_μ, and so $\lambda = f \circ j \circ f^{-1}$ is of the form (3.2).

The point $[\mu] \in T_S$ determines the conformal mapping $f_\mu | H$ uniquely, but not $f_\mu | H'$ and hence not λ. In 4.8 we shall see that for each G and $[\mu]$, there are Lipschitz-continuous reflections (3.2). We also remark that by Theorem 4.5 (to be proved in section 4), there are reflections λ which are real-analytic in the complement of $f_\mu(\mathbb{R})$.

3.6. Quasisymmetric Functions Compatible with a Group

The mapping f^μ induces an isomorphism of the group G onto the Fuchsian group

$$G^\mu = \{g^\mu = f^\mu \circ g \circ (f^\mu)^{-1} | g \in G\}.$$

In particular, $(f^\mu | \mathbb{R}) \circ g \circ (f^\mu)^{-1} | \mathbb{R}$ agrees with the restriction to $\mathbb{R}$ of a Möbius transformation.

Let us consider again the space X of normalized quasisymmetric functions, which we defined in III.1.1 and studied in III.3. We relate X to the group G as follows: $X(G)$ is the subset of X for whose functions h the composition $h \circ g \circ h^{-1}$ is the restriction to $\mathbb{R}$ of a Möbius transformation for every $g \in G$. We let $X(G)$ inherit the metric of X, i.e., the distance in $X(G)$ is the restriction to $X(G)$ of the distance function ρ on X. Then $(X(G), \rho)$ is a metric space, and if G is trivial we get back our previous space (X, ρ).

Theorem III.3.1 states that the mapping $[\mu] \to f^\mu | \mathbb{R}$ is a homeomorphism of the universal Teichmüller space onto X. In order to generalize this theorem to T_S and $X(G)$, we need the following result.

Theorem 3.4. *Every quasisymmetric function* $h \in X(G)$ *has a quasiconformal extension f to the half-plane, such that the mapping $f \circ g \circ f^{-1}$ is conformal for every $g \in G$.*

The conformal mapping $f \circ g \circ f^{-1}$ is of course the restriction to the half-plane of the Möbius transformation which agrees with $h \circ g \circ h^{-1}$ on the real axis.

Theorem 3.4 is a deep result which has been established in steps. First, let G be the covering group of a compact surface. Then Theorem 3.4 follows if the classical result that $h \in X(G)$ can be extended to a homeomorphic self-mapping of the half-plane compatible with G is combined with Theorem 1.5. Kra [1] generalized this result by proving Theorem 3.4 for all finitely generated groups G. The author showed that Theorem 3.4 holds for all groups provided the quasisymmetry constant of h does not exceed $\sqrt{2}$.

In full generality, Theorem 3.4 is due to Tukia [1]. Tukia's proof is so long that to include it here would unbalance our presentation. For this reason, we content ourselves with a reference to Tukia's paper.

Quite recently, the author has learned of a forthcoming paper by Douady and Earle [1] which contains another proof of Theorem 3.4. This proof is more explicit than that of Tukia. It is possible to verify that, like the Beurling–Ahlfors extension, the Douady–Earle extension possesses the following three properties: 1° It is a diffeomorphism. 2° It is Lipschitz-continuous in the hyperbolic metric of the half-plane. 3° Its maximal dilatation has a bound which depends only on the quasisymmetry constant of the boundary function.

Thanks to these three properties, the Douady–Earle extension can be used in applications instead of the Beurling–Ahlfors extension. The additional property of the Douady–Earle extension of being compatible with the group action leads to important new results. The proof of the contractibility of the universal Teichmüller space in III.3.2 was based on the use of the Beurling–Ahlfors extension. Application of the Douady–Earle extension in its place makes it possible to solve a long outstanding problem (Douady–Earle [1]):

Every Teichmüller space is contractible.

In section 4 we shall see that Theorem 3.4, no matter how it is proved, reveals remarkable properties of Teichmüller space. Whenever possible, however, we shall establish such results by other methods, so as to render the proofs self-contained.

Using Theorem 3.4 we can now generalize Theorem III.3.1.

Theorem 3.5. *The mapping*

$$[\mu] \to f^\mu | \mathbb{R} \tag{3.4}$$

is a homeomorphism of (T_S, τ_S) *onto* $(X(G), \rho)$.

PROOF. By Theorem 3.1, the mapping (3.4) is well defined and injective. By Theorem 3.4, it is surjective.

By Theorem III.3.1, the mapping (3.4) is a homeomorphism of (T, τ) onto (X, ρ). Hence (3.4), which maps T_S bijectively onto $X(G)$, is a homeomorphism of (T_S, τ) onto $(X(G), \rho)$. From $\tau | T_S \leq \tau_S$ we thus conclude that (3.4) is a continuous mapping of (T_S, τ_S) onto $(X(G), \rho)$. The proof would be complete if we had an inequality in the opposite direction between $\tau | T_S$ and τ_S, to demonstrate that these two metrics are topologically equivalent. Such an inequality can be derived, for instance, by means of the right-hand inequality (5.10) in I.5.7. We content ourselves here with this remark, because we shall study the relationships between $\tau | T_S$ and τ_S in detail in the next section (Theorem 4.7). □

We remarked in 3.4 that if G is of the first kind, $f_\mu(\mathbb{R})$ is either a straight line or a far from smooth quasicircle. For the corresponding quasisymmetric functions $f^\mu | \mathbb{R}$ there is also a remarkable dichotomy. Suppose that G is a finitely generated covering group of the first kind (e.g., the covering group of a compact surface of genus > 1). If μ is a trivial complex dilatation, then $f^\mu | \mathbb{R}$ is the identity mapping. In all other cases $f^\mu | \mathbb{R}$ is a singular function (Mostow [1], Kuusalo [1]). The result holds for some more general groups G as well, but it remains an open question whether it is true for all covering groups of the first kind.

3.7. Unique Extremality and Teichmüller Metrics

Let F_h be the class of all quasiconformal self-mappings of the upper half-plane which agree with the quasisymmetric function h on the real axis. This class was introduced in I.5.7, where we noted that F_h always contains an extremal mapping with the smallest maximal dilatation in F_h. With the help of an example, it was shown that the extremal is not necessarily unique.

Let us now make the additional assumption that $h \in X(G)$ for a Fuchsian group G, i.e., that

$$h \circ g \circ h^{-1} = \theta(g) | \mathbb{R}, \tag{3.5}$$

where $\theta(g)$ is a Möbius transformation. We denote by $F_h(G)$ the subclass of F_h whose functions f satisfy the condition $f \circ g \circ f^{-1} = \theta(g) | H$. By Theorem 3.4, the class $F_h(G)$ is not empty, and again it contains an extremal mapping (cf. 2.2). If f_1 and f_2 are extremals in F_h and $F_h(G)$, respectively, with complex dilatations μ_i and maximal dilatations K_i, $i = 1, 2$, then

$$\tau(0, [\mu_1]) = \tfrac{1}{2} \log K_1, \qquad \tau_S(0, [\mu_2]) = \tfrac{1}{2} \log K_2. \tag{3.6}$$

The following observation establishes a connection between the extremals in F_h and $F_h(G)$. Along with f_1, the mapping $\theta(g) \circ f_1 \circ g^{-1}$ is also extremal in F_h. This can be verified immediately, in view of (3.5). Hence, if the extremal in F_h is unique, it follows that

$$f_1 \circ g \circ f_1^{-1} = \theta(g)|H, \tag{3.7}$$

i.e., that $f_1 \in F_h(G)$. Consequently, in this case we see directly that $F_h(G)$ is not empty, without resorting to Theorem 3.4. Of course, f_1 is extremal in $F_h(G)$, and by (3.6), we have $\tau(0, [\mu_1]) = \tau_S(0, [\mu_1])$.

We pointed out in 3.2 that there are cases in which $\tau_S(0, [\mu]) > \tau(0, [\mu])$. It follows that whenever this occurs, the class F_h contains more than one extremal for $h = f^\mu|\mathbb{R}$.

4. Teichmüller Space and Schwarzian Derivatives

4.1. Schwarzian Derivatives and Quadratic Differentials

Again let S be a Riemann surface and G the covering group of H' over S. If μ denotes a Beltrami differential for G, then by Theorem 3.1, the Teichmüller space T_S can be characterized as the set of conformal mappings $f_\mu|H$. Let us now form the Schwarzian derivative of $f_\mu|H$. By Theorem 3.3, the mapping $f_\mu \circ g \circ f_\mu^{-1}$ is a Möbius transformation for every $g \in G$. It follows that

$$S_{f_\mu|H} = S_{(f_\mu \circ g \circ f_\mu^{-1}) \circ f_\mu|H} = S_{f_\mu \circ g|H} = (S_{f_\mu|H} \circ g)g'^2. \tag{4.1}$$

We see that *the Schwarzian derivative of $S_{f_\mu|H}$ is a quadratic differential for the group G acting on H*. Its projection is a holomorphic quadratic differential on the mirror image of the surface S, i.e., on H/G.

What we can actually deduce from (4.1) by reading it from left to right and from right to left is the following result: Let f be a conformal mapping of the upper half-plane H. Then S_f is a quadratic differential for the group G if and only if $f \circ g \circ f^{-1}$ agrees with a Möbius transformation in $f(H)$ for every $g \in G$.

Assume, in addition, that f is a mapping f_μ with $[\mu] \in T$. We then have a third condition equivalent to the above two. For convenience of later reference we express all these conditions in a lemma.

Lemma 4.1. *The following three conditions are equivalent:*

$1°$ $S_{f_\mu|H}$ *is a quadratic differential for G;*
$2°$ $f_\mu \circ g \circ f_\mu^{-1}$ *agrees with a Möbius transformation in $f_\mu(H)$ for $g \in G$;*
$3°$ $f^\mu \circ g \circ (f^\mu)^{-1}$ *agrees with a Möbius transformation on $\mathbb{R}$ for $g \in G$.*

PROOF. As we already remarked, the equivalence of $1°$ and $2°$ follows directly from (4.1). If $2°$ holds, i.e., if $f_\mu \circ g \circ f_\mu^{-1} = w$ in $f_\mu(H)$, where w is a Möbius transformation, then $h = f^\mu \circ f_\mu^{-1} \circ w \circ f_\mu \circ (f^\mu)^{-1} \colon H' \to H'$ is a Möbius transformation which coincides with $f^\mu \circ g \circ (f^\mu)^{-1}$ on $\mathbb{R}$. Hence $3°$ follows from $2°$.

Conversely, assume that $3°$ holds, i.e., that $f^\mu \circ g \circ (f^\mu)^{-1} = h$ on $\mathbb{R}$, where h is a Möbius transformation. Set $w = f_\mu \circ (f^\mu)^{-1} \circ h \circ f^\mu \circ f_\mu^{-1}$ in the closure of $f_\mu(H')$, and $w = f_\mu \circ g \circ f_\mu^{-1}$ in $f_\mu(H)$. Then w is a homeomorphism of the plane and is conformal in $f_\mu(H)$ and $f_\mu(H')$. Hence w is a Möbius transformation, and so $2°$ follows from $3°$.   □

4.2. Spaces of Quadratic Differentials

Following the procedure in III.4, we now introduce the counterpart of the space Q. In what follows we regard G as acting in the upper half-plane H.

Let $Q(G)$ be the space consisting of all functions φ holomorphic in the upper half-plane H which are quadratic differentials for G and have a finite hyperbolic sup norm:

$$\|\varphi\| = \sup_{z \in H} 4y^2 |\varphi(z)| < \infty,$$

$z = x + iy$. Like the space Q defined in III.4.1, the space $Q(G)$ has a natural linear structure over the complex numbers.

The non-euclidean line element $|dz|/(2y)$ is invariant under all conformal self-mappings of H. It follows that, when $\varphi \in Q(G)$, $y^2|\varphi(z)|$ is invariant under G. In the definition of the norm, we can therefore replace H by an arbitrary Dirichlet region $N \subset H$ of G:

$$\|\varphi\| = \sup_{z \in N} 4y^2 |\varphi(z)|. \tag{4.2}$$

If S is a compact surface, the closure of N lies in H. In this case it follows from (4.2) that all holomorphic quadratic differentials for G have finite norm.

If G_1 is a subgroup of G_2, we have the inclusion $Q(G_2) \subset Q(G_1)$. In particular, all spaces $Q(G)$ are subsets of the Banach space Q which corresponds to the trivial group. From now on we write $Q = Q(1)$. Note that all spaces $Q(G)$ inherit a metric from $Q(1)$; in III.4 we introduced the notation q for this distance function.

Every $Q(G)$ is a closed subspace of $Q(1)$ and hence a Banach space. For consider functions $\varphi_n \in Q(G)$ which converge to φ in $Q(1)$. Given a $g \in G$, we then have $\varphi_n(z) \to \varphi(z)$, $\varphi_n(g(z)) \to \varphi(g(z))$, uniformly on every compact subset of H. It follows that $\varphi(g(z))g'(z)^2 = \lim \varphi_n(g(z))g'(z)^2 = \lim \varphi_n(z) = \varphi(z)$. Consequently, $\varphi \in Q(G)$.

4.3. Schwarzian Derivatives of Univalent Functions

All points of $Q(1)$, and hence of $Q(G)$, are Schwarzian derivatives of functions f meromorphic and locally injective in H. Let us consider the set

$$U(G) = \{\varphi = S_f \in Q(G) | f \text{ univalent in } H\}.$$

Between $U(G)$ and $U = U(1)$ we have the simple relation

$$U(G) = U(1) \cap Q(G).$$

We just saw that $Q(G)$ is closed in $Q(1)$. In III.4.4 we proved that $U(1)$ is closed in $Q(1)$. It follows that $U(G)$ *is a closed subset of* $Q(1)$.

Since $\|S_f\|_H \leq 6$ whenever f is univalent, the remark in 4.1 preceding Lemma 4.1 yields another characterization of $U(G)$:

The set $U(G)$ consists of the Schwarzian derivatives of those functions f univalent in H for which $f \circ g \circ f^{-1}$ agrees with a Möbius transformation in $f(H)$ for every $g \in G$. In other words, every f with $S_f \in U(G)$ induces an isomorphism of the group G onto the group $G_f = \{f \circ g \circ f^{-1} | g \in G\}$ of Möbius transformations acting on the domain $f(H)$.

Let $T(G)$ be the subset of $U(G)$ consisting of the Schwarzian derivatives $S_f \in U(G)$ of functions f which admit a quasiconformal extension to the plane with a complex dilatation that is a Beltrami differential for G. *The set* $T(G)$ *is the image of the Teichmüller space* T_S *under the mapping*

$$[\mu] \to S_{f_\mu|H}. \tag{4.3}$$

This is clear: the normalization of the mappings $f_\mu|H$, which fix 0, 1 and ∞, is unessential when we are considering Schwarzian derivatives.

In III.4 we proved that (4.3) is a homeomorphism of the universal Teichmüller space (T, τ) onto its image $T(1)$ in $(Q(1), q)$, and that $T(1)$ is an open subset of $Q(1)$. We shall now start proving a sequence of theorems which, in combination, assert that the restriction of (4.3) to T_S maps (T_S, τ_S) homeomorphically onto $T(G)$ and that $T(G)$ is an open subset of $Q(G)$. From $\tau|T_S \leq \tau_S$ we conclude immediately that (4.3) maps (T_S, τ_S) continuously into $Q(G)$.

Let us briefly return to the Möbius groups G_f induced by functions f with $S_f \in U(G)$. If $S_f \in U(G) \setminus T(G)$, then the G_f-invariant domain $f(H)$ is not a quasidisc. This follows from relation (4.4), which we shall establish in the next subsection. The groups G_f with S_f lying on the boundary of $T(G)$ are of particular interest and have been studied extensively by Maskit [1], Abikoff [3], and others. (It is not known whether the inclusion $\partial T(G) \subset U(G) \setminus T(G)$ is proper in cases where G is not trivial; cf. III.4.6.)

4.4. Connection between Teichmüller Spaces and the Universal Space

For the sets $T(G)$ we have the natural inclusion $T(G_2) \subset T(G_1)$ if G_1 is a subgroup of G_2. Hence, all sets $T(G)$ are subsets of $T(1)$.

The following fundamental result connects an arbitrary Teichmüller space in a simple manner with the universal Teichmüller space.

Theorem 4.1. *The Teichmüller spaces satisfy the relation*

$$T(G) = Q(G) \cap T(1). \tag{4.4}$$

PROOF. The inclusion $T(G) \subset Q(G) \cap T(1)$ follows directly from the definitions. We choose an arbitrary point $S_f \in Q(G) \cap T(1)$ and prove that $S_f \in T(G)$.

Let w be a conformal mapping of the lower half-plane H' onto the complement of the closure of $f(H)$, normalized so that $w^{-1}(f(\infty)) = \infty$. Since $S_f \in T(1)$, the boundary of $f(H)$ is a quasicircle. Hence, the function $h = w^{-1} \circ f$, defined on the real axis, is quasisymmetric; we may assume that h is normalized. Furthermore, for $g \in G$,

$$h \circ g = w^{-1} \circ f \circ g \circ f^{-1} \circ f = w^{-1} \circ (f \circ g \circ f^{-1}) \circ w \circ h. \tag{4.5}$$

Since $S_f \in U(G)$, the mapping $f \circ g \circ f^{-1}$ agrees with a Möbius transformation g_1 in $f(H)$. Then $w^{-1} \circ g_1 \circ w$, which maps H' onto itself, agrees with a Möbius transformation g_2 in H'. It follows from (4.5) that h induces an isomorphism of G onto the group $\{g_2 | g \in G\}$ of Möbius transformations acting on H'. In other words $h \in X(G)$.

Next we utilize Theorem 3.4. It follows that there is a quasiconformal extension φ of h to the lower half-plane which also fulfills the condition $\varphi \circ g = g_2 \circ \varphi$ for every $g \in G$. Then $f_1 = w \circ \varphi$ is a quasiconformal extension of f to the lower half-plane. For $g \in G$ we have

$$f_1 \circ g = w \circ g_2 \circ \varphi = g_1 \circ w \circ \varphi = g_1 \circ f_1.$$

This shows that $f_1 \circ g \circ f_1^{-1}$ agrees with a Möbius transformation in $f_1(H')$, and it follows that $S_f \in T(G)$.   □

In the sixties, Bers posed the problem whether (4.4) is true. The above proof is due to Lehto [2] who proved the conditional result that (4.4) holds if and only if Theorem 3.4 is true. After Theorem 3.4 was established, the relation (4.4) thus followed immediately.

Theorem 4.1 allows important conclusions:

Theorem 4.2. *The set $T(G)$ is closed in $T(1)$.*

PROOF. The relation (4.4) is equivalent to $T(G) = U(G) \cap T(1)$. Since $U(G)$ is closed in $Q(1)$, the theorem follows.   □

Theorem 4.3. *The set $T(G)$ is open in $Q(G)$.*

PROOF. This can be read from (4.4), since $T(1)$ is open in $Q(1)$.   □

Theorems 4.2 and 4.3 can be proved more directly, without the use of the relation (4.4). Such alternate proofs will be given in subsections 4.6 and 4.7. Bers [8] was the first to prove that $T(G)$ is open; see also Earle [1].

Remark. Let S be the extended plane punctured at three points. Then S has the half-plane as its universal covering surface. We noted in 2.7 that the Teichmüller space of S reduces to a single point. Consequently, $T(G)$ consists of zero only. Since $T(G)$ is an open subset of the linear space $Q(G)$, we deduce via the Teichmüller theory that in this case, $Q(G)$ does not contain points other than the origin.

4.5. Distance to the Boundary

By Theorem 4.3, the mapping (4.3) carries a neighborhood of the origin of T_S onto a ball of $Q(G)$ centered at the origin, but we can say more.

Theorem 4.4. *The ball*

$$B(0, 2) = \{\varphi \in Q(G) | \,\|\varphi\| < 2\}$$

lies in $T(G)$.

PROOF. In III.4.3 we remarked that $\{\varphi \in Q(1)|\,\|\varphi\| < 2\}$ lies in $T(1)$ (Theorem II.5.1). Hence, the theorem follows immediately from (4.4). □

There is another more direct way to show that $B(0, 2)$ is contained in $T(G)$. Let $\varphi \in B(0, 2)$ and set $\mu(\bar{z}) = -2y^2 \varphi(z)$. Since $\varphi(g(z))g'(z)^2 = \varphi(z)$ and $1/y = |g'(z)|/\text{Im}\,g(z)$, we see that μ is a Beltrami differential for G. Hence $[\mu] \in T_S$. From what was said in III.4.3 we know that the mapping

$$\varphi \to [z \to -2y^2 \varphi(\bar{z})] \tag{4.6}$$

is the inverse of $[\mu] \to s_\mu = S_{f_\mu}|_H$ in $B(0, 2)$.

Since $\|s_\mu\| \leq 6\beta([\mu], 0)$ (formula (III.4.1)), we conclude that the ball

$$\{[\mu] \in T_S | \beta([\mu], 0) < 1/3\} \tag{4.7}$$

is contained in the preimage of $B(0, 2)$.

Theorem II.5.1 says that the largest ball in $T(1)$ centered at the origin has the radius 2. The Schwarzian derivative of the logarithm is a boundary point of $T(1)$ with distance 2 from 0. In the upper half-plane the logarithm is compatible with the group consisting of all Möbius transformations of the form $z \to az$ or $z \to -a/z$, where a is a positive real number. It follows that if G is a properly discontinuous subgroup of this group, then $B(0, 2)$ is the largest ball in $T(G)$ with center at 0. For instance, this is the case in the Teichmüller space of an annulus (cf. IV.4.3). For an arbitrary G, the determination of the largest ball in $T(G)$ centered at 0 remains an open problem.

We shall use the mapping (4.6) in section 5 in studying the complex analytic structure of T_S. Here we draw the following conclusion.

Theorem 4.5. *Every point of the Teichmüller space* T_S *can be represented by a real analytic Beltrami differential and by a real analytic quasiconformal mapping.*

PROOF. Let a point $[\mu] = p \in T_S$ be given. Suppose first that p can be represented by a quasiconformal mapping whose maximal dilatation is < 2. Then

p lies in the set (4.7), and so p can be represented by $z \to -2y^2 \varphi(\bar{z})$, which is a real analytic complex dilatation. We also have an explicit expression for the corresponding mapping f^μ (cf. II.5.1 and II.5.2) from which it becomes apparent that f^μ is real analytic.

The general case is handled by induction. Assuming that the theorem is true if $\tau_S(p, 0) < r$, we show that it holds if $\tau_S(p, 0) < 2r$. Let f^μ be an extremal for the point p, with $\tau_S(p, 0) < 2r$. We write $f^\mu = f^{\mu_1} \circ f^{\mu_2}$, where $\mu_2(z)$ is the middle point of the line segment from 0 to $\mu(z)$ in the non-euclidean metric of the unit disc (cf. Theorem III.2.2). Then μ_2 is a Beltrami differential for G and μ_1 a Beltrami differential for G^{μ_2}. Moreover, $\tau_S([\mu_2], 0) < r$ and $\tau_{S'}([\mu_1], 0) < r$ with $S' = H'/G^{\mu_2}$. From this the theorem follows. □

Theorem 4.4 can be generalized. Every point $[\mu] \in T_S$ determines uniquely the quasidisc $A_\mu = f_\mu(H)$. Putting together a large part of our previous analysis, we obtain a lower bound for the distance to the boundary of the point s_μ of $T(G)$ in terms of the inner radius of univalence σ_I (defined in III.5.1) of A_μ.

Theorem 4.6. *For every* $s_\mu \in T(G)$, *the ball*

$$B(s_\mu, \sigma_I(A_\mu)) = \{\varphi \in Q(G) | q(s_\mu, \varphi) < \sigma_I(A_\mu)\}$$

is contained in $T(G)$.

PROOF. In III.5.3 we proved that $\{\varphi \in Q(1) | q(s_\mu, \varphi) < \sigma_I(A_\mu)\}$ lies in $T(1)$. Consequently, the theorem follows immediately from (4.4). □

We recall that in $T(1)$, the inner radius $\sigma_I(A_\mu)$ is precisely the distance from s_μ to the boundary (Theorem III.5.1).

4.6. Equivalence of Metrics

The topological equivalence of the metrics τ_S and $\tau | T_S$ can be proved with the aid of Schwarzian derivatives or of quasisymmetric functions. Neither method goes to the heart of the matter, but both give the desired result easily. The method based on the use of Schwarzian derivatives produces better constants. In fact, we can prove that the metrics are even uniformly equivalent, i.e., the identity mapping $(T_S, \tau | T_S) \to (T_S, \tau_S)$ and its inverse are uniformly continuous.

Theorem 4.7. *The metric* τ_S *of the Teichmüller space* T_S *is uniformly equivalent to the metric* $\tau | T_S$ *induced on* T_S *by the metric of the universal Teichmüller space. For every point* $p \in T_S$,

$$\limsup_{q \to p} \frac{\tau_S(p, q)}{\tau(p, q)} \leq 3. \tag{4.8}$$

More generally, on every bounded set $A \subset T_S$,

$$\tau_S(p, q) \le 3(1 + \text{dia } A)\tau(p, q), \tag{4.9}$$

where dia A *denotes the diameter of A in the τ_S-metric.*

PROOF. Let $p \in T_S$ and $\mu \in p$. We know that

$$q(s_\mu, 0) \le 6\beta(p, 0).$$

On the other hand, it follows from Theorem 4.4 and formula (4.6) that

$$q(s_\mu, 0) \ge 2\beta_S(p, 0).$$

Hence,

$$\beta_S(p, 0) \le 3\beta(p, 0). \tag{4.10}$$

In order to generalize this estimate for an arbitrary pair of points of T_S, we fix $p = [\mu] \in T_S$ and consider the mapping $\tilde{\alpha}_\mu$, defined by

$$f^{\tilde{\alpha}_\mu(v)} = f^v \circ (f^\mu)^{-1}.$$

The function $\tilde{\alpha}_\mu$ induces a well-defined mapping

$$[v] \to \alpha_\mu([v]) = [\tilde{\alpha}_\mu(v)], \tag{4.11}$$

which is an isometric bijection of the universal Teichmüller space T onto itself. It maps T_S onto the Teichmüller space T_{S^μ}, where $S^\mu = H'/G^\mu$, and it is an isometry of (T_S, τ_S) onto $(T_{S^\mu}, \tau_{S^\mu})$. The last assertion follows directly from the definition of the Teichmüller metric.

The mapping (4.11) takes the points $p = [\mu]$ and $q = [v]$ of T_S to the points 0 and $[\lambda] = \alpha_\mu([v])$ of T_{S^μ}. Hence, by (4.10),

$$\beta_S(p, q) = \beta_{S^\mu}(0, [\lambda]) \le 3\beta(0, [\lambda]) = 3\beta(p, q).$$

Since trivially $\beta \le \beta_S$, we have established the double inequality

$$\beta(p, q) \le \beta_S(p, q) \le 3\beta(p, q) \tag{4.12}$$

for all points p and q of T_S.

The inequalities (4.12) show that the metrics $\beta | T_S$ and β_S are uniformly equivalent. Since

$$\beta = \tanh \tau, \tag{4.13}$$

it follows that the Teichmüller metrics $\tau | T_S$ and τ_S are also uniformly equivalent.

From (4.12) and (4.13) we obtain

$$\tau_S \le \frac{1}{2} \log \frac{1 + 3\tau}{1 - 3\tau} \le \frac{3\tau}{1 - 3\tau},$$

if $\tau < 1/3$. This yields inequality (4.8). Also

$$\tau_S \leq \frac{3}{1 - 3a} \tau \tag{4.14}$$

if $\tau \leq a < 1/3$.

Let A be a set in T_S with a finite diameter dia A in the τ_S-metric, and p and q points of A. If $\tau(p, q) \leq a$, we just proved that (4.14) holds. If $\tau(p, q) > a$, then

$$\tau_S(p, q) \leq \text{dia } A < \frac{\text{dia } A}{a} \tau(p, q).$$

By choosing $a = \text{dia } A/3(1 + \text{dia } A)$ we obtain (4.9). $\quad\square$

We proved in 3.3 that the Teichmüller space (T_S, τ_S) is complete. If the result is combined with Theorem 4.7, we conclude that T_S is a closed subset of T. Since the same is true of their homeomorphic images, we have reproved Theorem 4.2: $T(G)$ is a closed subset of $T(1)$.

4.7. Bers Imbedding

Theorem 4.7 makes it possible to generalize Theorem III.4.1 immediately (Bers [8]).

Theorem 4.8. *The mapping*

$$[\mu] \to S_{f_\mu | H} \tag{4.15}$$

is a homeomorphism of (T_S, τ_S) *onto* $(T(G), q)$.

PROOF. By Theorem III.4.1, this mapping is a homeomorphism of $(T_S, \tau | T_S)$ onto $T(G)$. By Theorem 4.7, the metrics τ_S and $\tau | T_S$ are equivalent, and the theorem follows. $\quad\square$

In view of Theorems 4.3 and 4.8, we are again justified in calling the mapping (4.15) the Bers imbedding of the Teichmüller space.

In 4.5 we proved that $B(0, 2) \subset T(G)$ without utilizing relation (4.4). We shall now show, without resorting to (4.4), that every point s_μ of $T(G)$ has an open neighborhood in $Q(G)$ which is contained in $T(G)$.

To prove this, let us consider in $T(1)$ the mapping ψ defined by $\psi(s_v) = s_{\tilde{\alpha}_\mu(v)}$. If λ is the Bers imbedding of the universal Teichmüller space onto $T(1)$, then $\psi = \lambda \circ \alpha_\mu \circ \lambda^{-1}$. (Here $\tilde{\alpha}_\mu$ and α_μ are as in 4.6.) It follows that ψ is a homeomorphism of $T(1)$ onto itself. Furthermore, ψ maps $T(G)$ onto $T(G^\mu)$, and $Q(G) \cap T(1)$ is mapped onto $Q(G^\mu) \cap T(1)$ (Lemma 4.1).

Let V be an open ball in $Q(G^\mu)$ centered at the origin which is contained in $T(G^\mu)$. Write $V = Q(G^\mu) \cap T(1) \cap V_0$, where V_0 is a neighborhood of the origin in $T(1)$. Since $T(1)$ is open in $Q(1)$, the preimage $\psi^{-1}(V) = Q(G) \cap \psi^{-1}(V_0)$ is a neighborhood of s_μ in $Q(G)$. This neighborhood is contained in $\psi^{-1}(T(G^\mu)) = T(G)$.

We have proved without making use of (4.4) that $T(G)$ is closed in $T(1)$ and open in $Q(G)$. These results imply that $T(G)$ is open and closed in $Q(G) \cap T(1)$. Since $T(G)$ is connected, it follows that $T(G)$ is the component of $Q(G) \cap T(1)$ containing the origin. This result can be used instead of (4.4) in proving Theorem 4.6.

Remark. In III.4.5 we proved that $T(1) = \operatorname{int} U(1)$. It follows that $T(1) = \operatorname{int} \operatorname{cl} T(1)$, i.e., every neighborhood of every boundary point of $T(1)$ contains points which are in the complement of the closure of $T(1)$.

In the general case it is not known whether $T(G)$ agrees with the interior of $U(G)$. However, it is true that

$$T(G) = \operatorname{int} \operatorname{cl} T(G). \tag{4.16}$$

For finitely generated groups G of the first kind, this was proved by Abikoff [3] (who resorted to an unpublished result of Thurston). Quite recently, (4.16) was established by Bers [13] in the general case. In a note, Žuravlev [1] has announced the result that $T(G)$ is the component of $\operatorname{int} U(G)$ containing the origin. By aid of Žuravlev's result, Shiga [1] proved that if G is finitely generated and of the first kind, then indeed $T(G) = \operatorname{int} U(G)$.

4.8. Quasiconformal Extensions Compatible with a Group

We conclude this section by showing that for suitable reflections, the Ahlfors extension of Theorem II.4.1 is compatible with the action of a group.

Given a point S_{f_μ} of $T(G)$ and hence a quasicircle $f_\mu(\mathbb{R})$, we consider quasiconformal reflections $\lambda = f_\mu \circ j \circ f_\mu^{-1}$ for all μ's in the equivalence class. We show that there always exists a μ such that λ is Lipschitz-continuous.

Since f_μ is compatible with the group G, i.e., $f_\mu \circ g \circ f_\mu^{-1}$ is a Möbius transformation for every $g \in G$, it follows from Lemma 4.1 that $f^\mu | \mathbb{R}$ is compatible with G. We construct a quasiconformal mapping $\psi: H' \to H'$ with boundary values $f^\mu | \mathbb{R}$ by means of the Douady–Earle method (see 3.6). Then ψ is G-compatible.

Let us consider $f_\mu \circ (f^\mu)^{-1} \circ \psi$ in H'. It agrees with f_μ on $\mathbb{R}$, and it is G-compatible, because f_μ, f^μ, and ψ are G-compatible. We take $f_\mu \circ (f^\mu)^{-1} \circ \psi$ to be the extension of $f_\mu | H$ to the lower half-plane.

Now ψ is a diffeomorphism and Lipschitz-continuous in the hyperbolic metric of H'. We conclude exactly as in Lemma I.6.4 that with our choice of $f_\mu | H'$, the reflection $\lambda = f_\mu \circ j \circ f_\mu^{-1}$ is Lipschitz-continuous and, with the exception of $f_\mu(\mathbb{R})$, continuously differentiable. In 3.5 we proved that

$$\lambda \circ g_\mu = g_\mu \circ \lambda \tag{4.17}$$

for every element g_μ of the deformed group $G_\mu = \{g_\mu = f_\mu \circ g \circ f_\mu^{-1} | g \in G\}$.

Let f be a univalent function in $A_\mu = f_\mu(H)$ compatible with G_μ. Following the proof of Theorem II.4.1, we use the representation $f = w_1/w_2$, where w_1, w_2 is a normalized pair of solutions of the differential equation $w'' = -\frac{1}{2}S_f w$. We assume that f is holomorphic on ∂A_μ, and set for $z \in A_\mu$,

$$w(z) = \frac{w_1(z) + (\lambda(z) - z)w_1'(z)}{w_2(z) + (\lambda(z) - z)w_2'(z)}.$$

By Theorem II.4.1, there is a positive constant ε, which depends only on $[\mu]$, such that if $\|S_f\| < \varepsilon$, then $w \circ \lambda$ is a quasiconformal extension of f. (A modification with the help of a Möbius transformation is required, because A_μ is unbounded. The bound ε need not be the same as in Theorem II.4.1, because λ is constructed differently.) Assuming that $\|S_f\| < \varepsilon$, we shall prove, as a supplement to Theorem II.4.1:

The quasiconformal extension $w \circ \lambda$ of f is compatible with the group G_μ.

PROOF. The result follows by direct computation. First of all (cf. III.5.4),

$$\mu_w(z) = \frac{\partial w}{\partial \bar{w}} = \frac{\partial \lambda(z)}{\partial \lambda(z)} + \frac{(\lambda(z) - z)^2 S_f(z)}{2\bar{\partial}\lambda(z)}.$$

Making use of the identity $(g(z_1) - g(z_2))^2 = g'(z_1)g'(z_2)(z_1 - z_2)^2$, which holds for all Möbius transformations g, and because of (4.17), we obtain

$$(\lambda(g_\mu(z)) - g_\mu(z))^2 = (g_\mu(\lambda(z)) - g_\mu(z))^2 = g_\mu'(\lambda(z))g_\mu'(z)(\lambda(z) - z)^2.$$

Furthermore,

$$\bar{\partial}\lambda(g_\mu(z)) = \bar{\partial}\lambda(z)g_\mu'(\lambda(z))/\bar{g}_\mu'(z),$$

and by our hypothesis, $S_f(g_\mu(z)) = S_f(z)g_\mu'(z)^{-2}$. These formulas yield

$$\frac{(\lambda(g_\mu(z)) - g_\mu(z))^2 S_f(g_\mu(z))}{2\bar{\partial}\lambda(g_\mu(z))} = \frac{(\lambda(z) - z)^2 S_f(z)}{2\bar{\partial}\lambda(z)} \cdot \frac{\bar{g}_\mu'(z)}{g_\mu'(z)}.$$

From this and (4.17) it follows that

$$\mu_w \circ g_\mu = \mu_w \bar{g}_\mu'/g_\mu'. \tag{4.18}$$

By (4.17),

$$w \circ \lambda \circ g_\mu \circ \lambda^{-1} \circ w^{-1} = w \circ g_\mu \circ w^{-1}.$$

Combined with (4.18), this shows that $w \circ \lambda$ is compatible with G_μ. □

5. Complex Structures on Teichmüller Spaces

5.1. Holomorphic Functions in Banach Spaces

With the aid of the Bers imbedding we can introduce a natural complex analytic structure into the Teichmüller space T_S. The Teichmüller space thus becomes a complex analytic Banach manifold, a generalization of the notion of a complex analytic n-manifold defined in IV.1.2.

First of all, we shall recall the definition of holomorphic functions in Banach spaces. Let E and F be Banach spaces over the complex numbers, and $U \subset E$ an open set. A function $f: U \to F$ has a derivative at a point $x_0 \in U$ if there exists a continuous complex linear mapping $Df(x_0): E \to F$ such that

$$\lim_{h \to 0} \frac{\| f(x_0 + h) - f(x_0) - Df(x_0)(h) \|_F}{\| h \|_E} = 0.$$

The mapping $Df(x_0)$ is called the derivative of f at x_0. A function $f: U \to F$ which has a derivative at every point of U is said to be *holomorphic* in U.

The composition of holomorphic functions is holomorphic where defined. A holomorphic function $f: U \to f(U)$ is *biholomorphic* if it has a holomorphic inverse. In the case $E = \mathbb{C}^m$, $F = \mathbb{C}^n$, the above definition coincides with the usual notion of holomorphic functions.

In order to get a connection with ordinary complex-valued analytic functions, we introduce the dual F^* of F. The set F^* consists of all continuous complex linear mappings of F into $\mathbb{C}$. The norm

$$\| x^* \|_{F^*} = \sup\{|x^*(x)| \, | \, \| x \|_F \le 1\}$$

makes F^* a Banach space. A set $A \subset F^*$ is called total if $\alpha(x) = 0$ for every $\alpha \in A$ implies $x = 0$.

There are two characterizations which, taken together, make it possible to consider only complex-valued functions of a complex variable when it comes to checking whether a mapping between Banach spaces is holomorphic.

Lemma 5.1. *A function $f: U \to F$ is holomorphic if and only if it satisfies one of the following two conditions:*

(i) *For every $x \in U$ and $e \in E$, the function $w \to f(x + we)$ is a holomorphic function on an open neighborhood of the origin in $\mathbb{C}$ with values in F.*

(ii) *The function $f: U \to F$ is continuous and there exists a total subset A of the dual F^* such that, for every $\alpha \in A$, the function $\alpha \circ f: U \to \mathbb{C}$ is holomorphic.*

Conditions (i) and (ii) are given in Bourbaki [1], § 3.3.1.

5.2. Banach Manifolds

A complex Banach manifold M is a Hausdorff space with an open covering of sets each of which is homeomorphic to an open subset of a complex Banach space (not necessarily the same for the open sets of M). Suppose M has an atlas in which all parameter transformations are biholomorphic. A maximal atlas with this property is called a complex (analytic) structure on

M. A manifold *M* with a complex analytic structure is called a *complex analytic Banach manifold*.

Theorems 4.8 and 4.3 assert that the Teichmüller space T_S is actually globally homeomorphic to an open set in the complex Banach space $Q(G)$, which consists of the quadratic differentials with finite norm for the covering group *G*. Hence, by using this result, we could make the Teichmüller space a complex analytic Banach manifold.

We shall show that, in fact, a natural complex structure for T_S is obtained in this manner. However, the required verifications are easiest to carry out with the results at our disposal if we introduce the complex structure a little differently, using local parameter mappings.

A complex analytic structure can be given to a Teichmüller space in several ways. This is particularly true in the case of compact Riemann surfaces. On the other hand, the various approaches lead to isomorphic structures, so that we are free to speak of a canonical complex structure on a Teichmüller space. Here we shall be dealing with the general case, excluding only those Riemann surfaces which do not have a disc as a universal covering surface. That we arrive at a natural complex structure is seen from Theorems 5.2–5.6.

In section 9 we shall construct the "Teichmüller imbedding" which shows that the Teichmüller space of a Riemann surface of genus *p* is homeomorphic to $\mathbb{C}^{3p-3}$. However, the complex structure the Teichmüller spaces inherit from $\mathbb{C}^{3p-3}$ through this imbedding is not a natural one: Given two Riemann surfaces *S* and *S'* of genus *p*, the bijective isometry between T_S and $T_{S'}$ induced by a quasiconformal mapping of *S* onto *S'* is usually not biholomorphic with respect to these structures. Teichmüller was aware of this state of affairs, but in one of his last papers (Teichmüller [3]), he claimed to have proved the existence of the "right" complex structure. This paper is difficult to read, and today Teichmüller's reasoning on this point is not regarded as convincing.

The first complete proof for the existence of the complex structure in Teichmüller spaces of compact surfaces is due to Ahlfors [2]; see also Bers [2]. Subsequently Bers ([7], [8]) introduced complex structure into an arbitrary Teichmüller space by means of the mapping $\mu \to S_{f_\mu}$.

5.3. A Holomorphic Mapping between Banach Spaces

The introduction of the complex structure on the Teichmüller space T_S is based on the following result.

Theorem 5.1. *The function*

$$\mu \to \Lambda(\mu) = S_{f_\mu|H}, \tag{5.1}$$

which maps the open unit ball $B(G)$ *of the space of measurable* $(-1, 1)$-*differentials for G into the space* $Q(G)$ *of holomorphic quadratic differentials for G, is holomorphic.*

PROOF. We have already seen that $Q(G)$ is a Banach space. The ball $B(G)$ is an open subset of the Banach space $L^\infty(G)$ of measurable $(-1, 1)$-differentials for G with finite L^∞-norm. Fix $\mu, v \in B(G)$.

For $z \in H$ and $\varphi \in Q(G)$, we set $\alpha_z(\varphi) = \varphi(z)$. Then $A = \{\alpha_z | z \in H\}$ is a total set in the dual of $Q(G)$. We apply condition (ii) of Lemma 5.1 to the function

$$w \to s(w) = S_{f_{\mu+wv}|H}. \tag{5.2}$$

The set U in condition (ii) is now the neighborhood $\{w \,|\, |w| < (1 - \|\mu\|_\infty)/\|v\|_\infty\}$ of the origin in the complex plane. Then $\mu + wv \in B(G)$. Furthermore, let $F = Q(G)$ and $\alpha = \alpha_z$.

By Corollary II.3.1 and the Remark following it, the function

$$w \to \alpha_z \circ s = S_{f_{\mu+wv}|H}(z)$$

is holomorphic in U for every $z \in H$. By formula (III.4.4), the function s is continuous in U. Hence, by condition (ii) of Lemma 5.1, the function (5.2) is holomorphic in U. Using this fact we conclude from condition (i) of Lemma 5.1 that (5.1) is holomorphic in $B(G)$. $\qquad\square$

Even though the definition of the complex structure on T_S with the aid of the holomorphic mapping (5.1) requires a number of auxiliary mappings, as illustrated by the diagram in Fig. 14, the idea is quite simple. The ball $B(0, 2)$ of $Q(G)$ plays a distinguished role, because it follows from what was said in 4.5 that (5.1) has a holomorphic section there. In the preimage of $B(0, 2)$ in T_S we take $[v] \to S_{f_v|H}$ as a local parameter. Near an arbitrary point $[\mu]$ of T_S a local parameter is obtained if we first map T_S isometrically onto T_{S^μ} in such a way that $[\mu]$ is moved to the origin. An easy verification shows that the parameter transformations are biholomorphic. In 5.4 and 5.5 we shall explain all this in detail.

5.4. An Atlas on the Teichmüller Space

Let us now introduce the auxiliary mappings which are needed in the definition of the complex analytic structure on T_S. For a given $\mu \in B(G)$, we again write $G^\mu = \{f^\mu \circ g \circ (f^\mu)^{-1} | g \in G\}$. As in 4.6, we consider the mapping $\tilde{\alpha}_\mu: B(G) \to B(G^\mu)$, defined by

$$f^{\tilde{\alpha}_\mu(v)} = f^v \circ (f^\mu)^{-1},$$

or, in more explicit terms, by

$$\tilde{\alpha}_\mu(v) = \left(\frac{v - \mu}{1 - \bar{\mu}v}\left(\frac{\partial f^\mu}{|\partial f^\mu|}\right)^2\right) \circ (f^\mu)^{-1}. \tag{5.3}$$

The function $\tilde{\alpha}_\mu$ maps $B(G)$ bijectively onto $B(G^\mu)$. (Application of $\tilde{\alpha}_\mu$ simply means transforming the Riemann surface S quasiconformally to the sur-

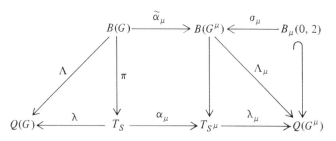

Figure 14

face $S^\mu = H'/G^\mu$.) From (5.3) and the definition in 5.1 it follows that $\tilde{\alpha}_\mu$ is biholomorphic.

In 4.6 we noticed that the induced mapping α_μ, defined by $\alpha_\mu([v]) = [\tilde{\alpha}_\mu(v)]$, is a bijective isometry of T_S onto T_{S^μ}.

For $\mu \in B(G)$, $v \in B(G^\mu)$, we write

$$\Lambda_\mu(v) = S_{f_v|H}. \tag{5.4}$$

By Theorem 5.1, this mapping of $B(G^\mu)$ into $Q(G^\mu)$ is holomorphic. From the proof of Theorem 4.4 we know that in the ball $B_\mu(0, 2) = \{\varphi \in Q(G^\mu) \mid \|\varphi\| < 2\}$, the mapping (5.4) has a section $\sigma_\mu: B_\mu(0, 2) \to B(G^\mu)$, defined by

$$\sigma_\mu(\varphi)(\bar{z}) = -2y^2 \varphi(z), \qquad z \in H. \tag{5.5}$$

From (5.5) and the definition in 5.1 we see that σ_μ is holomorphic.

Let π denote the canonical projection of $B(G)$ onto T_S, and λ and λ_μ the Bers imbeddings of T_S and T_{S^μ}, respectively, into $Q(G)$ and $Q(G^\mu)$. The commutative diagram in Fig. 14 illustrates the mappings introduced here.

The collection

$$\{V_\mu = (\pi \circ \tilde{\alpha}_\mu^{-1} \circ \sigma_\mu)(B_\mu(0, 2)) \mid \mu \in B(G)\} \tag{5.6}$$

is an open covering of T_S. In fact, V_μ is the preimage of $B_\mu(0, 2)$ under the homeomorphism

$$h_\mu = \lambda_\mu \circ \alpha_\mu \tag{5.7}$$

of T_S onto $T(G^\mu)$.

5.5. Complex Analytic Structure

We shall now prove that the restriction mappings $h_\mu|V_\mu$ define a complex analytic structure for the Teichmüller space T_S of the Riemann surface S and that this structure has the expected good properties. The proof does not cover the case of a torus; the complex structure of the Teichmüller space of a torus will be introduced in section 6.

Theorem 5.2. *The atlas*

$$\{(V_\mu, h_\mu) | \mu \in B(G)\} \tag{5.8}$$

defines a complex analytic structure on the Teichmüller space T_S. The Bers imbedding $[\mu] \to S_{f_\mu | H}$ of T_S into $Q(G)$ is holomorphic with respect to this structure.

PROOF. Assuming that V_μ and h_μ are defined by (5.6) and (5.7), we choose two elements μ_1 and μ_2 of $B(G)$ such that $V_{\mu_1} \cap V_{\mu_2}$ is not empty. In $h_{\mu_1}(V_{\mu_1} \cap V_{\mu_2})$ we have

$$h_{\mu_2} \circ h_{\mu_1}^{-1} = \Lambda_{\mu_2} \circ \tilde{\alpha}_{\mu_2} \circ \tilde{\alpha}_{\mu_1}^{-1} \circ \sigma_{\mu_1}.$$

All mappings on the right-hand side are holomorphic, as we have seen. Consequently, as a composition of holomorphic functions, $h_{\mu_2} \circ h_{\mu_1}^{-1}$ is holomorphic. By changing the roles of μ_1 and μ_2 we conclude that $h_{\mu_2} \circ h_{\mu_1}^{-1}$ is biholomorphic. Hence, (5.8) defines a complex analytic structure for T_S.

It is not difficult to see that the complex structure we obtained by means of the atlas (5.8) is independent of the representation H'/G we used for the Riemann surface S. Theorem 5.5 expresses an even stronger result.

In order to complete the proof of the theorem, we still have to show that the Bers imbedding $\lambda: T_S \to Q(G)$ is holomorphic, i.e., that $\lambda \circ h_\mu^{-1}$ is holomorphic in $B_\mu(0, 2)$. Now

$$\lambda \circ h_\mu^{-1} = \Lambda \circ \tilde{\alpha}_\mu^{-1} \circ \sigma_\mu.$$

Since all mappings on the right are holomorphic, their composition $\lambda \circ h_\mu^{-1}$ is holomorphic. □

We shall now establish further results (Theorems 5.3–5.6) which show that (5.8) defines a natural structure.

Theorem 5.3. *The canonical projection*

$$\pi: B(G) \to T_S$$

is holomorphic, and it has local holomorphic sections everywhere in T_S.

PROOF. First of all, we have

$$h_\mu \circ \pi = \Lambda_\mu \circ \tilde{\alpha}_\mu.$$

Since Λ_μ and $\tilde{\alpha}_\mu$ are holomorphic, it follows that π is holomorphic.

Next, let us consider the mapping

$$\psi_\mu = \tilde{\alpha}_\mu^{-1} \circ \sigma_\mu \circ h_\mu$$

of V_μ into $B(G)$. All functions on the right-hand side are holomorphic. Therefore, their composition ψ_μ is holomorphic. From the definitions we infer that

$$\pi \circ \psi_\mu = \text{identity mapping of } V_\mu. \tag{5.9}$$

Consequently, ψ_μ is the desired section. □

Since $\psi_\mu(\pi(\mu)) = \mu$, we conclude that π is an open mapping. We remark that Theorem 5.3 determines the complex analytic structure of T_S uniquely.

Theorem 5.4. *The Bers imbedding* $\lambda: T_S \to T(G)$ *is biholomorphic.*

PROOF. Suppose first that $Q(G)$ is finite dimensional. (By IV.5.5, this is the case if S is compact; cf. also 9.7.) We then conclude directly from Theorem 5.2 that λ is biholomorphic using the theorem by which a holomorphic bijection is always biholomorphic in finite dimensional manifolds (Narasimhan [1], p. 86).

In the general case, we fix a point $s_\mu \in T(G)$ and consider Schwarzians $s_v \in T(G)$ lying close to s_μ. Then $f = f_v \circ f_\mu^{-1}$ has a small Schwarzian derivative in the quasidisc $f_\mu(H)$. One proves that f has a quasiconformal extension whose complex dilatation is in $B(G_\mu)$ and depends holomorphically on s_v (Bers [9], Theorem 6; cf. the remark at the end of II.5.1). The conclusion is that Λ has a local holomorphic section at s_μ. Since $\Lambda = \lambda \circ \pi$, we deduce from Theorem 5.3 that λ is biholomorphic. $\qquad\square$

5.6. Complex Structure under Quasiconformal Mappings

Theorem 2.6 states that the Teichmüller spaces of quasiconformally equivalent Riemann surfaces are isomorphic in the sense that there exists a bijective isometry between the Teichmüller spaces. We can now strengthen this result and show that such an isometry can be chosen to be biholomorphic.

Theorem 5.5. *Quasiconformally equivalent Riemann surfaces have isometrically and biholomorphically isomorphic Teichmüller spaces.*

PROOF. Let $S = H'/G$ and $S' = H'/G'$ be quasiconformally equivalent Riemann surfaces. We consider a lift of a quasiconformal mapping of S onto S' to a self-mapping of the lower half-plane. There is no loss of generality in assuming that the lift is of the form f^μ. We have $\mu \in B(G)$.

The mapping $\tilde\alpha_\mu$ induced by μ is now a biholomorphic mapping of $B(G)$ onto $B(G')$. It induces the mapping α_μ of T_S onto $T_{S'}$. The theorem follows when we prove that α_μ is biholomorphic.

If $\pi': B(G') \to T_{S'}$ is the canonical projection, we have

$$\alpha_\mu \circ \pi = \pi' \circ \tilde\alpha_\mu. \tag{5.10}$$

Consider an arbitrary element $v \in B(G)$. In view of (5.9) and (5.10), we have in V_v,

$$\alpha_\mu = \alpha_\mu \circ \pi \circ \psi_v = \pi' \circ \tilde\alpha_\mu \circ \psi_v.$$

Again, on the right all functions are holomorphic, and so α_μ is holomorphic. By changing the roles of S and S' we conclude that α_μ is biholomorphic. $\qquad\square$

To be quite precise, we have not yet proved Theorem 5.5 for tori, which have the complex plane rather than the half-plane as universal covering surface. In the next section we shall show that the Teichmüller space of every torus is biholomorphically equivalent to the upper half-plane.

Theorem 5.5 can be applied to the modular group Mod(S) of T_S, which was introduced in 2.7.

Theorem 5.6. *The elements of the modular group* Mod(S) *are biholomorphic automorphisms of the Teichmüller space* T_S.

PROOF. By Theorem 5.5, a quasiconformal mapping between the Riemann surfaces S and S' induces a biholomorphic isomorphism $T_S \to T_{S'}$. The elements of the modular group are such isomorphisms induced by quasiconformal self-mappings of S.   □

We proved in 1.5 that two compact Riemann surfaces are quasiconformally equivalent as soon as they are homeomorphic. Moreover, we learned in IV.5.2 that they are homeomorphic if and only if they have the same genus. Therefore, in view of Theorem 5.5, *the genus completely determines the Teichmüller space of a compact surface.* The abstract Teichmüller space corresponding to surfaces of genus p is denoted by T_p. Starting from a specific Riemann surface S, as we have done, means fixing the origin in T_p.

In IV.5.5 we saw that the space of holomorphic quadratic differentials of a compact Riemann surface of genus $p > 1$ has finite dimension $3p - 3$. Therefore, $Q(G)$ can be identified with $\mathbb{C}^{3p-3}$, and T_p becomes a complex analytic $(3p - 3)$-manifold. We shall consider this question in greater detail in section 9 by using another imbedding of T_S into $Q(G)$. In section 6 we shall prove that T_1 is a complex 1-manifold.

6. Teichmüller Space of a Torus

6.1. Covering Group of a Torus

In sections 3–5 we considered Riemann surfaces having the half-plane as a universal covering surface. Let us now assume that S is a Riemann surface which has the complex plane $\mathbb{C}$ as its universal covering surface. If the covering group G of $\mathbb{C}$ over S is trivial or cyclic, then all quasiconformal images of S are conformally equivalent (cf. IV.4.1). Leaving aside these cases, which are uninteresting from the point of view of the theory of Teichmüller spaces, we assume that S is a torus, i.e., a compact surface of genus 1. By the remark at the end of 2.6, the Teichmüller spaces of different tori are all isometrically bijective.

The covering group G of $\mathbb{C}$ over a torus consists of translations

$$z \rightarrow z + m\omega_1 + n\omega_2,$$

where ω_1 and ω_2 are complex numbers, $\text{Im}(\omega_1/\omega_2) \neq 0$, and m and n run through all integers $\mathbb{Z}$. We call the pair (ω_1, ω_2) a base of G; the base is said to be normalized if $\text{Im}(\omega_1/\omega_2) > 0$.

Lemma 6.1. *Let (ω_1, ω_2) be a base of G. Then (ω_1', ω_2') is a base of G if and only if*

$$\omega_1' = a\omega_1 + b\omega_2, \qquad \omega_2' = c\omega_1 + d\omega_2, \tag{6.1}$$

where a, b, c, d are integers and

$$ad - bc = \pm 1. \tag{6.2}$$

If (ω_1, ω_2) is normalized, then (ω_1', ω_2') is normalized if and only if $ad - bc = 1$.

PROOF. The validity of (6.1) with integral coefficients is clearly a necessary condition. It becomes sufficient if (6.1) can be solved with respect to ω_1 and ω_2 so that ω_1 and ω_2 are linear combinations of ω_1' and ω_2' with coefficients in $\mathbb{Z}$. This occurs if and only if (6.2) holds.

From (6.1) it follows that

$$\text{Im}\left(\frac{\omega_1'}{\omega_2'}\right) = \frac{ad - bc}{|c\omega_1/\omega_2 + d|^2} \text{Im}\left(\frac{\omega_1}{\omega_2}\right). \tag{6.3}$$

Assuming that (6.2) is true we see that $\text{Im}(\omega_1/\omega_2)$ and $\text{Im}(\omega_1'/\omega_2')$ are simultaneously positive if and only if $ad - bc = 1$. □

Consider the *elliptic modular group* Γ which consists of the restrictions to H of all Möbius transformations

$$z \rightarrow \frac{az + b}{cz + d}$$

where a, b, c, d are real integers and $ad - bc = 1$. By (6.3) the group Γ acts on H.

The group Γ contains elliptic transformations. These are of two types. Either they are conjugates of the mapping $z \rightarrow -1/z$, which has the fixed point i in H. These transformations are of period 2. Or they are conjugates of $z \rightarrow -1/(z + 1)$, which has the fixed point $-1/2 + i\sqrt{3}/2$ in H, and are then of period 3. Even though Γ is not fixed point free, it is not difficult to prove that Γ acts properly discontinuously on H. (For more details about the properties of Γ we refer to Lehner [1], pp. 87, 99, and 139.) By Theorem IV.3.2 and the remark following it, the quotient H/Γ is a Riemann surface.

Let $S = \mathbb{C}/G$ be a torus and (ω_1, ω_2) a normalized base of G. We conclude from Lemma 6.1 that if (ω_1', ω_2') is a pair of non-zero complex numbers, then $(\lambda\omega_1', \lambda\omega_2')$ is a normalized base of G for a $\lambda \neq 0$ if and only if ω_1'/ω_2' is equivalent to ω_1/ω_2 under the modular group Γ. Hence every torus can be represented as a point of the Riemann surface H/Γ.

6.2. Generation of Group Isomorphisms

Theorem 1.3 says that if S is a Riemann surface which admits a disc as its universal covering surface and for which the covering group is not elementary, then the only conformal deformation of S onto itself is the identity mapping. For tori the situation is quite different:

Lemma 6.2. *Let S be a torus and p and q arbitrary points of S. Then there is a conformal mapping $f: S \to S$ homotopic to the identity such that $f(p) = q$.*

PROOF. Let $\pi: \mathbb{C} \to S = \mathbb{C}/G$ be the canonical projection and $z \in \pi^{-1}\{p\}$, $w \in \pi^{-1}\{q\}$. A translation commutes with every $g \in G$. Therefore, the mapping $\zeta \to \zeta + t(w - z)$ can be projected to a conformal mapping $f_t: S \to S$ for every $t, 0 \leq t \leq 1$. As t varies from 0 to 1, we obtain a homotopy from the identity mapping to $f_1 = f$, and $f(p) = q$. $\qquad\square$

Lemma 6.2 tells us that in studying the Riemann space R_S and the Teichmüller space T_S of a torus we can restrict ourselves to mappings $\varphi: S \to S'$ which are normalized by a condition $\varphi(p) = p'$, $p \in S$, $p' \in S'$. If $\pi: \mathbb{C} \to S$ and $\pi': \mathbb{C} \to S'$ are the canonical projections and we take $p = \pi(0)$, $p' = \pi'(0)$, then φ has a unique lift $f: \mathbb{C} \to \mathbb{C}$ with the property $f(0) = 0$. We call such a lift f normalized. A normalized f induces an isomorphism of G onto G' under which the transformation $z \to z + m\omega_1 + n\omega_2$ maps to $z \to z + mf(\omega_1) + nf(\omega_2)$.

In the case of tori it is easy to show that every isomorphism between G and G' can be generated in this manner.

Theorem 6.1. *Let $S = \mathbb{C}/G$ and $S' = \mathbb{C}/G'$ be tori and $\theta: G \to G'$ an isomorphism. Then there is a homeomorphism of S onto S' which induces θ.*

PROOF. Let (ω_1, ω_2) be a base of G and suppose that $(\omega_1, \omega_2) \to (\omega_1', \omega_2')$ under θ. Consider the affine transformation α which fixes 0 and maps ω_i to ω_i', $i = 1, 2$. Then α determines θ, and it projects to a homeomorphism of S onto S'. $\qquad\square$

Let $\varphi: S \to S'$ be a homeomorphism with a normalized lift f such that $f(\omega_i) = \omega_i'$, $i = 1, 2$. If $h = f \circ \alpha^{-1}$, then $h(z + \omega_i') = h(z) + \omega_i'$. From this we conclude that h is sense-preserving (Theorem IV.3.5). Hence f and α are simultaneously sense-preserving.

Let $\tau = \omega_1/\omega_2$, $\tau' = \omega_1'/\omega_2'$. If $\tau' \neq \bar{\tau}$, then

$$\alpha(z) = \lambda(z + \mu\bar{z}). \qquad (6.4)$$

Direct calculation shows that

$$|\mu| = \left| \frac{\tau' - \tau}{\tau' - \bar{\tau}} \right|. \qquad (6.5)$$

Suppose that (ω_1, ω_2) is a normalized base of G, i.e., that $\operatorname{Im} \tau > 0$. By (6.5), we then have $\operatorname{Im} \tau' > 0$ if and only if $|\mu| < 1$. But $|\mu| < 1$ is equivalent to α being sense-preserving. It follows that φ is *sense-preserving if and only if* $\operatorname{Im}(f(\omega_1)/f(\omega_2)) > 0$. If $\tau' = \bar{\tau}$, then φ is sense-reversing and $\operatorname{Im}(f(\omega_1)/f(\omega_2)) < 0$.

If $|\mu| < 1$, then (6.4) defines a quasiconformal mapping. Since the projection of a quasiconformal α is quasiconformal, it follows that all tori are quasiconformally equivalent. We also conclude from the proof of Theorem 6.1, by use of Theorem IV.3.5, that to every sense-preserving homeomorphism φ between two tori S and S' there is a quasiconformal mapping of S onto S' which is homotopic to φ. (This is a new proof of Theorem 1.5 for tori.)

6.3. Conformal Equivalence of Tori

Let $S = \mathbb{C}/G$ and $S' = \mathbb{C}/G'$ be two tori and (ω_1, ω_2) a base of G. We first show that S and S' are conformally equivalent if and only if there is a complex number $\lambda \neq 0$ such that $(\lambda\omega_1, \lambda\omega_2)$ is a base of G'.

In fact, if S and S' are conformally equivalent, then by Lemma 6.2 there is a conformal map of S onto S' whose lift $f: \mathbb{C} \to \mathbb{C}$ is normalized. Since f is a conformal self-mapping of $\mathbb{C}$ with $f(0) = 0$, we have $f(z) = \lambda z$. Then $f(\omega_i) = \lambda\omega_i$, and so $(\lambda\omega_1, \lambda\omega_2)$ is a base of G'. Conversely, if $(\lambda\omega_1, \lambda\omega_2)$ is a base of G', then the projection of $z \to \lambda z$ is a conformal mapping of S onto S'.

Theorem 6.2. *Let* $S = \mathbb{C}/G$ *and* $S' = \mathbb{C}/G'$ *be tori and* (ω_1, ω_2) *and* (ω_1', ω_2') *normalized bases of* G *and* G'. *Then* S *and* S' *are conformally equivalent if and only if the points* ω_1/ω_2 *and* ω_1'/ω_2' *are equivalent under the elliptic modular group.*

PROOF. We just showed that S and S' are conformally equivalent if and only if there is a $\lambda \neq 0$ such that $(\lambda\omega_1, \lambda\omega_2)$ is a base of G'. From what we said at the end of 6.1 it follows that this is the case if and only if the points ω_1/ω_2 and ω_1'/ω_2' are equivalent under the elliptic modular group. $\square$

Theorem 6.2 provides a model for the Riemann space R_S of tori. We conclude from it that R_S can be mapped bijectively onto the quotient of the upper half-plane H by the modular group Γ. The mapping $\chi: R_S \to H/\Gamma$ is obtained if we fix a normalized base (ω_1, ω_2) for the covering group of $\mathbb{C}$ over S and set $\chi(p) = [f(\omega_1)/f(\omega_2)]$ for $p \in R_S$, $f \in p$.

Repeated application of the reflection principle shows the existence of a "modular" function J, which is holomorphic in H, maps H onto $\mathbb{C}$, and is automorphic with respect to the group Γ. By using J we can identify H/Γ with the complex plane (Lehner [1], pp. 4–5).

Let f be a conformal self-mapping of S. The lift of f to $\mathbb{C}$ is then of the form $z \to \lambda z + \text{constant}$. If $\lambda = 1$ or $\lambda = -1$, then conversely, the projection of $z \to \pm z + \text{constant}$ is a conformal self-mapping of S. These are the trivial conformal self-mappings of a torus.

In certain exceptional cases, a torus admits other conformal self-mappings. We deduce from the preceding considerations that $z \to \lambda z$ projects to a conformal self-mapping of S for some $\lambda \neq \pm 1$ if and only if $\omega_1/\omega_2 = h(\omega_1/\omega_2)$ for a transformation $h \in \Gamma$ different from the identity mapping. In other words, ω_1/ω_2 must be a fixed point of an elliptic transformation $h \in \Gamma$. As stated before, there are two possibilities. First, h is a transformation of period 2, i.e., a conjugate of $z \to -1/z$. Then (ω_1, ω_2) can be chosen such that the fundamental parallelogram with the vertices 0, ω_1, $\omega_1 + \omega_2$, ω_2 is a square, and $\lambda = \pm i$. Second, h is of period 3, i.e., a conjugate of $z \to -1/(z + 1)$. In this case there is a fundamental parallelogram which again has sides of equal length but the angles are $\pi/3$ and $2\pi/3$, and $\lambda = \pm 1/2 \pm i\sqrt{3}/2$. These are the only cases in which a torus admits non-trivial conformal self-mappings.

The rest of this section is concerned with showing that the Teichmüller space T_S of a torus is isomorphic to the hyperbolic upper half-plane (or to the unit disc). It follows, in particular, that the Teichmüller space of a torus is branched over the Riemann space at points which correspond to the fixed points of Γ or, what is the same, which correspond to symmetric tori admitting non-trivial conformal self-mappings.

6.4. Extremal Mappings of Tori

We obtain a key to the properties of the Teichmüller space of a torus by studying mappings which are extremal in a homotopy class. The following basic result is due to Teichmüller ([1], p. 31).

Theorem 6.3. *In each homotopy class of sense-preserving homeomorphisms between tori there is a mapping whose lifts are affine and which is extremal, i.e., which has smallest maximal dilatation. Having affine lifts or being extremal determines the mapping uniquely up to conformal mappings homotopic to the identity.*

PROOF. Let $S = \mathbb{C}/G$ and $S' = \mathbb{C}/G'$ be tori, and consider mappings in a given homotopy class of sense-preserving homeomorphisms of S onto S'. By Lemma 6.2 and the fact that conformal self-mappings of $\mathbb{C}$ are affine, we may restrict ourselves to mappings with normalized lifts. Let F denote the class of these normalized lifts.

Fix a normalized base (ω_1, ω_2) of G. There exists a normalized base (ω_1', ω_2') of G' such that F is the class of self-homeomorphisms f of $\mathbb{C}$ satisfying $f(0) = 0$ and

$$f(z + m\omega_1 + n\omega_2) = f(z) + m\omega_1' + n\omega_2' \qquad (6.6)$$

for all $z \in \mathbb{C}$ and $m, n \in \mathbb{Z}$. This follows from Theorem IV.3.5, in view of the fact that the identity is the only inner automorphism of G'. Clearly, there is a unique affine mapping w in F.

Let $f \in F$ be K-quasiconformal, and set $f_k(z) = f(kz)/k$, $k = 1, 2, \ldots$. Then every $f_k \in F$ is K-quasiconformal. From (6.6) we conclude that for $k \to \infty$, the mappings f_k converge to w, uniformly in the euclidean metric. By Theorem I.2.2, the maximal dilatation of w is at most K. It follows that the projection of w is extremal in the given homotopy class.

We now prove that there are no other extremals in this homotopy class. There is no loss of generality in assuming that w is a stretching in the direction of the real axis, or more precisely, that $w(x + iy) = Kx + iy$, $K \geq 1$. In fact, this normalization can be obtained by suitable conformal self-mappings of $\mathbb{C}$, which do not change the maximal dilatation. Note that K is the maximal dilatation of w.

Let f be an extremal mapping competing with w. We prove that $f = w$. Since $f - w$ has the periods ω_1 and ω_2, there is a positive number L such that

$$|f(z) - w(z)| \leq L \qquad (6.7)$$

for every $z \in \mathbb{C}$. Since f is absolutely continuous on lines, we infer from (6.7) that for every $r > 0$,

$$\int_0^r |f_x(x + iy)| \, dx \geq \left| \int_0^r f_x(x + iy) \, dx \right| \geq Kr - 2L$$

for almost all $y \in [0, r]$. If $Q = \{(x, y)|0 \leq x \leq r, 0 \leq y \leq r\}$, it follows that

$$\iint_Q |f_x| \, dx \, dy \geq Kr^2 - 2Lr.$$

As $r \to \infty$, the number of period parallelograms meeting Q is of the order of magnitude $r^2(1 + o(1))/m(P)$, where $m(P)$ is the area of the period parallelogram P. Hence,

$$Kr^2 - 2Lr \leq \frac{r^2(1 + o(1))}{m(P)} \iint_P |f_x| \, dx \, dy.$$

Letting $r \to \infty$ we obtain

$$Km(P) \leq \iint_P |f_x| \, dx \, dy.$$

As an extremal, f is K-quasiconformal. Therefore, $|f_x|^2 \leq KJ$ a.e., where J

is the Jacobian of f (cf. I.3.4). Application of Schwarz's inequality yields

$$K^2(m(P))^2 \leq m(P) \iint_P |f_x|^2 \, dx \, dy \leq Km(P)m(f(P)). \tag{6.8}$$

Because $f(P)$ and $w(P)$ are fundamental domains of G' with boundaries of measure zero, $m(f(P)) = m(w(P)) = Km(P)$. We see that equality holds everywhere in (6.8), and so

$$|f_x(z)|^2 = KJ(z) \tag{6.9}$$

almost everywhere.

We write $f_x = \partial f + \bar{\partial} f$, $J = |\partial f|^2 - |\bar{\partial} f|^2$. From (6.9) we first conclude that $|\partial f + \bar{\partial} f| = |\partial f| + |\bar{\partial} f|$ a.e., and then that the complex dilatation $\bar{\partial} f / \partial f$ equals a.e. $(K - 1)/(K + 1)$. Thus f is affine, hence $f = w$, and the theorem is proved. $\square$

6.5. Distance of Group Isomorphisms from the Identity

Let us represent the torus $S = \mathbb{C}/G$ as the point ω_1/ω_2 of the upper half-plane, where as before (ω_1, ω_2) is a normalized base of G. We recall that if we change (ω_1, ω_2) to another normalized base (η_1, η_2) of G, then $\eta_1 = a\omega_1 + b\omega_2$ and $\eta_2 = c\omega_1 + d\omega_2$ for $a, b, c, d \in \mathbb{Z}$ with $ad - bc = 1$. This implies that η_1/η_2 is the image of ω_1/ω_2 under the Möbius transformation $z \to h(z) = (az + b)/(cz + d)$ belonging to the elliptic modular group Γ.

Let $S' = \mathbb{C}/G'$ be another torus and $\theta: G \to G'$ a given isomorphism. If θ is induced by a sense-preserving homeomorphism $\varphi: S \to S'$ with a normalized lift f, the numbers $f(\omega_1)$ and $f(\omega_2)$ do not depend on the particular choice of φ (and hence of f). Also, if (η_1, η_2) and h are as above, it follows that $f(\eta_1)/f(\eta_2) = h(f(\omega_1)/f(\omega_2))$ for the same h; this is seen from $f(m\omega_1 + n\omega_2) = mf(\omega_1) + nf(\omega_2)$, $m, n \in \mathbb{Z}$. We conclude that the hyperbolic distance in the upper half-plane between the points ω_1/ω_2 and $f(\omega_1)/f(\omega_2)$ depends solely on the isomorphism θ, not on the base used for G nor on the choice of the generator f. We denote this distance by δ_θ. It measures how much θ deviates from the isomorphisms induced by conformal mappings.

Lemma 6.3. *Let $\theta: G \to G'$ be an isomorphism generated by a normalized K-quasiconformal mapping f. Then*

$$\delta_\theta \leq \tfrac{1}{2} \log K. \tag{6.10}$$

Equality holds if and only if f is the affine transformation generating θ.

PROOF. Let w be the affine normalized mapping which generates θ. If $w(z) = \lambda(z + \mu \bar{z})$, we see from (6.5) that $|\mu| = |\tau' - \tau|/|\tau' - \bar{\tau}|$, where $\tau = \omega_1/\omega_2$, $\tau' = w(\omega_1)/w(\omega_2)$. Hence, if K is the maximal dilatation of w,

$$\log K = \log \frac{1 + |\mu|}{1 - |\mu|} = \log \frac{|\tau' - \bar{\tau}| + |\tau' - \tau|}{|\tau' - \bar{\tau}| - |\tau' - \tau|} = 2\delta_\theta.$$

Consequently, (6.10) holds as an equality for the affine mapping. Inequality (6.10) and the "only if" part of Lemma 6.3 follow from Theorem 6.3. □

6.6. Representation of the Teichmüller Space of a Torus

Let us start again from a fixed torus $S = \mathbb{C}/G$ and from a normalized base (ω_1, ω_2) of G. Let p be a point of the Teichmüller space T_S. In considering representatives of p we may restrict ourselves to mappings φ of S which have a normalized lift f; this follows from Lemma 6.2. We take such a $\varphi \in p$ and set

$$\psi(p) = f(\omega_1)/f(\omega_2).$$

We show that ψ is a well defined mapping of T_S. Let us consider another mapping $\varphi' \in p$ with the normalized lift f'. Then there is a conformal map $\sigma: \varphi'(S) \to \varphi(S)$ with a normalized lift such that $\sigma \circ \varphi'$ is homotopic to φ. The lifts of φ and $\sigma \circ \varphi'$ induce the same group isomorphism. But the lift of σ is of the form $z \to \lambda z$, and so $f(\omega_i) = \lambda f'(\omega_i)$, $i = 1, 2$. This shows that $f(\omega_1)/f(\omega_2)$ does not depend on the choice of $\varphi \in p$. From the discussion after Theorem 6.1 it follows that the image point $\psi(p)$ lies in the upper half-plane H.

The change of (ω_1, ω_2) to another normalized base of G means that ψ is to be replaced by $h \circ \psi$, where h is an element of the elliptic modular group. This follows from what was said in 6.5 before Lemma 6.3.

Theorem 6.4. *The mapping $\psi: T_S \to H$, defined by*

$$\psi([\varphi]) = f(\omega_1)/f(\omega_2), \qquad (6.11)$$

where f is the normalized lift of φ, is a bijective isometry of T_S onto the upper half-plane furnished with the hyperbolic metric.

PROOF. The mapping ψ is injective: If $\psi(p_1) = \psi(p_2)$, there are mappings $\varphi_i \in p_i$, $i = 1, 2$, whose normalized lifts f_i satisfy the equations $f_1(\omega_i) = \lambda f_2(\omega_i)$, $i = 1, 2$. If $z \to \lambda z = s(z)$, then f_1 and $s \circ f_2$ determine the same group isomorphism. It follows that their projections φ_1 and $\sigma \circ \varphi_2$ are homotopic. Here σ, as the projection of s, is conformal, and so $p_1 = p_2$.

The mapping ψ is surjective: Given a point $z \in H$, we choose an arbitrary non-zero complex number ω_2'. After this we set $\omega_1' = z\omega_2'$. By Theorem 6.1, there is a homeomorphism φ of S whose lift f is normalized and has the properties $f(\omega_i) = \omega_i'$, $i = 1, 2$. From the discussion following Theorem 6.1 it follows that φ is sense-preserving. It determines a point $p \in T_S$, and we see that $\psi(p) = z$.

The mapping ψ is an isometry: Given two points $p_i \in T_S$, $i = 1, 2$, consider their representatives φ_i with the normalized lifts f_i. Let $\varphi: \varphi_1(S) \to \varphi_2(S)$ be the extremal in the class of quasiconformal mappings homotopic to $\varphi_2 \circ \varphi_1^{-1}$ and having normalized lifts. For the Teichmüller distance τ_S we then have $\tau_S(p_1, p_2) = \frac{1}{2} \log K$, where K is the maximal dilatation of φ. On the other hand, if f is the normalized lift of φ, it follows from Lemma 6.3 that $\frac{1}{2} \log K$ is equal to the hyperbolic distance of the points $f_1(\omega_1)/f_1(\omega_2)$ and $f(f_1(\omega_1))/f(f_1(\omega_2))$. But $f \circ f_1$ induces the same group isomorphism as f_2, and so $f(f_1(\omega_i)) = f_2(\omega_i), i = 1, 2$. Thus the Teichmüller distance $\tau_S(p_1, p_2)$ coincides with the hyperbolic distance between $\psi(p_1)$ and $\psi(p_2)$, and the theorem is proved.   □

We know that every equivalence class $[\varphi]$ has a representative whose lift f is an affine transformation $\zeta \to \zeta + z\bar{\zeta}$; here the (constant) complex dilatation z is of absolute value <1 and depends only on $[\varphi]$. This fact makes it possible to express (6.11) in an explicit form. If we denote the point $[\varphi]$ by $[z]$, we obtain

$$\psi([z]) = \frac{\omega_1 + \bar{\omega}_1 z}{\omega_2 + \bar{\omega}_2 z}. \tag{6.12}$$

The expression (6.12) leads to a simple representation of the Teichmüller space of a torus.

Theorem 6.5. *The mapping*

$$[z] \to z \tag{6.13}$$

is a bijective isometry of the Teichmüller space T_S onto the hyperbolic unit disc D.

PROOF. The theorem follows immediately from the fact that (6.12) is a bijective isometry of T_S onto the hyperbolic upper half-plane.   □

6.7. Complex Structure of the Teichmüller Space of Torus

By Theorems 6.4 and 6.5, the Teichmüller space of a torus is a complex 1-manifold. We now introduce a complex analytic structure for T_S by means of the mapping (6.13). The use of (6.11) would of course lead to the same structure, because $z \to \psi([z])$ is a conformal mapping.

It follows that the Teichmüller spaces of tori are not only isometrically but also biholomorphically equivalent. We show that a quasiconformal mapping always induces such an equivalence. Recall that any two tori are quasiconformally equivalent.

Theorem 6.6. *Let S and S' be tori and $w: S \to S'$ a quasiconformal mapping. Then the bijective isometry $[\varphi] \to [\varphi \circ w^{-1}]$ of T_S onto $T_{S'}$ is biholomorphic.*

PROOF. The isometry depends only on the homotopy class of w. Therefore, by Theorem 6.3, there is no loss of generality in assuming that w has a lift $\zeta \to \lambda(\zeta + \mu\bar{\zeta})$.

We may assume that the lift of φ is $\zeta \to \zeta + z\bar{\zeta}$. Then $\varphi \circ w^{-1}$ has the complex dilatation $e^{i\theta}(z - \mu)/(1 - \bar{\mu}z)$, where $\theta = 2 \arg \lambda$. It follows that $[\varphi] \to [\varphi \circ w^{-1}]$ induces the conformal self-mapping $z \to e^{i\theta}(z - \mu)/(1 - \bar{\mu}z)$ of the unit disc. $\square$

By Theorem 6.6, which completes the proof of Theorem 5.5, we can speak of the complex analytic structure of the abstract Teichmüller space T_1 of tori.

It is of particular interest that under the mapping (6.13), which defines the complex analytic structure for T_1, the Teichmüller metric agrees with the hyperbolic metric of the unit disc. In Teichmüller spaces T_p, $p > 1$, connections between the metric and the complex analytic structure will be studied in 9.5 and 9.6.

Another interesting result is the concrete model we obtain for the modular group Mod(S) of T_S. It follows from our considerations that two points p_1 and p_2 of T_S are equivalent under Mod(S) if and only if their images $\psi(p_1)$ and $\psi(p_2)$ in H are equivalent under the elliptic modular group Γ. In fact, $h \to \psi^{-1} \circ h \circ \psi$ is an isomorphism of Γ onto Mod(S). (Cf., also, the isomorphisms $R_S \simeq H/\Gamma$ and $T_S \simeq H$ with Theorem 2.7, which says that $R_S \simeq T_S/\text{Mod}(S)$.)

The isomorphism Mod(S) $\simeq \Gamma$ exhibits the discontinuous nature of Mod(S). The discontinuity of Mod(S) is true of all compact surfaces S but more difficult to prove if the genus of S is > 1 (Kravetz [1], Abikoff [2]).

For every Möbius transformation g which maps H onto itself, $\psi^{-1} \circ g \circ \psi$ is a biholomorphic self-mapping of T_S. It follows that the modular group is a proper subgroup of the full group of biholomorphic automorphisms of the Teichmüller space of a torus. In T_p, $p > 1$, the situation is different, as we shall see in 9.6.

7. Extremal Mappings of Riemann Surfaces

7.1. Dual Banach Spaces

Having discussed the Teichmüller space of a compact Riemann surface of genus 1, we shall now focus our attention on compact surfaces whose genus is greater than 1. Their Teichmüller spaces are not as accessible as the space of a torus, because we no longer have a simple explicit representation for the covering group.

In order to get an insight into the properties of Teichmüller spaces of compact surfaces, we shall study in this section and in section 8 the extremal quasiconformal mappings whose maximal dilatations determine the distance between points in the Teichmüller space. This is equivalent to considering mappings which are extremal for given boundary values. Some preliminary remarks were made in I.5.7 and in 2.2 and 3.7 of this chapter.

A good part of our considerations can be done in the general setting. The special case of compact surfaces will not be taken up until the last subsection 7.8.

Let S be a Riemann surface whose universal covering surface is conformally equivalent to a disc. In this section, we take the universal covering surface to be the unit disc D. As before, G denotes the covering group of D over S.

We shall derive a *necessary* condition satisfied by the complex dilatation of the extremal mapping. The condition will be in terms of notions related to L^p-spaces. First, we introduce the space $L^\infty(G)$ of all bounded measurable $(-1, 1)$-differentials of G. As we have noted before $L^\infty(G)$ is a complex Banach space in which the Beltrami differentials of G form the open unit ball.

Next we consider quadratic differentials of G, i.e., measurable functions φ on D which satisfy the condition $(\varphi \circ g)g'^2 = \varphi$ for $g \in G$. We define the L^1-norm

$$\|\varphi\| = \int_N |\varphi|,$$

where N is a Dirichlet region for G. Since $|\varphi|$ is a $(1, 1)$-differential of G (cf. IV.1.4), this norm is independent of N. The linear space $L^1(G)$ of quadratic differentials of G with a finite L^1-norm is also a Banach space.

If $\mu \in L^\infty(G)$ and $\varphi \in L^1(G)$, the integral of the product $\mu\varphi$ is well defined on S (cf. IV.1.4). We write

$$\lambda_\mu(\varphi) = \int_N \mu\varphi.$$

Because $\{\mu|N|\mu \in L^\infty(G)\}$ and $\{\varphi|N|\varphi \in L^1(G)\}$ coincide with the L^∞- and L^1-spaces of N, it follows from standard results on L^p-spaces that the mapping $\mu \to \lambda_\mu$ is an isomorphism of $L^\infty(G)$ onto the dual of $L^1(G)$ (cf. Dunford–Schwartz [1], p. 289).

7.2. Space of Integrable Holomorphic Quadratic Differentials

Let $A(G)$ denote the subspace of $L^1(G)$ whose functions are holomorphic. *The set $A(G)$ is closed in $L^1(G)$.* For if $D_r = D(z_0, r) = \{z||z - z_0| < r\}$ is contained in D and $\varphi \in A(G)$, then by the mean value theorem for analytic functions,

$$|\varphi(z_0)| = \frac{1}{\pi r^2}\left|\int_{D_r} \varphi\right| \le \frac{1}{\pi r^2}\int_{D_r} |\varphi|. \tag{7.1}$$

It follows, since $\{g(N)|g \in G\}$ is locally finite in D, that if $\varphi_n \in A(G)$ and $\varphi_n \to \varphi$

in $L^1(G)$, then (choosing φ suitably from its equivalence class) $\varphi_n \to \varphi$ locally uniformly in D. Hence, the limit φ is holomorphic. We infer that $A(G)$ is a Banach space.

Let Φ be a bounded set in $A(G)$. From (7.1) we see that the functions $\varphi \in \Phi$ are locally uniformly bounded in absolute value in D. Consequently, Φ is a normal family, and so every sequence of functions $\varphi_n \in \Phi$ contains a subsequence (φ_{n_i}) which converges locally uniformly in D to a limit function φ. It follows from Fatou's lemma that

$$\|\varphi\| \leq \liminf_{i \to \infty} \|\varphi_{n_i}\|. \tag{7.2}$$

The functions φ_{n_i} are said to converge *weakly* to φ in $A(G)$, i.e., the norms of φ_{n_i} are uniformly bounded and $\lim \varphi_{n_i} = \varphi$ locally uniformly in D.

Convergence of a sequence in $A(G)$ implies its weak convergence, whereas the converse is not always true.

In 4.2 we introduced the Banach space $Q(G)$ consisting of holomorphic quadratic differentials of G for which the hyperbolic sup-norm

$$\|\varphi\|_q = \sup_{z \in N} |\varphi(z)|/\eta(z)^2$$

is finite. If S is compact, the spaces $A(G)$ and $Q(G)$ have the same elements: In this case the closure of N lies in D so that every holomorphic quadratic differential of G has a finite L^1-norm and a finite hyperbolic sup-norm.

In certain other cases we can readily compare the spaces $A(G)$ and $Q(G)$. Writing $|\varphi| = \eta^2(|\varphi|/\eta^2)$ and integrating over N, we obtain the inequality

$$\int_N |\varphi| \leq |S|_h \|\varphi\|_q,$$

where $|S|_h$ denotes the hyperbolic area of S (cf. IV.3.6). Hence, if this area is finite, then $Q(G) \subset A(G)$.

On the other hand, it follows from (7.1) that

$$\pi(1 - |z|)^2 |\varphi(z)| \leq \int_D |\varphi|.$$

Hence, if G is trivial, then $A(G) \subset Q(G)$. The example $\varphi(z) = (1 - z)^{-2}$ shows that $A(G)$ is properly contained in $Q(G)$.

7.3. Poincaré Theta Series

There is a classical method for producing quadratic differentials from analytic functions. If f is holomorphic in D, then

$$\Theta f = \sum_{g \in G} (f \circ g) g'^2$$

is called a *Poincaré theta series* of f.

Let $A(1)$ denote the space $A(G)$ in case G is the trivial group, i.e., $A(1)$ consists of functions holomorphic and integrable in D.

Theorem 7.1. *The mapping* $f \to \Theta f$ *is a continuous, linear surjection of* $A(1)$ *onto* $A(G)$ *of norm* ≤ 1.

PROOF. Let $f \in A(1)$, $z_0 \in D$, and $r = (1 - |z_0|)/3$. There exist G-equivalent Dirichlet regions $N_1, \ldots, N_k$ such that $D(z_0, 2r) \subset \bar{N}_1 \cup \cdots \cup \bar{N}_k$. For every $z \in D(z_0, r)$ we have

$$\pi r^2 \sum_{g \in G} |f(g(z))||g'(z)|^2 \leq \sum_{g \in G} \int_{D_r(z)} |f \circ g||g'|^2$$

$$\leq \sum_{g \in G} \sum_{j=1}^{k} \int_{N_j} |f \circ g||g'|^2 = \sum_{j=1}^{k} \sum_{g \in G} \int_{g(N_j)} |f| = k \int_D |f| < \infty,$$

where (7.1) yields the first inequality. It follows that $\sum_{g \in G} |f \circ g||g'|^2$ is locally uniformly convergent in D.

From this we conclude that $\Theta f \in A(G)$. Also,

$$\|\Theta f\| = \int_N |\Theta f| \leq \sum_{g \in G} \int_N |f \circ g||g'|^2 = \sum_{g \in G} \int_{g(N)} |f| = \|f\|.$$

This implies continuity of the mapping $f \to \Theta f$, since its linearity is clear. To prove surjectivity requires more analysis, and we refer to Lehner [2]. $\square$

If $\mu \in L^\infty$ and $f \in L^1$ in D, then

$$\int_D \mu f = \sum_{g \in G} \int_{g(N)} \mu f = \sum_{g \in G} \int_N (\mu \circ g)(f \circ g)|g'|^2.$$

Hence, for $\mu \in L^\infty(G)$ and $f \in A(1)$, we arrive at the relation

$$\int_D \mu f = \int_N \mu \Theta f, \tag{7.3}$$

which will have applications later.

7.4. Infinitesimally Trivial Differentials

Let $N(G)$ denote the subset of $L^\infty(G)$ which is orthogonal to $A(G)$, i.e., for whose elements μ,

$$\int_N \mu \varphi = 0$$

for every $\varphi \in A(G)$. Clearly $N(G)$ is a closed linear subspace of $L^\infty(G)$. Differentials which belong to $N(G)$ are called *infinitesimally trivial*.

Let us extend all functions $\mu \in L^\infty(G)$ to the plane by setting $\mu(z) = 0$ outside D. We denote by f_μ the quasiconformal mapping of the plane which has complex dilatation μ and is so normalized that

$$\lim_{z \to \infty} (f_\mu(z) - z) = 0.$$

Then μ determines f_μ uniquely, and f_μ is conformal in $E = \{z||z| > 1\}$. As before, we use the notation $s_\mu = S_{f_\mu|E}$ for the Schwarzian derivative.

In section 4 we studied extensively the mapping $[\mu] \to s_\mu$ of the Teichmüller space into the space $Q(G)$ of quadratic differentials; the different choice for the universal covering surface and for the normalization of f_μ which we have made here is of course unessential. We called a differential μ trivial if it determines the origin of T_S. Trivial differentials μ can also be characterized by the property that the Schwarzian derivative s_μ vanishes identically. Infinite-simally trivial differentials admit a similar characterization. Note that if $\mu \in L^\infty(G)$ and w is a complex number, then $w\mu$ is a complex dilatation whenever $|w| < 1/\|\mu\|_\infty$.

Theorem 7.2. *A differential μ is infinitesimally trivial if and only if*

$$\lim_{w \to 0} \frac{s_{w\mu}(z)}{w} = 0$$

for every $z \in E$.

PROOF. In I.4.4 and II.3.2 we established the representation formula

$$f_{w\mu}(z) = z + \sum_{n=1}^{\infty} a_n(z)w^n,$$

with

$$a_1(z) = T\mu(z) = -\frac{1}{\pi} \int \int_D \frac{\mu(\zeta)}{\zeta - z} \, d\xi \, d\eta.$$

From this we obtain by straightforward computation

$$\lim_{w \to 0} \frac{s_{w\mu}(z)}{w} = a_1'''(z) = -\frac{6}{\pi} \int \int_D \frac{\mu(\zeta)}{(\zeta - z)^4} \, d\xi \, d\eta,$$

because we are now dealing with points z for which $|z| > 1$. Hence

$$\lim_{w \to 0} \frac{s_{w\mu}(z)}{w} = \sum_{n=0}^{\infty} b_n z^{-(n+4)},$$

where

$$b_n = -\frac{(n + 1)(n + 2)(n + 3)}{\pi} \int \int_D \mu(\zeta) \zeta^n \, d\xi \, d\eta.$$

Therefore, the theorem follows if we prove that

$$\int \int_D \mu(\zeta) \zeta^n \, d\xi \, d\eta = 0, \qquad n = 0, 1, 2, \ldots, \tag{7.4}$$

if and only if $\mu \in N(G)$.

If Θ_n is the Θ-series for $f(z) = z^n$, then by formula (7.3),

$$\int\int_D \mu(\zeta)\zeta^n \, d\xi \, d\eta = \int\int_N \mu(\zeta)\Theta_n(\zeta) \, d\xi \, d\eta.$$

By Theorem 7.1, the function Θ_n is an element of $A(G)$. Hence, if $\mu \in N(G)$, then (7.4) follows.

Conversely, if (7.4) holds, it follows from an approximation theorem of Carleman [1] that μ is orthogonal in D to all functions $f \in A(1)$. Formula (7.3) then shows that μ is orthogonal in N to all functions Θf with f in $A(1)$. By Theorem 7.1, every element of $A(G)$ is of this form, and so $\mu \in N(G)$.   □

A slight modification of the above proof gives the following result: *A differential μ is infinitesimally trivial if and only if $T\mu$ vanishes identically in E.*

In section 5 we studied the function $\mu \to \Lambda(\mu) = s_\mu$ in the unit ball $B(G)$ of $L^\infty(G)$ and proved that it is holomorphic. It follows that Λ has a derivative $D\Lambda(\mu)$ for every $\mu \in B(G)$ (cf. the definition in 5.1). Direct computation shows that

$$D\Lambda(0)(\mu) = \lim_{w\to 0} \frac{s_{w\mu}}{w}.$$

Hence, Theorem 7.2 says that *the set of infinitesimally trivial differentials of G is the kernel of the mapping $D\Lambda(0)$.*

In this section we shall use the class $N(G)$ for studying extremal mappings, but infinitesimally trivial complex dilatations are also met in connection with other problems in the theory of Teichmüller spaces. Their importance was noticed already by Teichmüller, and they were utilized by Ahlfors and Bers in their studies regarding the complex structure of Teichmüller spaces. For information about the class $N(G)$ in the Teichmüller theory we refer to the surveys Royden [2], Earle [2], and Kra [2].

7.5. Mappings with Infinitesimally Trivial Dilatations

If the complex dilatation of f_μ is in $N(G)$, we can improve the estimate $\|s_\mu\|_q \leq 6\|\mu\|_\infty$ derived in II.3.3.

Theorem 7.3. *If f_μ has an infinitesimally trivial complex dilatation, then*

$$\|S_{f_\mu|E}\|_q \leq 6\|\mu\|_\infty^2. \tag{7.5}$$

PROOF. Fix a point $z \in E$ and consider the holomorphic function

$$w \to \psi(w) = (|z|^2 - 1)^2 S_{f_{w\mu/\|\mu\|_\infty}}(z)$$

in the unit disc. It follows from Theorem 7.2 that ψ has a zero of order at least 2 at the origin. Application of Schwarz's lemma to ψ, which is bounded in absolute value by 6, yields therefore

$$|\psi(w)| \le 6|w|^2.$$

Setting $w = \|\mu\|_\infty$, we get back our function f_μ, and the estimate (7.5) follows.

$\square$

Equality can hold in (7.5). To prove this consider the mapping f defined by $f(z) = z + k^2/z$ for $|z| \ge 1$ and $f(z) = z + 2k(|z| - 1) + k^2\bar{z}$ for $|z| < 1$, $0 \le k < 1$. Then $f = f_\mu$ is a quasiconformal mapping of the plane with $\mu(z) = 0$ in E and $\mu(z) = kz/|z|$ in D. We see that

$$\iint_D \mu(z)z^n \, dx \, dy = k \int_0^1 \int_0^{2\pi} (re^{i\varphi})^{n+1} \, dr \, d\varphi = 0, \qquad n = 0, 1, \dots.$$

It follows that μ is in $N(G)$ for the trivial group. Furthermore, $\|\mu\|_\infty = k$ and

$$\lim_{z \to \infty} (|z|^2 - 1)^2 |S_f(z)| = 6k^2.$$

We shall now apply Theorem 7.3 to prove an auxiliary result about infinitesimally trivial dilatations, which will come into use in what follows.

Lemma 7.1. *Let $v \in N(G)$ and $\|v\|_\infty < 2$. Then, for $0 \le t \le 1/4$, there is a $\sigma(t) \in [tv]$ such that $\|\sigma(t)\|_\infty \le 12t^2$.*

PROOF. By Theorem 7.3,

$$\|S_{f_{tv}|E}\|_q \le 24t^2.$$

For $0 \le t \le 1/4$, we have $24t^2 < 2$. Hence by formula (4.6), there is a complex dilatation $\sigma(t) \in [tv]$ for which

$$\|\sigma(t)\|_\infty = \tfrac{1}{2}\|S_{f_{tv}|E}\|_q \le 12t^2.$$

$\square$

7.6. Complex Dilatations of Extremal Mappings

Fix a point $p \in T_S$ and consider the family $\{f^\mu | \mu \in p\}$; here f^μ fixes the points $1, i, -1$. In other words, we consider all quasiconformal self-mappings of D whose complex dilatations are in $L^\infty(G)$ and which agree with an f^μ on ∂D. As has been noted repeatedly, this family contains one or more extremal mappings, i.e., mappings with the smallest maximal dilatation (Theorem 2.1).

The proof of the following lemma uses ideas applied by Krushkal [1] to a special case and later elaborated by Reich and Strebel [1]; see also Reich [1].

Lemma 7.2. *Let f^μ be extremal in its equivalence class. If $\mu - \kappa \in N(G)$, then $\|\mu\|_\infty \le \|\kappa\|_\infty$.*

PROOF. Set $\|\mu\|_\infty = k$, $\|\kappa\|_\infty = k_1$. We assume that $k_1 < k$ and prove that f^μ cannot then be extremal.

Writing $v = \mu - \kappa$, we first prove that for all sufficiently small positive values of t,

$$f^\lambda = f^\mu \circ (f^{tv})^{-1}$$

has a smaller maximal dilatation than f^μ. Using Lemma 7.1 we shall then correct the boundary values of f^λ and still keep the maximal dilatation smaller than that of f^μ.

Direct calculation gives

$$|\lambda(\zeta)| = \left| \frac{\mu(z) - tv(z)}{1 - t\mu(z)\bar{v}(z)} \right| = |\mu(z)| - \frac{1 - |\mu(z)|^2}{|\mu(z)|} \operatorname{Re}(\mu(z)\bar{v}(z))t + O(t^2). \quad (7.6)$$

Here $\zeta = f^{tv}(z)$, and the remainder term $O(t^2)$ is uniformly bounded in z.

Write $E_1 = \{z \in D \,|\, |\mu(z)| < (k + k_1)/2\}$, $E_2 = D \backslash E_1$. In E_1, $|\mu(z)|$ is strictly less than k. By (7.6), there are positive numbers δ_1 and t_1, such that for $z \in E_1$, $|\lambda(\zeta)| < k - \delta_1 t$ if $t < t_1$. In E_2,

$$\frac{1 - |\mu|^2}{|\mu|} \operatorname{Re}(\mu\bar{v}) \geq (1 - |\mu|^2)(|\mu| - |\kappa|) \geq \frac{1}{2}(1 - k^2)(k - k_1).$$

From this and (7.6) we deduce that there are positive numbers δ_2 and t_2, such that for $z \in E_2$, $|\lambda(\zeta)| < k - \delta_2 t$ if $t < t_2$. If $\delta = \min(\delta_1, \delta_2)$, $t_0 = \min(t_1, t_2)$, we thus have

$$|\lambda(\zeta)| < k - \delta t$$

for $t < t_0$, at every point $\zeta \in D$.

The mapping f^λ need not agree with f^μ on the boundary. But if $\sigma(t)$ is as in Lemma 7.1, then $f^\tau = f^\lambda \circ f^{\sigma(t)}$ has the same boundary values as f^μ. We have

$$\|\tau\|_\infty \leq \frac{\|\lambda\|_\infty + \|\sigma(t)\|_\infty}{1 - \|\sigma(t)\|_\infty} < \frac{k - \delta t + 12t^2}{1 - 12t^2}.$$

Hence, for t small enough, $\|\tau\|_\infty < k = \|\mu\|_\infty$, and the lemma is proved. $\square$

With the aid of this lemma, the desired characterization of extremal complex dilatations can be readily given. Let us assume that $\mu \in L^\infty(G)$ and $\varphi \in A(G)$. We write

$$l_\mu(\varphi) = \int_N \mu\varphi$$

and denote by

$$\|\mu\|_G^* = \sup\{|l_\mu(\varphi)| \,|\, \|\varphi\| = 1\}$$

the norm of μ as an element of the dual space of $A(G)$. If G is the trivial group, this "dual" norm is denoted by $\|\mu\|^*$. Obviously, $\|\mu\|_G^* \leq \|\mu\|_\infty$.

We are now ready to prove the main result about extremal complex dilatations (Hamilton [1], Krushkal [1]).

Theorem 7.4. *If f^μ is extremal in its equivalence class, then*

$$\|\mu\|_G^* = \|\mu\|_\infty. \tag{7.7}$$

PROOF. By the Hahn–Banach extension theorem (Dunford–Schwartz [1], p. 63), there is a linear functional λ in $L^1(G)$ with $\lambda|A(G) = l_\mu$ and $\|\lambda\| = \|\mu\|_G^*$. From what was said in 7.1 we deduce the existence of a $\kappa \in L^\infty(G)$, such that $\lambda(\varphi) = \int \kappa\varphi$ and $\|\lambda\| = \|\kappa\|_\infty$. Since $\|\mu\|_G^* \leq \|\mu\|_\infty$, we have to prove that $\|\mu\|_\infty \leq \|\kappa\|_\infty$.

From $\lambda|A(G) = l_\mu$ we conclude that

$$\int_N (\kappa - \mu)\varphi = 0$$

for $\varphi \in A(G)$. Hence $\mu - \kappa \in N(G)$, and by Lemma 7.2, $\|\mu\|_\infty \leq \|\kappa\|_\infty$. □

The necessary condition (7.7) for μ to be extremal is also sufficient. Reich and Strebel [2], with the aid of their "main inequality", proved that this is so for the trivial group. Later Strebel ([2],[5]) generalized the result, first to finitely generated groups and then to all cases.

Theorem 7.4 allows the following conclusion (Kra [2]): *Let f^μ be extremal for its boundary values among all quasiconformal mappings. If the operator $\Theta: A(1) \to A(G)$ is of norm < 1, then $\mu \, (\neq 0)$ is not in $L^\infty(G)$.*

PROOF. Suppose that $\mu \in L^\infty(G)$. By formula (7.3),

$$\|\mu\|^* = \sup\left\{\left|\int_N \mu\Theta\varphi\right|\,\Big|\,\|\varphi\| = 1\right\} \leq \|\mu\|_\infty\|\Theta\| < \|\mu\|_\infty.$$

This contradicts Theorem 7.4, since f^μ is extremal. □

In particular, the extremal μ in $L^\infty(G)$ is not extremal in L^∞. In terms of Teichmüller distances, the result can be expressed as follows: If $S = D/G$, then $(\tau|T_S)(0, [\mu]) < \tau_S(0, [\mu])$ (cf. 3.2 and 3.7).

7.7. Teichmüller Mappings

Let f^μ be an extremal mapping for the point $[\mu] \in T_S$. By Theorem 7.4, there is a sequence of functions $\varphi_n \in A(G)$ with $\|\varphi_n\| = 1$, such that

$$\lim_{n\to\infty} \int_N \mu\varphi_n = \|\mu\|_\infty.$$

Such a sequence (φ_n) is called a *Hamilton sequence* for μ.

A Hamilton sequence always contains a subsequence (φ_{n_i}) which converges weakly in $A(G)$ to a function $\varphi \in A(G)$ (cf. 7.2). It may happen that φ vanishes identically, in which case the sequence (φ_{n_i}) is said to be *degenerate*.

If no Hamilton sequence is degenerate, we can draw a remarkable conclusion about the extremal mapping (Strebel [3], Reich [1]).

Theorem 7.5. *Let μ be an extremal complex dilatation for which no Hamilton sequence is degenerate. If $\|\mu\|_\infty > 0$, then every weakly convergent Hamilton sequence tends in $L^1(G)$ to the same holomorphic quadratic differential φ, and*

$$\mu = \|\mu\|_\infty \frac{\bar\varphi}{|\varphi|}. \tag{7.8}$$

PROOF. Let (φ_n) be a weakly convergent Hamilton sequence for μ, and $\varphi(z) = \lim \varphi_n(z)$. In 7.2 we showed that $\varphi \in A(G)$ and $\|\varphi\| \le 1$ (formula (7.2)). By the triangle inequality

$$\|\varphi_n - \varphi\| \ge 1 - \|\varphi\|. \tag{7.9}$$

Given an $\varepsilon > 0$, let F be a compact subset of a Dirichlet region N of G, such that

$$\int_{N \setminus F} |\varphi| < \varepsilon. \tag{7.10}$$

Since $\varphi_n(z) \to \varphi(z)$ uniformly on F, there is an n_0 such that for $n > n_0$,

$$\int_F (|\varphi_n - \varphi| - (|\varphi_n| - |\varphi|)) < \varepsilon.$$

From $|\varphi_n - \varphi| - (|\varphi_n| - |\varphi|) \le |\varphi_n| + |\varphi| - (|\varphi_n| - |\varphi|) = 2|\varphi|$ and from (7.10) we conclude that

$$\int_{N \setminus F} (|\varphi_n - \varphi| - (|\varphi_n| - |\varphi|)) < 2\varepsilon.$$

Hence,

$$\int_N |\varphi_n - \varphi| < 1 - \|\varphi\| + 3\varepsilon$$

for $n > n_0$. From this and (7.9) it follows that

$$\lim_{n \to \infty} \|\varphi_n - \varphi\| = 1 - \|\varphi\|. \tag{7.11}$$

Here $\|\varphi\| = 1$. In order to prove this, we suppose that $\|\varphi\| < 1$, and form the sequence $((\varphi_n - \varphi)/\|\varphi_n - \varphi\|)$. Trivially,

$$\left| \int_N \mu \frac{\varphi_n - \varphi}{\|\varphi_n - \varphi\|} \right| \le \|\mu\|_\infty.$$

Moreover,

$$\mathrm{Re} \int_N \mu \frac{\varphi_n - \varphi}{\|\varphi_n - \varphi\|} \ge \frac{1}{\|\varphi_n - \varphi\|} \left(\mathrm{Re} \int_N \mu \varphi_n - \|\mu\|_\infty \|\varphi\| \right).$$

By (7.11), the right-hand expression tends to $\|\mu\|_\infty$ as $n \to \infty$. Hence, $((\varphi_n - \varphi)/\|\varphi_n - \varphi\|)$ is a degenerate Hamilton sequence for μ, which contradicts the hypothesis. It follows that $\|\varphi\| = 1$, and so by (7.11), $\varphi_n \to \varphi$ in $L^1(G)$.

We conclude that

$$\left| \|\mu\|_\infty - \int_N \mu\varphi \right| = \left| \lim_{n\to\infty} \int_N \mu(\varphi_n - \varphi) \right| \leq \|\mu\|_\infty \lim_{n\to\infty} \|\varphi_n - \varphi\| = 0.$$

Consequently,

$$0 = \int_N (\|\mu\|_\infty |\varphi| - \mu\varphi) = \int_N \left(\|\mu\|_\infty - \mu\frac{|\varphi|}{\bar\varphi} \right) |\varphi|.$$

This is possible only if the bracketed expression in the right-hand integral is zero, and (7.8) follows.

Finally, let ψ be the limit of another Hamilton sequence for μ. Then we deduce from (7.8) that $\varphi/\psi = |\varphi/\psi|$. Hence, the meromorphic function φ/ψ is a positive constant. From $\|\varphi\| = \|\psi\| = 1$ it follows that $\varphi = \psi$. $\square$

A quasiconformal mapping which is conformal or whose complex dilatation is of the form (7.8) is said to be a *Teichmüller mapping*.

In 2.3 we constructed geodesic lines in T_S. Suppose that an extremal for $[\mu] \in T_S$ is a Teichmüller mapping with μ determined by (7.8). From formula (2.4) we see that a geodesic ray from the origin through $[\mu]$ to the boundary of T_S allows the simple representation

$$t \to [t\bar\varphi/|\varphi|], \qquad 0 \leq t < 1.$$

In particular, in T_S the complex dilatation $t\bar\varphi/|\varphi|$ is extremal for the boundary values of $f^{t\bar\varphi/|\varphi|}$ for every t, $0 \leq t < 1$.

7.8. Extremal Mappings of Compact Surfaces

Theorem 7.5 is still implicit, but there are special cases in which it is possible to deduce that Hamilton sequences cannot degenerate. In particular, this conclusion can be drawn if the Riemann surface S is compact. We prefer to express the result in terms of the mappings of S, and recall that studying the family $\{f^\mu | \mu \in p\}$ is equivalent to studying a homotopy class of quasiconformal mappings of S (cf. 3.1).

Theorem 7.6. *On a compact Riemann surface, a quasiconformal mapping with the smallest maximal dilatation in its homotopy class is a Teichmüller mapping.*

PROOF. Let (φ_n) be a weakly convergent Hamilton sequence with limit φ. For a compact Riemann surface of genus > 1, the Dirichlet regions N are relatively compact in D (IV.5.1). Therefore

$$1 = \lim \int_N |\varphi_n| = \int_N |\varphi|.$$

Hence, (φ_n) is not degenerate and the result we wanted to prove follows from Theorem 7.5.   □

Since we assumed in this section that the Riemann surface S has a disc as its universal covering surface, our reasoning does not apply as such to a torus. But this case was thoroughly handled in section 6. It follows from Theorem 6.3 that also in the case of a torus the extremals are Teichmüller mappings.

Theorem 7.6 is *Teichmüller's existence theorem*. The first proof, which is based on a continuity argument, is in Teichmüller [2]; see also Ahlfors [1] and Bers [3]. Apart from technical details, the reasoning we used to prove Theorem 7.6 is due to Hamilton [1].

Extremal mappings of compact Riemann surfaces will be studied more closely in the following section, where Theorem 7.6 will be complemented by another famous result of Teichmüller: On a compact Riemann surface, a Teichmüller mapping is always the unique extremal in its homotopy class.

Our proof of this uniqueness theorem will be based on the original reasoning of Teichmüller [1], with due regard to the clarifications later made in it by Bers [3]. It would also be possible to apply the reasoning which yields the converse of Theorem 7.4. Strebel [5] showed that such a method also establishes the uniqueness of the extremal.

8. Uniqueness of Extremal Mappings of Compact Surfaces

8.1. Teichmüller Mappings and Quadratic Differentials

Having classified in IV.4.1 all Riemann surfaces which admit the plane as a universal covering surface, we know that if a compact Riemann surface is not the sphere or a torus, i.e., if its genus is > 1, then it has a disc as its universal covering surface. The covering group is finitely generated, and its Dirichlet regions are relatively compact on the universal covering surface (IV.5.1).

In section 4 we proved that $[\mu] \rightarrow S_{f_\mu|H}$ is an imbedding of the Teichmüller space T_S into the space of holomorphic quadratic differentials. Let us now assume that S is a compact Riemann surface with genus $p > 1$. Then there is another way to associate a holomorphic quadratic differential with each point of T_S. This leads to an imbedding which is simpler than the one given by the mapping $[\mu] \rightarrow S_{f_\mu|H}$, in that the image of T_S is the open unit ball.

In order to associate quadratic differentials with points of the Teichmüller space of a compact surface, we return to the extremal problem treated in section 7. We proved that in every homotopy class of sense-preserving homeomorphisms between two compact Riemann surfaces, a mapping with the smallest maximal dilatation is always a Teichmüller mapping (Theorem 7.6). More precisely, an extremal mapping f of a surface S is either conformal or its complex dilatation is of the form

$$k\bar{\varphi}/|\varphi|, \tag{8.1}$$

where $0 < k < 1$ and φ is a holomorphic quadratic differential of S.

In this section we shall prove that every Teichmüller mapping is the unique extremal in its homotopy class, i.e., that each homotopy class contains exactly one Teichmüller mapping. This makes it possible to define an injective mapping of the Teichmüller space into the space of quadratic differentials.

Given a holomorphic quadratic differential φ on S, we say that the pair (φ, k) determines the Teichmüller mapping f which has the complex dilatation (8.1). If f is determined by another pair (φ_1, k_1), we clearly have $k_1 = k$. Thus $\bar{\varphi}_1/|\varphi_1| = \bar{\varphi}/|\varphi|$, and it follows that φ_1/φ is a positive constant. We see that a non-conformal Teichmüller mapping determines the associated quadratic differential up to a positive multiplicative constant.

We note that the absolute value of the complex dilatation of a Teichmüller mapping is constant and that a mapping determined by the pair (φ, k) has the maximal dilatation $K = (1 + k)/(1 - k)$. The complex dilatation (8.1) is defined at every point of S, except for the finitely many zeros of φ.

8.2. Local Representation of Teichmüller Mappings

In the case of a torus, the properties of extremal mappings can be determined without great difficulty. This is due to the fact that on a torus, all holomorphic quadratic differentials are constants. The induced metric is therefore euclidean, and the universal covering surface, the complex plane, with its explicit covering group offers a convenient framework for studying extremal mappings. We proved that on $\mathbb{C}$ the extremals are globally affine, and that, up to translations, they are unique in any homotopy class.

If the Riemann surface S has genus $p > 1$, the situation is much more complicated. Holomorphic quadratic differentials now have $4p - 4$ zeros, which means that the induced metric has singularities. Global coordinates cannot be used for the study of Teichmüller mappings. However, there is a certain analogy with the case of a torus. It turns out that in a sense which we shall now make precise, every Teichmüller mapping is locally affine.

The local behavior of a Teichmüller mapping f becomes clear when we introduce a suitable quadratic differential on the surface S' onto which f maps S (Teichmüller [1]).

Theorem 8.1. *Let* $f: S \to S'$ *be a Teichmüller mapping determined by a pair* (φ, k). *Then there exists a unique holomorphic quadratic differential* ψ *on* S' *with the following properties:* $1°$ *If* φ *has a zero of order* $n \geq 0$ *at* p, *then* ψ *has a zero of the same order at* $f(p)$. $2°$ *If* ζ *is a natural parameter of* φ *at a regular point* p, *then the mapping* f *has the representation*

$$\zeta' \circ f = \frac{\zeta + k\bar{\zeta}}{1 - k} \tag{8.2}$$

in a neighborhood of p *in terms of a natural parameter* ζ' *of* ψ *at* $f(p)$.

PROOF. Let p be an arbitrary point of S. (We actually prove a little more than what is stated in the theorem.) Assume that φ has a zero of order n at p, $n \geq 0$ ($n = 0$ means of course that $\varphi(p) \neq 0$), and denote by ζ a natural parameter of φ at p. We define a mapping ζ' in a neighborhood of $f(p)$ by setting

$$\zeta' \circ f = \left(\frac{\zeta^{(n+2)/2} + k\bar{\zeta}^{(n+2)/2}}{1 - k} \right)^{2/(n+2)}. \tag{8.3}$$

If p is a regular point, (8.3) reduces to the simple form (8.2). We prove that ζ' is a local parameter on the Riemann surface S'.

Let z' be an arbitrary local parameter on S' in a neighborhood of the point $f(p)$. Then $z' \circ f$ has the complex dilatation $k\bar{\varphi}/|\varphi|$. Since $\varphi(z) dz^2 = ((n + 2)/2)^2 \zeta^n d\zeta^2$ (formula (6.1) in IV.6) and since complex dilatation is a $(-1, 1)$-differential, we see that the complex dilatation of $z' \circ f$ can also be expressed in the form $k(\bar{\zeta}/|\zeta|)^n$.

On the other hand, we compute from (8.3) that the mapping $\zeta' \circ f$ has the complex dilatation $k(\bar{\zeta}/|\zeta|)^n$. It follows that $\zeta' \circ (z')^{-1}$ is conformal, and we conclude that ζ' is a local parameter of S'.

In order to construct the differential ψ, we consider a point $p_1 \neq p$ which is so close to p that p_1 is a regular point and lies in the domain of ζ. A natural parameter at p_1 is obtained if we integrate $\varphi(z)^{1/2} dz = ((n + 2)/2)\zeta^{n/2} d\zeta$. It follows that $\zeta_1 = \zeta^{(n+2)/2}$ is a natural parameter at p_1. Then $\zeta_1' = (\zeta_1 + k\bar{\zeta}_1)/(1 - k)$ is a local parameter of S' at $f(p_1)$, and $4 d\zeta_1'^2 = (n + 2)^2 \zeta'^n d\zeta'^2$. We conclude that

$$\psi(z') dz'^2 = ((n + 2)/2)^2 \zeta'^n d\zeta'^2$$

defines a holomorphic quadratic differential ψ on S'. We see that ζ' is a natural parameter of ψ at $f(p)$. It has a zero of order n at $f(p)$. □

If $\zeta = \xi + i\eta$, $\zeta' \circ f = \xi' + i\eta'$, the mapping (8.2) assumes the form

$$\xi' + i\eta' = K\xi + i\eta \tag{8.4}$$

with $K = (1 + k)/(1 - k)$. Hence, formula (8.2) can be expressed as follows: *In a neighborhood of a regular point of* φ, *the associated Teichmüller mapping is a conformal transformation, followed by a fixed stretching in the direction of the positive real axis, followed by another conformal mapping.*

For the Teichmüller mapping $f: S \to S'$, we call φ its *initial* and ψ its *terminal* differential.

8.3. Stretching Function and the Jacobian

The stretching (8.4) is a characteristic property of a Teichmüller mapping. We shall now define a stretching function for an arbitrary quasiconformal mapping of S.

Let φ be a holomorphic quadratic differential on the Riemann surface S. On S we use the metric induced by φ (cf. IV.7 where such a metric was studied). In addition to S, we shall consider in the following the image S' of S under a Teichmüller mapping with φ as initial differential. On S' we use the metric induced by the terminal quadratic differential ψ of this Teichmüller mapping.

Let $f: S \to S'$ be a quasiconformal mapping and $z \to z' = w(z)$ its representation in a neighborhood of a regular point $p \in S$. We wish to define a stretching function λ_f of f on S, so that if $z = x + iy$ and z' are natural parameters, we have $\lambda_f = |w_x|$. This is achieved if we set

$$\lambda_f(p) = \left| \frac{\partial w(z)}{\varphi(z)^{1/2}} + \frac{\bar{\partial} w(z)}{\bar{\varphi}(z)^{1/2}} \right| |\psi(w(z))|^{1/2}. \tag{8.5}$$

We first note that since φ and ψ are quadratic differentials, λ_f is a Borel function on S. If z and z' are natural parameters, $\varphi = \psi \equiv 1$, and so $\lambda_f = |\partial w + \bar{\partial} w| = |w_x|$.

Let α be a horizontal arc on S. The differential $\lambda_f |\varphi|^{1/2}$ can be integrated along α (with the possible exception of a family of arcs α whose union has the area zero), and from the representation $\lambda_f = |w_x|$ it follows that

$$\int_\alpha \lambda_f |\varphi|^{1/2} = l(f(\alpha)). \tag{8.6}$$

In later integrations we also need the Jacobian with respect to the φ- and ψ-metrics. It is defined by the formula

$$J_f(p) = (|\partial w(z)|^2 - |\bar{\partial} w(z)|^2) \left| \frac{\psi(w(z))}{\varphi(z)} \right|.$$

We see that J_f is a function on S. In natural parameters, J_f becomes the ordinary Jacobian $|\partial w|^2 - |\bar{\partial} w|^2$. Therefore,

$$\int_S J_f |\varphi| = |S'|,$$

where $|S'|$ denotes the ψ-area of S'.

8.4. Average Stretching

The following result about average stretching is an essential step on the road leading to the extremal properties of Teichmüller mappings.

Lemma 8.1. *Let $f: S \to S$ be a quasiconformal mapping homotopic to the identity and φ a holomorphic quadratic differential on S. Define λ_f by (8.5) for φ and $\psi = \varphi$. Then*

$$\int_S \lambda_f |\varphi| \geq |S|.$$

PROOF. We first define a one-dimensional average of $\lambda = \lambda_f$. Let α be a subarc of a horizontal trajectory with midpoint p and of length $2a$. We set

$$\lambda_a(p) = \frac{1}{2a} \int_\alpha \lambda |\varphi|^{1/2}. \tag{8.7}$$

Assume, for a moment, that φ has an oriented trajectory structure (cf. IV.6.3). Let S_0 be the union of the non-critical horizontal trajectories of φ. Since there are only finitely many critical horizontal trajectories on S, the set $S \backslash S_0$ has the area zero.

We define a flow on S_0. Let p be a point on a horizontal trajectory $\alpha \subset S_0$, and t a real number. Let $\chi(p, t)$ denote the point on α which has the distance $|t|$ from p and lies in the positive direction from p if $t > 0$, in the negative direction if $t < 0$. Then $\chi: S_0 \times \mathbb{R} \to S_0$ is a continuous mapping such that $p \to \chi(p, t)$ is a conformal and isometric bijection for every t.

We conclude that $\lambda^t = \lambda \circ \chi(\cdot, t)$ is a Borel-measurable function on S_0, and

$$\int_S \lambda^t |\varphi| = \int_{S_0} \lambda^t |\varphi| = \int_{S_0} \lambda |\varphi| = \int_S \lambda |\varphi|.$$

Hence,

$$\int_S \lambda |\varphi| = \frac{1}{2a} \int_{-a}^{a} \left(\int_S \lambda^t |\varphi| \right) dt = \int_S \left(\frac{1}{2a} \int_{-a}^{a} \lambda^t \, dt \right) |\varphi|,$$

where Fubini's theorem justifies the last step. Here

$$\frac{1}{2a} \int_{-a}^{a} \lambda^t(p) \, dt = \lambda_a(p).$$

We have thus proved that

$$\int_S \lambda |\varphi| = \int_S \lambda_a |\varphi|, \tag{8.8}$$

provided that φ has an orientable trajectory system.

If the trajectory system of φ is not orientable, we construct as in IV.6.3 a two-sheeted covering $\tilde{S}$ of S on which the trajectories of the lift $\tilde{\varphi}$ of φ (these are the lifts of the trajectories of φ) can be oriented. If $\tilde{\lambda}$ and $\tilde{\lambda}_a$ denote the lifts of λ and λ_a on $\tilde{S}$, then it is clear from the construction of $\tilde{S}$ that

$$\int_S \lambda|\varphi| = \tfrac{1}{2}\int_{\tilde{S}} \tilde{\lambda}|\tilde{\varphi}| = \tfrac{1}{2}\int_{\tilde{S}} \tilde{\lambda}_a|\tilde{\varphi}| = \int_S \lambda_a|\varphi|.$$

It follows that (8.8) is always valid.

From (8.7), (8.6), and Lemma IV.7.4 we deduce that

$$\lambda_a(p) \geq 1 - M/a$$

almost everywhere, where the number M does not depend on α. Hence, by (8.8)

$$\int_S \lambda|\varphi| \geq (1 - M/a)|S|. \tag{8.9}$$

On every non-critical trajectory, no matter whether it is a spiral or a closed curve, we can take a as large as we please. Letting $a \to \infty$, we obtain the lemma from (8.9). $\qquad\square$

8.5. Teichmüller's Uniqueness Theorem

We have now completed the preparations needed for our proof of Teichmüller's basic result about the uniqueness of extremal quasiconformal mappings of compact Riemann surfaces.

Theorem 8.2. *Let S be a compact Riemann surface of genus > 1. In every homotopy class of sense-preserving homeomorphisms of S onto another Riemann surface S', there is exactly one Teichmüller mapping, and its maximal dilatation is uniquely smallest.*

PROOF. By Theorem 1.5, every homotopy class of sense-preserving homeomorphisms of S onto S' contains quasiconformal mappings. By Theorem 2.1, each class contains a mapping with the smallest maximal dilatation. By Theorem 7.6, every such extremal is a Teichmüller mapping. Consequently, it remains to be proved that every Teichmüller mapping is the unique extremal in its homotopy class. Taken together the results then imply that each homotopy class contains exactly one Teichmüller mapping.

Suppose first that $f_0: S \to S'$ is a conformal mapping. If $f: S \to S'$ is also conformal and homotopic to f_0, then $f^{-1} \circ f_0: S \to S$ is conformal and homotopic to the identity. By Theorem 1.3, $f = f_0$, and so f_0 is a unique extremal.

Henceforth we assume that $f_0: S \to S'$ is a Teichmüller mapping defined by the pair (φ, k_0), $k_0 > 0$. Let $f: S \to S'$ be a quasiconformal mapping homotopic to f_0. If K_0 and K denote the maximal dilatations of f_0 and f, we prove first that

$$K \geq K_0. \tag{8.10}$$

The quasiconformal mapping $h = f \circ f_0^{-1} \colon S' \to S'$ is homotopic to the identity. We apply Lemma 8.1 to h, using on S' the metric induced by the terminal quadratic differential ψ of f_0. It follows, by Schwarz's inequality, that

$$\int_{S'} \lambda_h^2 |\psi| \geq |S'|. \tag{8.11}$$

In order to estimate the left-hand integral we consider a regular point p of S with the additional property that $f(p)$ is a regular point of S' for ψ. These conditions exclude only finitely many points p of S. Let $\zeta = \xi + i\eta$ be a natural parameter at p, and ζ_0 and ζ_1 natural parameters at the points $f_0(p)$ and $f(p)$, respectively, such that in terms of ζ and ζ_0, the mapping f_0 has the representation

$$\zeta_0 = \frac{\zeta + k_0 \bar{\zeta}}{1 - k_0} = K_0 \xi + i\eta, \qquad k_0 = \frac{K_0 - 1}{K_0 + 1}. \tag{8.12}$$

We see that λ_{f_0} and J_{f_0} are both equal to the constant K_0.

Let $\zeta_1 = w(\zeta_0)$ be a representation of h. (The various mappings are illustrated in Fig. 15.) By (8.12), f has the representation

$$\zeta_1 = w\left(\frac{\zeta + k_0 \bar{\zeta}}{1 - k_0}\right).$$

Differentiation yields $\partial \zeta_1(\zeta) + \bar{\partial} \zeta_1(\zeta) = K_0(\partial w(\zeta_0) + \bar{\partial} w(\zeta_0))$. Because $f = h \circ f_0$, we conclude that

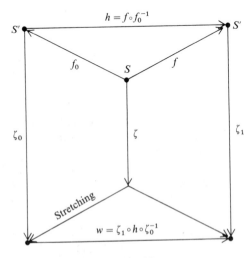

Figure 15

$$\lambda_f = K_0(\lambda_h \circ f_0)$$

almost everywhere.

Reverting to the integral in (8.11) we now obtain

$$\int_{S'} \lambda_h^2 |\psi| = \int_S (\lambda_h \circ f_0)^2 J_{f_0} |\varphi| = K_0^{-1} \int_S \lambda_f^2 |\varphi|.$$

Next we take into account the fact that f is K-quasiconformal. We express the dilatation condition in the natural parameters ζ and ζ_1 and deduce that

$$\lambda_f^2 \leq K J_f \qquad \text{a.e.} \tag{8.13}$$

Thus

$$\int_{S'} \lambda_h^2 |\psi| \leq (K/K_0) \int_S J_f |\varphi| = (K/K_0)|S'|.$$

In conjunction with (8.11), this yields (8.10).

Assume now that equality holds in (8.10). The above reasoning shows that (8.13) must then hold as an equality a.e., so that

$$\lambda_f^2 = K_0 J_f \qquad \text{a.e.} \tag{8.14}$$

We show that (8.14) is possible only if f and f_0 satisfy the same Beltrami equation.

We use again the natural parameters ζ, ζ_0 and ζ_1. From (8.12) we first see that

$$\bar{\partial}\zeta_0 = k_0 \partial \zeta_0.$$

In terms of ζ_1, equation (8.14) assumes the form

$$|\partial \zeta_1 + \bar{\partial}\zeta_1|^2 = K_0(|\partial \zeta_1|^2 - |\bar{\partial}\zeta_1|^2) \qquad \text{a.e.}$$

On the other hand, $\zeta \to \zeta_1$ is K_0-quasiconformal, so that

$$|\partial \zeta_1| + |\bar{\partial}\zeta_1| \leq K_0(|\partial \zeta_1| - |\bar{\partial}\zeta_1|) \qquad \text{a.e.}$$

These relations imply that $|\partial \zeta_1 + \bar{\partial}\zeta_1| = |\partial \zeta_1| + |\bar{\partial}\zeta_1|$, and hence $\arg \partial \zeta_1 = \arg \bar{\partial}\zeta_1$. Further, $|\partial \zeta_1| + |\bar{\partial}\zeta_1| = K_0(|\partial \zeta_1| - |\bar{\partial}\zeta_1|)$, whence $|\bar{\partial}\zeta_1| = k_0|\partial \zeta_1|$. It follows that ζ_1 also satisfies the equation

$$\bar{\partial}\zeta_1 = k_0 \partial \zeta_1.$$

We conclude that the mapping $\zeta_0 \to \zeta_1$ is conformal, which means that $f \circ f_0^{-1}$ is conformal. Since $f \circ f_0^{-1}$ is homotopic to the identity, Theorem 1.3 tells that $f \circ f_0^{-1}$ is the identity mapping. The unique extremality of f_0 has thus been proved. $\qquad\square$

As we already said at the end of section 7, the above proof of Theorem 8.2 follows essentially the modification Bers ([3] and the appendix in [12]) gave of the original proof of Teichmüller [1].

9. Teichmüller Spaces of Compact Surfaces

9.1. Teichmüller Imbedding

Let S be a compact Riemann surface of genus > 1. Theorem 8.2 provides a new way to map the Teichmüller space of S into the space Q_S of holomorphic quadratic differentials of S. By Theorem 8.2, every point $p \in T_S$ contains Teichmüller mappings and they all have the same complex dilatation. If $p = 0$, these mappings are conformal. For all other points, the mappings are determined by a pair (φ, k), where $0 < k < 1$ and $\varphi \in Q_S$ is unique up to a positive multiplicative constant.

By Theorem IV.5.5, the linear space Q_S is finite dimensional. By a fundamental theorem of linear algebra, all norm-induced metrics on Q_S are topologically equivalent. Of the many natural metrics available, we choose here the one which is induced by the hyperbolic sup norm: $\|\varphi\| = \sup |\varphi|/\eta^2$.

In representing Teichmüller mappings by pairs (φ, k), we assume henceforth that φ is normalized by the requirement $\|\varphi\| = 1$. For the point $p \in T_S$, $p \neq 0$, we then have the unique representation

$$p = [k\bar{\varphi}/|\varphi|].$$

We use this representation for $p = 0$ also. In this case $k = 0$, and φ is not uniquely defined, but it is of course immaterial how φ is chosen.

Theorem 9.1. *The mapping*

$$[k\bar{\varphi}/|\varphi|] \to k\varphi, \tag{9.1}$$

where $\varphi \in Q_S$, $\|\varphi\| = 1$, and $0 \leq k < 1$, is a homeomorphism of T_S onto the open unit ball of Q_S.

PROOF. By our previous remarks, (9.1) is a well defined injection of T_S into the open unit ball of Q_S. It is clearly also a surjection onto this ball.

In order to prove that (9.1) is continuous, we assume that $p_n = [k_n \bar{\varphi}_n/|\varphi_n|]$ converges to $p = [k\bar{\varphi}/|\varphi|]$ in T_S. We use in T_S its β-metric (see III.2.2), which is topologically equivalent to the Teichmüller metric τ_S of T_S. By Theorem 8.2, we have $\beta_S(p_n, 0) = k_n$, $\beta_S(p, 0) = k$. Hence, by the triangle inequality $|k_n - k| \leq \beta_S(p_n, p)$. It follows that $k_n \to k$. If $k = 0$, this implies that $\|k_n \varphi_n - k\varphi\| \to 0$.

Assuming that $k > 0$ we show that $\varphi_n \to \varphi$ in Q_S. Let us lift the differentials φ_n and φ to the universal covering surface H. In H, the analytic functions φ_n, $n = 1, 2, \ldots$, constitute a normal family. If $\|\varphi_n - \varphi\|$ does not tend to 0 as $n \to \infty$, there is a subsequence (φ_{n_i}) and a differential $\psi \in Q_S$, $\psi \neq \varphi$, $\|\psi\| = 1$, such that $\|\varphi_{n_i} - \psi\| \to 0$. Then $k_{n_i} \bar{\varphi}_{n_i}/|\varphi_{n_i}| \to k\bar{\psi}/|\psi|$ a.e. By Lemma 3.1, $[k\bar{\varphi}/|\varphi|] = [k\bar{\psi}/|\psi|]$. Hence $\varphi = \psi$, which is a contradiction, and so

$\lim \| \varphi_n - \varphi \| = 0$. From

$$\| k_n \varphi_n - k\varphi \| \le |k_n - k| + \| \varphi_n - \varphi \|$$

we finally see that $k_n \varphi_n \to k\varphi$, i.e., that (9.1) is continuous.

The continuity of the inverse of (9.1) follows from the invariance of domain. The space Q_S is finite dimensional and T_S is homeomorphic with an open subset of Q_S (Theorems 4.3 and 4.8). Hence (9.1), as a continuous injection of T_S into Q_S, is a homeomorphism of T_S onto its image.   □

We call (9.1) the *Teichmüller imbedding.*

9.2. Teichmüller Space as a Ball of the Euclidean Space

In proving Theorem 9.1 we chose the hyperbolic sup norm in Q_S for technical reasons only. Since Q_S is a $(3p - 3)$-dimensional linear space over the complex numbers, we can fix a base $\varphi_1, \varphi_2, \ldots, \varphi_{3p-3}$ in Q_S. Then an arbitrary $\varphi \in Q_S$ has a representation

$$\varphi = \sum_{i=1}^{3p-3} z_i \varphi_i$$

with complex coefficients $z_i = x_i + iy_i$. The quadratic differential φ can thus be identified with the point $(z_1, z_2, \ldots, z_{3p-3})$ of $\mathbb{C}^{3p-3}$ or with the point $(x_1, \ldots, x_{3p-3}, y_1, \ldots, y_{3p-3})$ of the euclidean space $\mathbb{R}^{6p-6}$. We now introduce the euclidean norm

$$\| \varphi \| = \left(\sum_{i=1}^{3p-3} (x_i^2 + y_i^2) \right)^{1/2}$$

in Q_S.

Using this new norm, we can rephrase Theorem 9.1 as follows:

Theorem 9.2. *The mapping*

$$[k\bar{\varphi}/|\varphi|] \to (kx_1, \ldots, kx_{3p-3}, ky_1, \ldots, ky_{3p-3}), \qquad (9.2)$$

where $\varphi \in Q_S$, $\| \varphi \| = 1$, and $0 \le k < 1$, is a homeomorphism of T_S onto the open unit ball of the euclidean space $\mathbb{R}^{6p-6}$.

The validity of this statement follows immediately from Theorem 9.1, by virtue of the fact that all norms in Q_S define the same topology.

We conclude, in particular, that *the Teichmüller space of a compact Riemann surface of genus $p > 1$ is homeomorphic to $\mathbb{R}^{6p-6}$.* This is an old result of Fricke who proved it without the use of quasiconformal mappings by suitably parametrizing covering groups over compact surfaces. A modernized version of Fricke's proof has been given by Keen [1].

If $p = 1$, we proved that the Teichmüller space is isomorphic to the hyperbolic upper half-plane (Theorem 6.4). Hence, for $p = 1$, the space T_S is homeomorphic to $\mathbb{R}^2$.

The Bers imbedding of Theorem 4.8 also allows the conclusion that for $p > 1$, the space T_S is homeomorphic to a subdomain of $\mathbb{R}^{6p-6}$. However, not much is known about this domain. Now we can immediately draw the following conclusion: The Teichmüller space of a compact Riemann surface is contractible. This special result has, of course, lost some of its interest in light of the Douady–Earle result that every Teichmüller space is contractible (see 3.6).

9.3. Straight Lines in Teichmüller Space

Let S be a compact Riemann surface of genus $p > 1$. Then every point of the Teichmüller space of S can be represented by the complex dilatation of a unique Teichmüller mapping. If $q = [k\bar{\varphi}/|\varphi|] \in T_S$, then $\tau_S(0, q) = \frac{1}{2}\log((1 + k)/(1 - k))$.

In 2.3 we studied geodesic lines in Teichmüller space. Since extremal mappings are unique, the remark made in 7.7 can be complemented. We deduce from formula (2.4), by using Theorem 8.2, that

$$t \to [t\bar{\varphi}/|\varphi|], \qquad -1 < t < 1,$$

is a geodesic line in T_S which passes through the origin and the point q. Here

$$\tau_S(0, [t\bar{\varphi}/|\varphi|]) = \frac{1}{2}\log\frac{1 + |t|}{1 - |t|}.$$

Consequently, by changing the parameter, we obtain the mapping

$$x \to [(\tanh x)\bar{\varphi}/|\varphi|], \qquad -\infty < x < \infty,$$

which is an isometry of the real axis into T_S. We call such a path in T_S a *straight line*. It follows that every point of T_S lies on a straight line through the origin.

In order to study geodesics through two arbitrary points of T_S, we recall that all Teichmüller spaces of Riemann surfaces of genus p are isomorphic (see 5.6). The space T_S is a model of the abstract space T_p.

Let q_1 and q_2 be two points of T_S. We can map T_S isometrically onto another Teichmüller space such that the image of one of the given points lies at the origin. In view of what we proved about straight lines through the origin, we obtain the following result: *In the Teichmüller space T_p any two points lie on a straight line.*

This result holds in the case $p = 1$ also, because T_p is then isomorphic to the hyperbolic upper half-plane.

9.4. Composition of Teichmüller Mappings

For an application in 9.5 we need a result about the composition of Teichmüller mappings. As before, S is a compact Riemann surface of genus >1.
Let $f: S \to S'$ be a Teichmüller mapping determined by a pair (φ, k). Then
$f^{-1}: S' \to S$ is a Teichmüller mapping determined by the pair $(-\psi, k)$, where
ψ is the terminal differential of f. For it follows from formula (8.2) that near
a regular point of ψ,

$$\zeta \circ f^{-1} = \frac{\zeta' - k\bar{\zeta}'}{1 + k}, \tag{9.3}$$

where ζ and ζ' are the natural parameters associated with φ and ψ. From
this we conclude that the complex dilatation of f^{-1} is $-k\bar{\psi}/|\psi|$, by making
use of the fact that $d\zeta'^2 = \psi(z')\,dz'^2$ and that the complex dilatation is a
$(-1, 1)$-differential.

Let us multiply (9.3) by $iK = i(1 + k)/(1 - k)$. Comparison with (8.2) then
yields the result that the terminal differential of f^{-1} is $-K^2\varphi$.

For composed mappings we prove what will be needed later.

Lemma 9.1. *Let $f_1: S \to S_1$ and $f_2: S \to S_2$ be Teichmüller mappings determined
by (φ_1, k_1) and (φ_2, k_2) such that φ_2/φ_1 is a constant. Then $f_2 \circ f_1^{-1}$ is a Teichmüller mapping. Up to constants, the initial differential of $f_2 \circ f_1^{-1}$ agrees with
that of f_1^{-1}, and the terminal differential with that of f_2.*

PROOF. Let $p \in S$ be a regular point for φ_1 and φ_2. We denote by ζ and ζ^*
natural parameters for φ_1 and φ_2 at p vanishing at p, and by ζ_1 and ζ_2 the
corresponding natural parameters for the terminal differentials of f_1 and f_2
at $f_1(p)$ and $f_2(p)$.

Let us consider the projection mapping $\zeta_1 \to \zeta_2$ of $f_2 \circ f_1^{-1}$. It can be de-
composed into mappings $\zeta_1 \to \zeta$, $\zeta \to \zeta^*$, and $\zeta^* \to \zeta_2$ (see Fig. 16). Here
$\zeta_1 \to \zeta$ and $\zeta^* \to \zeta_2$ are stretchings. Furthermore, $\zeta \to \zeta^*$ is conformal and
$d\zeta^*/d\zeta = \sqrt{\varphi_2/\varphi_1}$. Since φ_2/φ_1 is constant, we see that ζ^*/ζ is constant. It
follows that $\zeta_1 \to \zeta_2$ is an affine transformation. Explicitly, we write $\varphi_2/\varphi_1 =
a^2 e^{-i\theta}, a > 0, 0 \le \theta < 2\pi$. Then we can choose $\zeta^* = a e^{-i\theta/2}\zeta$.

The stretchings $\zeta_1 \to \zeta$ and $\zeta^* \to \zeta_2$ can be determined from (9.3) and (8.2).
It follows that

$$\zeta_2 = \frac{ae^{-i\theta/2}}{(1 + k_1)(1 - k_2)}[(1 - k_1 k_2 e^{i\theta})\zeta_1 + (k_2 e^{i\theta} - k_1)\bar{\zeta}_1]. \tag{9.4}$$

Since ζ_1 is the natural parameter of the terminal differential of f_1, we see
from (9.4) that $f_2 \circ f_1^{-1}$ is a Teichmüller mapping whose initial differential is
a constant times the initial differential of f_1^{-1}. The result about the terminal
differential of $f_2 \circ f_1^{-1}$ is obtained if we interchange the roles of f_1 and f_2.

$\square$

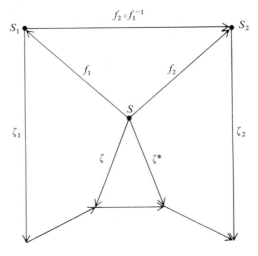

Figure 16

For the complex dilatation of $\zeta_1 \to \zeta_2$ we obtain the expression

$$\frac{k_2 e^{i\theta} - k_1}{1 - k_1 k_2 e^{i\theta}}. \tag{9.5}$$

Let K_1 and K_2 be the maximal dilatations of f_1 and f_2. As regards the maximal dilatation of $f_2 \circ f_1^{-1}$, we use (9.5) to make precise what seems clear geometrically. Suppose first that the initial differential φ_2 of f_2 is a positive constant times the terminal differential $-K_1^2 \varphi_1$ of f_1^{-1}, i.e., that $\theta = \pi$. We then conclude from (9.5), by simple computation, that $f_2 \circ f_1^{-1}$ has the maximal dilatation $K_1 K_2$. The other extreme occurs if $\varphi_2/(-K_1^2 \varphi_1)$ is a negative constant, i.e., if $\theta = 0$. It then follows from (9.5) that $f_2 \circ f_1^{-1}$ has the maximal dilatation $\max(K_1/K_2, K_2/K_1)$.

The result of Lemma 9.1 can also be expressed as follows: If $f_1 : S \to S_1$ and $f_2 : S_1 \to S'$ are Teichmüller mappings, and the ratio of the terminal differential of f_1 and the initial differential of f_2 is constant, then $f_2 \circ f_1$ is a Teichmüller mapping.

9.5. Teichmüller Discs

In 6.6 and 6.7 we proved that there exists a holomorphic isometry between the Teichmüller space of a torus and the hyperbolic unit disc. We shall now show that the same is true in a local sense in every Teichmüller space T_p.

Let S be a compact Riemann surface whose genus is > 1. We fix a holomorphic quadratic differential φ of S (up to a multiplicative positive constant) and consider the mapping

$$z \to [z\bar{\varphi}/|\varphi|] \tag{9.6}$$

of the unit disc D into the Teichmüller space T_S. By Theorem 8.2, this mapping is injective. The mapping $z \to z\bar{\varphi}/|\varphi|$ of D into the set $B(G)$ of Beltrami differentials of S is holomorphic. As a composition of $z \to z\bar{\varphi}/|\varphi|$ and the canonical projection π, *the mapping (9.6) is holomorphic.*

Let

$$\Delta = \{[z\bar{\varphi}/|\varphi|] \mid z \in D\}$$

denote the image of D under (9.6). The subset Δ of T_S is called the *Teichmüller disc* determined by φ.

With varying φ, every Δ clearly contains the origin of T_S. But otherwise two Teichmüller discs are disjoint if they are determined by quadratic differentials whose quotient is not constant. This follows from the definition of Δ, in view of the uniqueness result of Theorem 8.2.

Theorem 9.3. *The mapping $z \to [z\bar{\varphi}/|\varphi|]$ is an isometry of the hyperbolic unit disc onto the Teichmüller disc determined by φ.*

PROOF. For distances from the origin, isometry is clear:

$$\tau_S(0, [z\bar{\varphi}/|\varphi|]) = \frac{1}{2}\log\frac{1 + |z|}{1 - |z|} = h(0, z),$$

where h denotes the hyperbolic distance in the unit disc.

In the general case, let f_1 and f_2 be the Teichmüller mappings with complex dilatations $z_1\bar{\varphi}/|\varphi|$ and $z_2\bar{\varphi}/|\varphi|$. By Lemma 9.1, the composition $f_2 \circ f_1^{-1}$ is also a Teichmüller mapping. From (9.5) (or from formula (I.4.4)) we see that its complex dilatation has the absolute value $|(z_1 - z_2)/(1 - \bar{z}_1 z_2)|$. Therefore, by Theorem 8.2,

$$\tau_S([z_1\bar{\varphi}/|\varphi|], [z_2\bar{\varphi}/|\varphi|]) = \frac{1}{2}\log\frac{|1 - \bar{z}_1 z_2| + |z_1 - z_2|}{|1 - \bar{z}_1 z_2| - |z_1 - z_2|} = h(z_1, z_2). \quad \square$$

9.6. Complex Structure and Teichmüller Metric

The results concerning Teichmüller discs admit far-reaching generalizations. In order to explain this, we first generalize the notion of hyperbolic distance.

Let f be a holomorphic function in the unit disc D with values lying in D. By using suitable Möbius transformations, we see that Schwarz's lemma can be expressed in the form

$$h(f(z_1), f(z_2)) \le h(z_1, z_2). \tag{9.7}$$

From the way the hyperbolic metric was introduced to other domains by use of conformal mappings, we conclude that (9.7) holds if f is an analytic mapping of D into an arbitrary simply connected domain with more than one boundary point, and also, if f maps D holomorphically into a Riemann surface which admits D as its universal covering surface.

Property (9.7) gives rise to one more generalization. Let X be a complex analytic manifold. There exists a largest pseudometric d on X such that

$$d(f(z_1), f(z_2)) \leq h(z_1, z_2)$$

for all holomorphic mappings $f: D \to X$ and for all z_1, z_2 in D. If d is a metric, it is called the *hyperbolic* (or *Kobayashi*) *metric* of X (Kobayashi [1]).

The following result of Royden [1] connects the Teichmüller metric with the complex analytic structure of T_p.

Theorem 9.4. *In the space T_p, $p \geq 1$, the hyperbolic metric and the Teichmüller metric are the same.*

PROOF. For $p = 1$, the theorem follows from (9.7) and the fact that T_p can then be identified with the hyperbolic unit disc (see 6.7).

For $p > 1$, let us consider a model T_S of T_p. In studying the distances between two points q_1 and q_2 of T_S, we may assume, by replacing T_S by another model of T_p, that q_1 lies at the origin. Let $q_2 = [k\bar{\varphi}/|\varphi|]$.

Let us apply again the holomorphic mapping (9.6) of the unit disc into T_S. By Theorem 9.3,

$$\tau_S(q_1, q_2) = h(z_1, z_2), \tag{9.8}$$

where z_1 and z_2 are the preimages of q_1 and q_2 under (9.6). Here the holomorphic mapping of D into T_S depends on the given points. Therefore, we can only infer from (9.8) that the Teichmüller metric τ_S is greater than or equal to the hyperbolic metric of T_S.

Equality is in fact more difficult to establish; for the proof we refer to Royden [1].   □

Theorem 9.4 says that the Teichmüller metric can be recovered from the complex analytic structure of T_p. It follows that the *Teichmüller metric is invariant under biholomorphic self-mappings of T_p*.

Royden [1] also proved that in T_p, $p > 1$, *the modular group is the full group of biholomorphic automorphisms of T_p*.

We showed in 2.7 that two points $[f_1]$ and $[f_2]$ of T_S are equivalent under the modular group if and only if the Riemann surfaces $f_1(S)$ and $f_2(S)$ are conformally equivalent (Theorem 2.7). Hence, if S is a compact Riemann surface of genus greater than 1, then by Royden's result, the points $[f_1]$ and $[f_2]$ of T_S are equivalent under biholomorphic mappings of T_S if and only if the surfaces $f_1(S)$ and $f_2(S)$ are conformally equivalent.

Royden's result also shows that the Teichmüller imbedding (9.2) is not holomorphic. For if it were, it would be biholomorphic. The group G of biholomorphic self-mappings of the unit ball of $\mathbb{C}^{3p-3}$ would then be isomorphic to the modular group of T_p. However, the modular group is known to be properly discontinuous, whereas G (which contains all unitary mappings) is not.

9.7. Surfaces of Finite Type

A Riemann surface S_0 is of *finite type* if S_0 is contained in a compact Riemann surface S and $S \backslash S_0$ consists of finitely many points and of finitely many disjoint closed parametric discs of S.

Let S be of genus p, and consider the case in which $S \backslash S_0$ is the set of n points. We say that S_0 is then of type (p, n). The points of $S \backslash S_0$ are called punctures.

It follows from our previous analysis of covering groups that a Riemann surface of type (p, n) admits a disc as its covering surface if and only if $2p - 2 + n > 0$. Under this condition, which leaves out only the tori and the extended plane with no more than two punctures, the Teichmüller theory of surfaces of type (p, n) is largely analogous with the theory of compact surfaces of genus > 1. In particular, the following basic result is true:

Theorem 9.5. *The Teichmüller space of a Riemann surface of type (p, n), $2p - 2 + n > 0$, is homeomorphic to the euclidean space $\mathbb{R}^{6p-6+2n}$.*

The proof can be reduced to the case of a compact surface. The idea is to construct suitable coverings of S branched at the points of $S \backslash S_0$. The procedure is explained in Ahlfors [1], pp. 20–23.

We have emphasized many times the fact that the Teichmüller spaces of compact surfaces with the same genus are all isomorphic. Similarly, the Teichmüller spaces of all Riemann surfaces of type (p, n) are isomorphic, because such surfaces are all quasiconformally equivalent.

Finally, let S_0 be a Riemann surface of finite type such that $S \backslash S_0$ contains discs. A Teichmüller theory analogous to the theory of compact surfaces can again be developed for such surfaces. However, in this case the notion of reduced Teichmüller space (see 2.1) must be used. For an exposition of the theory of Teichmüller spaces of Riemann surfaces of finite type we refer to Abikoff [2].

Bibliography

Abikoff, W. [1] Some remarks on Kleinian groups. Advances in the Theory of Riemann Surfaces. *Annals of Math. Studies* **66**. Princeton Univ. Press (1971), 1–5.

———— [2] The Real Analytic Theory of Teichmüller Space. *Lecture Notes in Math.* **820**. Springer-Verlag (1980).

———— [3] A geometric property of Bers' embedding of the Teichmüller space. Riemann Surfaces and Related Topics. *Annals of Math. Studies* **97**. Princeton Univ. Press (1981), 3–5.

Ahlfors, L. V. [1] On quasiconformal mappings. *Journal d'Analyse Math.* **3** (1953/54), 1–58.

———— [2] The complex analytic structure of the space of closed Riemann surfaces. *Analytic Functions*. Princeton Univ. Press (1960), 45–66.

———— [3] Some remarks on Teichmüller's space of Riemann surfaces. *Annals of Math.* (2) **74** (1961), 171–191.

———— [4] Quasiconformal reflections. *Acta Math.* **109** (1963), 291–301.

———— [5] *Lectures on Quasiconformal Mappings*. Van Nostrand (1966).

Ahlfors, L. V. and Bers, L. [1] Riemann's mapping theorem for variable metrics. *Annals of Math.* (2) **72** (1960), 385–404.

Ahlfors, L. V. and Sario, L. [1] *Riemann Surfaces*. Princeton Univ. Press (1960).

Ahlfors, L. V. and Weill, G. [1] A uniqueness theorem for Beltrami equations. *Proc. Amer. Math. Soc.* **13** (1962), 975–978.

Astala, K. and Gehring, F. W. [1] Injectivity, the BMO norm and the universal Teichmüller space. To appear in *Journal d'Analyse Math.*

Bers, L. [1] On a theorem of Mori and the definition of quasiconformality. *Trans. Amer. Math. Soc.* **84** (1957), 78–84.

———— [2] Spaces of Riemann surfaces. *Proc. International Congress of Mathematicians*, Edinburgh 1958. Cambridge Univ. Press (1960), 349–361.

———— [3] Quasiconformal mappings and Teichmüller's theorem. *Analytic Functions*. Princeton Univ. Press (1960), 89–119.

———— [4] Simultaneous uniformization. *Bull. Amer. Math. Soc.* **66** (1960), 94–97.

———— [5] Uniformization by Beltrami equations. *Comm. Pure Appl. Math.* **14** (1961), 215–228.

———— [6] Correction to "Spaces of Riemann surfaces as bounded domains". *Bull. Amer. Math. Soc.* **67** (1961), 465–466.

————— [7] On moduli of Riemann surfaces. *Lectures at Forschungsinstitut für Mathematik.* Eidgenössische Technische Hochschule, Zürich (1964), mimeographed.

————— [8] Automorphic forms and general Teichmüller spaces. *Proc. Conf. Complex Analysis,* Minneapolis 1964. Springer-Verlag (1965), 109–113.

————— [9] A non-standard integral equation with applications to quasiconformal mappings. *Acta Math.* **116** (1966), 113–134.

————— [10] Uniformization, moduli, and Kleinian groups. *Bull. London Math. Soc.* **4** (1972), 257–300.

————— [11] Quasiconformal mappings, with applications to differential equations, function theory and topology. *Bull. Amer. Math. Soc.* **83** (1977), 1083–1100.

————— [12] A new proof of a fundamental inequality for quasiconformal mappings. *Journal d'Analyse Math.* **36** (1979), 15–30.

————— [13] On a theorem of Abikoff. *Ann. Acad. Sci. Fenn. A I Math.* **10** (1985), 83–87.

Beurling, A. and Ahlfors, L. V. [1] The boundary correspondence under quasiconformal mappings. *Acta Math.* **96** (1956), 125–142.

Bourbaki, N. [1] *Variétés différentielles et analytiques.* Eléments de mathématique XXXIII. Fascicule de résultats–Paragraphes 1 à 7. Hermann (1967).

Calvis, D. [1] The inner radius of univalence of normal circular triangles and regular polygons. *Complex Variables Theory Appl.* **4** (1985), 295–304.

Carleman, T. [1] Über die Approximation analytischer Funktionen durch lineare Aggregate von vorgegebenen Potenzen. *Ark. Mat. Astr. Fys.* **17**, no. 9 (1923), 1–30.

Douady, A. and Earle, C. J. [1] Conformally natural extension of homeomorphisms of the circle. To appear in *Acta Math.*

Dunford, N. and Schwartz, J. T. [1] *Linear operators,* Part I. Interscience Publishers (1958).

Duren, P. L. [1] *Univalent Functions.* Springer-Verlag (1983).

Earle, C. J. [1] The Teichmüller space of an arbitrary Fuchsian group. *Bull. Amer. Math. Soc.* **70** (1964), 699–701.

————— [2] Teichmüller theory. *Discrete Groups and Automorphic Functions,* ed. W. J. Harvey. Academic Press (1977), 143–162.

Earle, C. J. and Eells, J. Jr. [1] On the differential geometry of Teichmüller spaces. *Journal d'Analyse Math.* **19** (1967), 35–52.

Flinn, B. (Brown) [1] Jordan domains and the universal Teichmüller space. *Trans. Amer. Math. Soc.* **282** (1984), 603–610.

Ford, L. R. [1] *Automorphic Functions,* 2nd ed. Chelsea, New York (1951).

Fricke, R. and Klein, F. [1] *Vorlesungen über die Theorie der automorphen Functionen.* Teubner (1926).

Gardiner, F. P. [1] An analysis of the group operation in universal Teichmüller space. *Trans. Amer. Math. Soc.* **132** (1968), 471–486.

Gauss, C. F. [1] Allgemeine Auflösung der Aufgabe die Theile einer gegebenen Fläche auf einer andern gegebenen Fläche so abzubilden, dass die Abbildung dem Abgebildeten in den kleinsten Theilen ähnlich wird. *Astronomische Abhandlungen,* vol. 3 (ed. H. C. Schumacher), Altona (1825), or *Carl Friedrich Gauss Werke,* vol. IV, Göttingen (1880), 189–216.

Gehring, F. W. [1] Quasiconformal mappings which hold the real axis pointwise fixed. *Math. Essays Dedicated to A. J. Macintyre.* Ohio Univ. Press, Athens (1970), 145–148.

————— [2] Univalent functions and the Schwarzian derivative. *Comment. Math. Helv.* **52** (1977), 561–572.

————— [3] Spirals and the universal Teichmüller space. *Acta Math.* **141** (1978), 99–113.

————— [4] Characteristic properties of quasidisks. *Sém. Math. Sup.* **84**. Presses Univ. Montréal (1982).

Gehring, F. W. and Osgood, B. G. [1] Uniform domains and the quasi-hyperbolic metric. *Journal d'Analyse Math.* **36** (1979), 50–74.

Gehring, F. W. and Pommerenke, Ch. [1] On the Nehari univalence criterion and quasicircles. *Comment. Math. Helv.* **59** (1984), 226–242.

Gehring, F. W. and Väisälä, J. [1] Hausdorff dimension and quasiconformal mappings. *J. London Math. Soc.* (2) **6** (1973), 504–512.

Grötzsch, H. [1] Über die Verzerrung bei schlichten nichtkonformen Abbildungen und über eine damit zusammenhängende Erweiterung des Picardschen Satzes. *Ber. Verh. Sächs. Akad. Wiss. Leipzig* **80** (1928), 503–507.

Hamilton, R. S. [1] Extremal quasiconformal mappings with prescribed boundary values. *Trans. Amer. Math. Soc.* **138** (1969), 399–406.

Keen, L. [1] On Fricke moduli. Advances in the Theory of Riemann Surfaces. *Annals of Math. Studies* **66**. Princeton Univ. Press (1971), 205–224.

Klein, F. [1] *Gesammelte mathematische Abhandlungen*, Vol. III. Julius Springer (1923).

Kobayashi, S. [1] *Hyperbolic Manifolds and Holomorphic Mappings*. Marcel Dekker (1970).

Kra, I. [1] On Teichmüller spaces of finitely generated Fuchsian groups. *Amer. J. Math.* **91** (1969), 67–74.

——— [2] Canonical mappings between Teichmüller spaces. *Bull. Amer. Math. Soc. N. S.* **4** (1981), 143–179.

Kraus, W. [1] Über den Zusammenhang einiger Charakteristiken eines einfach zusammenhängenden Bereiches mit der Kreisabbildung. *Mitt. Math. Sem. Giessen* **21** (1932), 1–28.

Kravetz, S. [1] On the geometry of Teichmüller spaces and the structure of their modular groups. *Ann. Acad. Sci. Fenn. A I Math.* **278** (1959), 1–35.

Kruschkal, S. L. and Kühnau, R. [1] Quasikonforme Abbildungen—neue Methoden und Anwendungen. *Teubner-Texte zur Math.* **54**, Leipzig (1983).

Kruškal', S. L. [1] Teichmüller's theorem on extremal quasi-conformal mappings. *Siberian Math. J.* **8** (1967), 231–244. Translated from *Sibirsk. Mat. Ž.* **8** (1967), 313–332.

Kühnau, R. [1] Wertannahmeprobleme bei quasikonformen Abbildungen mit ortsabhängiger Dilatationsbeschränkung. *Math. Nachr.* **40** (1969), 1–11.

——— [2] Verzerrungssätze und Koeffizientenbedingungen vom Grunskyschen Typ für quasikonforme Abbildungen. *Math. Nachr.* **48** (1971), 77–105.

Kuusalo, T. [1] Boundary mappings of geometric isomorphisms of Fuchsian groups. *Ann. Acad. Sci. Fenn. A I Math.* **545** (1973), 1–7.

Lehner, J. [1] *Discontinuous Groups and Automorphic Functions*. Amer. Math. Soc. (1964).

——— [2] Automorphic forms. *Discrete Groups and Automorphic Functions*, ed. W. J. Harvey. Academic Press (1977), 73–120.

Lehtinen, M. [1] A real-analytic quasiconformal extension of a quasisymmetric function. *Ann. Acad. Sci. Fenn. A I Math.* **3** (1977), 207–213.

——— [2] On the inner radius of univalency for non-circular domains. *Ann. Acad. Sci. Fenn. A I Math.* **5** (1980), 45–47.

——— [3] Remarks on the maximal dilatation of the Beurling–Ahlfors extension. *Ann. Acad. Sci. Fenn. A I Math.* **9** (1984), 133–139.

——— [4] Estimates of the inner radius of univalency of domains bounded by conic sections. *Ann. Acad. Sci. Fenn. A I Math.* **10** (1985), 349–353.

——— [5] Angles and the inner radius of univalency. To appear in *Ann. Acad. Sci. Fenn. A I Math.* **11**.

Lehto, O. [1] Schlicht functions with a quasiconformal extension. *Ann. Acad. Sci. Fenn. A I Math.* **500** (1971), 1–10.

——— [2] Group isomorphisms induced by quasiconformal mappings. *Contributions to Analysis.* A collection of papers dedicated to Lipman Bers. Academic Press (1974), 241–244.

———— [3] Quasiconformal mappings and singular integrals. Istituto Nazionale di Alta Matematica. *Symposia Mathematica XVIII.* Academic Press (1976), 429–453.

———— [4] Quasiconformal homeomorphisms and Beltrami equations. *Discrete Groups and Automorphic Functions,* ed. W. J. Harvey. Academic Press (1977), 121–142.

———— [5] On univalent functions with quasiconformal extensions over the boundary. *Journal d'Analyse Math.* **30** (1976), 349–354.

———— [6] Domain constants associated with Schwarzian derivative. *Comment. Math. Helv.* **52** (1977), 603–610.

———— [7] Remarks on Nehari's theorem about the Schwarzian derivative and schlicht functions. *Journal d'Analyse Math.* **36** (1979), 184–190.

Lehto, O. and Tammi, O. [1] Schwarzian derivative in domains of bounded boundary rotation. *Ann. Acad. Sci. Fenn. A I Math.* **4** (1978/79), 253–257.

Lehto, O. and Virtanen, K. I. [1] *Quasiconformal Mappings in the Plane.* Springer-Verlag (1973).

Martio, O. and Sarvas, J. [1] Injectivity theorems in plane and space. *Ann. Acad. Sci. Fenn. A I Math.* **4** (1978/79), 383–401.

Maskit, B. [1] On boundaries of Teichmüller spaces and on kleinian groups: II. *Annals of Math.* (*2*) **91** (1970), 607–639.

Mori, A. [1] On quasi-conformality and pseudo-analyticity. *Trans. Amer. Math. Soc.* **84** (1957), 56–77.

Morrey, C. B. [1] On the solutions of quasi-linear elliptic partial differential equations. *Trans. Amer. Math. Soc.* **43** (1938), 126–166.

Mostow, G. D. [1] Strong rigidity of locally symmetric spaces. *Annals of Math. Studies* **78**. Princeton Univ. Press (1973).

Narasimhan, R. [1] *Several Complex Variables.* Univ. of Chicago Press (1971).

Nehari, Z. [1] The Schwarzian derivative and schlicht functions. *Bull. Amer. Math. Soc.* **55** (1949), 545–551.

———— [2] *Conformal Mapping.* McGraw-Hill (1952).

Nevanlinna, R. [1] *Uniformisierung.* Springer-Verlag (1953).

Newman, M. H. A. [1] *Elements of the Topology of Plane Sets of Points.* Cambridge Univ. Press (1961).

Paatero, V. [1] Über die konforme Abbildung von Gebieten deren Ränder von beschränkter Drehung sind. *Ann. Acad. Sci. Fenn. Ser. A XXXIII:* **9** (1931), 1–79.

Pfluger, A. [1] Quasikonforme Abbildungen und logarithmische Kapazität. *Ann. Inst. Fourier (Grenoble)* **2** (1951), 69–80.

———— [2] Über die Konstruktion Riemannscher Flächen durch Verheftung. *J. Indian Math. Soc.* **24** (1961), 401–412.

Pommerenke, Chr. [1] *Univalent Functions.* Vandenhoeck & Ruprecht (1975).

Reich, E. [1] On criteria for unique extremality of Teichmüller mappings. *Ann. Acad. Sci. Fenn. A I Math.* **6** (1981), 289–301.

Reich, E. and Strebel, K. [1] On the extremality of certain Teichmüller mappings. *Comment. Math. Helv.* **45** (1970), 353–362.

———— [2] Extremal quasiconformal mappings with given boundary values. *Contributions to Analysis.* A collection of papers dedicated to Lipman Bers. Academic Press (1974), 375–391.

Riemann, B. [1] Grundlagen für eine allgemeine Theorie der Functionen einer veränderlichen complexen Grösse. Inauguraldissertation, Göttingen 1851. *Gesammelte mathematische Werke und wissenschaftlicher Nachlass,* 2nd ed. Teubner, Leipzig (1892), 3–48.

Royden, H. L. [1] Automorphisms and isometries of Teichmüller space. *Advances in* the Theory of Riemann Surfaces. *Annals of Math. Studies* **66**. Princeton Univ. Press (1971), 369–383.

——— [2] Intrinsic metrics on Teichmüller space. *Proc. International Congress of Mathematicians*, Vancouver 1974, vol. 2. Canadian Math. Congress (1975), 217–221.

Schiffer, M. [1] A variational method for univalent quasiconformal mappings. *Duke Math. J.* **33** (1966), 395–411.

Schiffer, M. and Schober, G. [1] Coefficient problems and generalized Grunsky inequalities for schlicht functions with quasiconformal extensions. *Arch. Rational Mech. Anal.* **60** (1976), 205–228.

Schober, G. [1] Univalent Functions—Selected Topics. *Lecture Notes in Math.* **478**, Springer-Verlag (1975).

Schwarz, H. A. [1] Über einige Abbildungsaufgaben. *J. für reine und angewandte Math.* **70** (1869), 105–120, or *Gesammelte Mathematische Abhandlungen*, Vol. II, Julius Springer, Berlin (1890), 65–83.

Shiga, H. [1] Characterization of quasi-disks and Teichmüller spaces. *Tôhoku Math. J.* (2) **37** (1985), 541–552.

Springer, G. [1] *Introduction to Riemann Surfaces*. Addison-Wesley (1957).

Stein, E. M. [1] *Singular Integrals and Differentiability Properties of Functions*. Princeton Univ. Press (1970).

Strebel, K. [1] Zur Frage der Eindeutigkeit extremaler quasikonformer Abbildungen des Einheitskreises. *Comment. Math. Helv.* **36** (1962), 306–323.

——— [2] On the trajectory structure of quadratic differentials. Discontinuous Groups and Riemann Surfaces. *Annals of Math. Studies* **79**. Princeton Univ. Press (1974), 419–438.

——— [3] On the existence of extremal Teichmueller mappings. *Journal d'Analyse Math.* **30** (1976), 464–480.

——— [4] On lifts of extremal quasiconformal mappings. *Journal d'Analyse Math.* **31** (1977), 191–203.

——— [5] On quasiconformal mappings of open Riemann surfaces. *Comment. Math. Helv.* **53** (1978), 301–321.

——— [6] *Quadratic Differentials*. Springer-Verlag (1984).

Teichmüller, O. [1] Extremale quasikonforme Abbildungen und quadratische Differentiale. *Abh. Preuss. Akad. Wiss., math.-naturw. Kl.* **22** (1939), 1–197, or *Gesammelte Abhandlungen—Collected Papers*, ed. by L. V. Ahlfors and F. W. Gehring, Springer-Verlag (1982), 335–531.

——— [2] Bestimmung der extremalen quasikonformen Abbildungen bei geschlossenen orientierten Riemannschen Flächen. *Abh. Preuss. Akad. Wiss., math.-naturw. Kl.* **4** (1943), 1–42, or *Gesammelte Abhandlungen—Collected Papers*, ed. by L. V. Ahlfors and F. W. Gehring, Springer-Verlag (1982), 635–676.

——— [3] Veränderliche Riemannsche Flächen. *Deutsche Math.* **7** (1944), 344–359, or *Gesammelte Abhandlungen—Collected Papers*, ed. by L. V. Ahlfors and F. W. Gehring, Springer-Verlag (1982), 712–727.

Thurston, W. P. [1] Zippers and schlicht functions. To appear.

Tukia, P. [1] Quasiconformal extension of quasisymmetric mappings compatible with a Möbius group. *Acta Math.* **154** (1985), 153–193.

Vekua, I. N. [1] *Generalized Analytic Functions*. Pergamon Press, Oxford (1962).

Weyl, H. [1] *Die Idee der Riemannschen Fläche*. Teubner, Leipzig u. Berlin (1913).

Žuravlev, I. V. [1] Univalent functions and Teichmüller spaces. *Soviet Math. Dokl.* **21** (1980), 252–255. Translated from *Dokl. Akad. Nauk SSSR* **250** (1980), 1047–1050.

Index

Abikoff, W. 2, 192, 198, 204, 221, 247
Absolutely continuous on lines 20
Ahlfors, L.V. 1, 3, 8, 12, 16, 26, 33,
    34, 35, 36, 40, 49, 72, 81, 88, 116,
    173, 182, 207, 226, 232, 247
Ahlfors–Sario 3, 128, 135, 136, 137,
    138, 140, 141, 142, 143, 147
Arc condition 45
Area theorem 58, 78
Astala, K. 117

Beltrami differential 133
Beltrami equation 24
    existence theorem 28, 177
    uniqueness theorem 24, 177
Bers, L. 1, 2, 23, 24, 72, 88, 89, 98,
    99, 111, 116, 144, 182, 191, 199,
    203, 204, 207, 211, 226, 232, 239
Bers imbedding 114, 203
Beurling, A. 8, 16, 33, 34, 36
Beurling–Ahlfors extension 33
Bieberbach 58
Biholomorphic 206
Bojarski 27
Bordered (Riemann) surface 131, 132
Boundary rotation 62
Bourbaki, N. 206

Calderón 26
Calderón–Zygmund inequality 26
Calvis, D. 126
Canonical mapping of a quadrilateral 7
Carleman, T. 226
Class
    C_0^∞ 25
    S 59
    S_k, $S_k(\infty)$ 79
    Σ 59
    Σ_k 73
Complex analytic structure 130
Complex dilatation 23, 177
    extremal 105
    trivial 191
Conformal structure 130
    lifted 142
    projected 142
Contractible space 109
Cover transformation 138
Covering group 138
    transitive 138
Covering surface 137
    branched 137
    normal 136
    smooth 135
    universal 137
    unlimited 135

Deformation equivalent 185
Differential 132
  Abelian 132
  Beltrami 133
  for a group 148
  infinitesimally trivial 224
  (m,n)- 132
  quadratic 132
  trivial 191
Dilatation quotient 19
Dirichlet region 151
Distance
  between domains 67
  between quasisymmetric functions 108
  from a disc 61
  hyperbolic 5
  Teichmüller 103, 104, 183, 184
  β 105
  δ 61, 67
  q 103
  ρ 108
  τ 103, 104, 183, 184
  τ_s 190
Distortion function λ 15
Divisor 159
Douady, A. 194
Double of a Riemann surface 155
Dunford–Schwartz 26, 222, 229
Duren, P.L. 77

Earle, C.J. 2, 34, 109, 194, 199, 226
Eells, J., Jr. 34, 109
Elementary group 152
Equivalent
  Beltrami differentials 184
  complex dilatations 97
  quasiconformal mappings 97, 182

Flinn, B. 117
Ford, L.R. 152
Fricke, R. 156, 241
Fuchsian group 150
Fundamental domain 149
Fundamental group 136

Gardiner, F.P. 102

Gauss, C.F. 24, 128, 135, 144
Gehring, F.W. 21, 23, 38, 41, 49, 84,
  90, 92, 94, 105, 116, 117
Generalized derivatives 21
Genus 158
Geodesic
  of quadratic differential 170
  of Teichmüller metric 105, 184
Golusin 79
Gronwall 58
Grothendieck 2
Grötzsch, H. 11, 20
  ring domain 11
Group (see also Möbius group)
  discrete 153
  elementary 152
  elliptic modular 213
  Fuchsian 150
  fundamental 136
  Kleinian 153
  limit point of 140
  limit set of 151
  modular 188
  of first (second) kind 154
  properly discontinuous 140
  quasi-Fuchsian 155
  of right translations 99
  set of discontinuity 152
  universal modular 99
Grunsky 79

Hamilton, R.S. 228, 232
  sequence 229
Hersch 13
Hilbert transformation 25
Hille 90
Holomorphic function 51
  in a Banach space 206
  on a Riemann surface 131
Homotopy
  lifting property 147
  of mappings 146
  modulo the boundary 180
Hyperbolic
  distance 5
  metric 5, 6, 148, 246
  sup norm 54

Induced group homomorphism 145
Inner radius of univalence 118
Isothermal coordinates 134

Jordan domain 7

Keen, L. 241
Klein, F. 135, 156
Kleinian group 153
Klein's method 156
Kobayashi, S. 246
Koebe 144
  function 59
  one-quarter theorem 6
K-quasiconformal 12, 176
Kra, I. 2, 194, 226, 229
Kraus, W. 60
Kravetz, S. 221
Krushkal, S.L. 79, 227, 228
Kühnau, R. 72, 74, 75, 78, 79
Kuusalo, T. 195

Lebesgue measurable mapping 130
Lehner, J. 3, 129, 149, 152, 153, 154,
      156, 213, 216, 224
Lehtinen, M. 34, 36, 37, 122, 124, 125,
      126, 127
Lehto, O. 17, 21, 23, 28, 65, 66, 71,
      72, 74, 76, 78, 79, 122, 199
Lehto–Virtanen 2, 4, 8, 9, 10, 11, 12,
      13, 15, 16, 18, 20, 21, 22, 23, 25,
      26, 27, 28, 29, 33, 38, 48, 101
Lift
  of a differential 148
  of a mapping 145
  of a path 135
Limit point of a group 140
Limit set of a group 151
Linearly locally connected 44
L^p–derivatives 21

Majorant principle 77
Manifold 129
  Banach 206
  complex analytic 130, 207

differentiable 129
  real 129
Martio, O. 49, 52
Maskit, B. 198
Maximal dilatation
  of a quasiconformal mapping 12
  of a quasisymmetric function 31
Meromorphic function 51, 131
Mirror image of a Riemann surface 155
Möbius group (*see also* Group)
  Dirichlet region 151
  invariant domain 155
  normal polygon 158
  quasiconformal deformation 191
Modular group
  elliptic 213
  of a Teichmüller space 188
  universal 99
Module
  of a path family 12
  of a quadrilateral 8, 9
  of a ring domain 10
Monodromy theorem 136
Mori, A. 21
Morrey, C.B. 24
Mostow, G.D. 195

Narasimhan, R. 211
Natural parameter 162
Nehari, Z. 6, 60, 90, 124
Nevanlinna, R. 143, 144, 158
Newman, M.H.A. 7, 15, 44, 85
Norm
  hyperbolic sup- 54
  of Schwarzian derivative 54
Normal family 13
Normal polygon 158

Ohtsuka 143
Osgood, B.G. 49
Outer radius of univalence 66

Paatero, V. 63
Parameter transformation 129
Pfluger, A. 12, 21, 23, 49, 101
Poincaré 144

density 6, 148
  metric 6
  theta series 223
Pommerenke, Ch. 79, 90, 92, 94

Quadratic differential 132
  critical point 161
  initial 235
  metric induced by 168
  natural parameter 162
  orientable 164
  regular point 161
  terminal 235
  trajectory 163
Quadrilateral 7
  conjugate 46
Quasicircle 38
Quasiconformal deformation 191
Quasiconformal mapping 12, 176
  analytic definition 22
  diffeomorphism 19
  extremal 37, 183
  geometric definition 12
  1-dimensional 31
  of a Riemann surface 176
Quasiconformal reflection 39
Quasidisc 38
Quasi-Fuchsian group 155
Quasisymmetric functions 31
  space of 108
Quasisymmetry constant 32

Reduced Teichmüller space 183
Reflection principle 16
Reich, E. 37, 227, 229, 230
Rengel's inequality 9
Riemann 130, 135
  mapping theorem 143
  space 182
  surface 130
Riemann–Roch theorem 160
Ring domain 10
Royden, H.L. 2, 226, 246

Sario, L. (see Ahlfors–Sario)
Sarvas, J. 49, 52

Schiffer, M. 79
Schober, G. 79
Schwartz, J.T. (see Dunford–Schwartz)
Schwarz, H.A. 52, 137
Schwarzian derivative 51
Schwarzian domain 83
Sewing problem 100
Shiga, H. 204
Sobolev space 21
Space
  $A(G)$ 222
  A (1) 223
  $L^\infty (G), L^1 (G)$ 222
  $N(G)$ 224
  Q 111
  $Q(G)$ 197
  $Q(1)$ 197
  R_s 182
  T 97
  $T(G)$ 198
  $T(1)$ 114
  T_p 212
  T_s 182
  U 115
  $U(G)$ 197
  $U(1)$ 198
  X 97, 108
  $X(G)$ 194
Springer, G. 3, 129, 151, 158, 159, 160
Stein, E.M. 26
Stoïlov 138
Straight line 163, 242
Strebel, K. 3, 21, 37, 129, 166, 167,
    169, 190, 227, 229, 230, 232
Study 63
Surface 130
  of finite type 247

Tammi, O. 65
Teichmüller 1, 11, 103, 163, 170, 174,
    181, 182, 207, 216, 226, 232, 233,
    239
  disc 245
  distance 103, 104, 183, 184
  existence theorem 232
  imbedding 241
  lemma 170
  mapping 231

  ring domain 11
  space 97, 182
  uniqueness theorem 237
Thurston, W.P. 2, 118, 204
Tienari 49
Torus 150
Total set 206
Trajectory of a quadratic differential
  critical 164
  cross-cut 167
  horizontal 163
  periodic 166
  spiral 167
Tukia, P. 110, 194

Uniform domain 41
Uniformly equivalent 201
Univalent function 52

Universal covering surface 137
Universal Teichmüller space 97

Väisälä, J. 13, 17, 38
Vekua, I.N. 26, 28
Virtanen, K.I. 17 (*see also* Lehto–
  Virtanen)

Weill, G. 88
Weyl, H. 130, 135

Yûjôbô 23

Žuravlev, I.V. 204
Zygmund 26

Graduate Texts in Mathematics

continued from page ii

48 SACHS/WU. General Relativity for Mathematicians.
49 GRUENBERG/WEIR. Linear Geometry. 2nd ed.
50 EDWARDS. Fermat's Last Theorem.
51 KLINGENBERG. A Course in Differential Geometry.
52 HARTSHORNE. Algebraic Geometry.
53 MANIN. A Course in Mathematical Logic.
54 GRAVER/WATKINS. Combinatorics with Emphasis on the Theory of Graphs.
55 BROWN/PEARCY. Introduction to Operator Theory I: Elements of Functional Analysis.
56 MASSEY. Algebraic Topology: An Introduction.
57 CROWELL/FOX. Introduction to Knot Theory.
58 KOBLITZ. *p*-adic Numbers, *p*-adic Analysis, and Zeta-Functions. 2nd ed.
59 LANG. Cyclotomic Fields.
60 ARNOLD. Mathematical Methods in Classical Mechanics.
61 WHITEHEAD. Elements of Homotopy Theory.
62 KARGAPOLOV/MERZLJAKOV. Fundamentals of the Theory of Groups.
63 BOLLABÁS. Graph Theory.
64 EDWARDS. Fourier Series. Vol. I. 2nd ed.
65 WELLS. Differential Analysis on Complex Manifolds. 2nd ed.
66 WATERHOUSE. Introduction to Affine Group Schemes.
67 SERRE. Local Fields.
68 WEIDMANN. Linear Operators in Hilbert Spaces.
69 LANG. Cyclotomic Fields II.
70 MASSEY. Singular Homology Theory.
71 FARKAS/KRA. Riemann Surfaces.
72 STILLWELL. Classical Topology and Combinatorial Group Theory.
73 HUNGERFORD. Algebra.
74 DAVENPORT. Multiplicative Number Theory. 2nd ed.
75 HOCHSCHILD. Basic Theory of Algebraic Groups and Lie Algebras.
76 IITAKA. Algebraic Geometry.
77 HECKE. Lectures on the Theory of Algebraic Numbers.
78 BURRIS/SANKAPPANAVAR. A Course in Universal Algebra.
79 WALTERS. An Introduction to Ergodic Theory.
80 ROBINSON. A Course in the Theory of Groups.
81 FORSTER. Lectures on Riemann Surfaces.
82 BOTT/TU. Differential Forms in Algebraic Topology.
83 WASHINGTON. Introduction to Cyclotomic Fields.
84 IRELAND/ROSEN. A Classical Introduction to Modern Number Theory.
85 EDWARDS. Fourier Series: Vol. II. 2nd ed.
86 VAN LINT. Introduction to Coding Theory.
87 BROWN. Cohomology of Groups.
88 PIERCE. Associative Algebras.
89 LANG. Introduction to Algebraic and Abelian Functions. 2nd ed.
90 BRØNDSTED. An Introduction to Convex Polytopes.
91 BEARDON. On the Geometry of Discrete Groups.
92 DIESTEL. Sequences and Series in Banach Spaces.

93 DUBROVIN/FOMENKO/NOVIKOV. Modern Geometry — Methods and Applications Vol. I.
94 WARNER. Foundations of Differentiable Manifolds and Lie Groups.
95 SHIRYAYEV. Probability, Statistics, and Random Processes.
96 CONWAY. A Course in Functional Analysis.
97 KOBLITZ. Introduction in Elliptic Curves and Modular Forms.
98 BRÖCKER/tom DIECK. Representations of Compact Lie Groups.
99 GROVE/BENSON. Finite Reflection Groups. 2nd ed.
100 BERG/CHRISTENSEN/RESSEL. Harmonic Analysis on Semigroups: Theory of Positive Definite and Related Functions.
101 EDWARDS. Galois Theory.
102 VARADARAJAN. Lie Groups, Lie Algebras and Their Representations.
103 LANG. Complex Analysis. 2nd. ed.
104 DUBROVIN/FOMENKO/NOVIKOV. Modern Geometry — Methods and Applications Vol. II.
105 LANG. $SL_2(\mathbf{R})$.
106 SILVERMAN. The Arithmetic of Elliptic Curves.
107 OLVER. Applications of Lie Groups to Differential Equations.
108 RANGE. Holomorphic Functions and Integral Representations in Several Complex Variables.
109 LEHTO. Univalent Functions and Teichmüller Spaces.